Communications
in Computer and Information Science 10

Joaquim Filipe Boris Shishkov
Markus Helfert (Eds.)

# Software and Data Technologies

First International Conference, ICSOFT 2006
Setúbal, Portugal, September 11-14, 2006
Revised Selected Papers

Springer

Volume Editors

Joaquim Filipe
Polytechnic Institute of Setúbal – INSTICC
Av. D. Manuel I, 27A - 2. Esq., 2910-595 Setúbal, Portugal
E-mail: j.filipe@est.ips.pt

Boris Shishkov
Interdisciplinary Institute for Collaboration and Research
on Enterprise Systems and Technology – IICREST
P.O. Box 104, 1618 Sofia, Bulgaria
E-mail: b.b.shishkov@ewi.utwente.nl

Markus Helfert
Dublin City University, School of Computing
Dublin 9, Ireland
E-mail: markus.helfert@computing.dcu.ie

Library of Congress Control Number: 2008931195

CR Subject Classification (1998): D.2, D.3, C.2.4, H.2, I.2.4

ISSN 1865-0929
ISBN-10 3-540-70619-4 Springer Berlin Heidelberg New York
ISBN-13 978-3-540-70619-9 Springer Berlin Heidelberg New York



Springer is a part of Springer Science+Business Media

springer.com


Printed in Germany

Typesetting: Camera-ready by author, data conversion by Scientific Publishing Services, Chennai, India
Printed on acid-free paper SPIN: 12441204 06/3180 5 4 3 2 1 0

# Preface

This book contains the best papers of the First International Conference on Software and Data Technologies (ICSOFT 2006), organized by the Institute for Systems and Technologies of Information, Communication and Control (*INSTICC*) in cooperation with the Object Management Group (*OMG*). Hosted by the School of Business of the Polytechnic Institute of Setubal, the conference was sponsored by Enterprise Ireland and the Polytechnic Institute of Setúbal.

The purpose of ICSOFT 2006 was to bring together researchers and practitioners interested in information technology and software development. The conference tracks were "Software Engineering", "Information Systems and Data Management", "Programming Languages", "Distributed and Parallel Systems" and "Knowledge Engineering."

Being crucial for the development of information systems, software and data technologies encompass a large number of research topics and applications: from implementation-related issues to more abstract theoretical aspects of software engineering; from databases and data-warehouses to management information systems and knowledge-base systems; next to that, distributed systems, pervasive computing, data quality and other related topics are included in the scope of this conference.

ICSOFT included in its program a panel to discuss the future of software development, composed by six distinguished world-class researchers. Furthermore, the conference program was enriched by a tutorial and six keynote lectures.

ICSOFT 2006 received 187 paper submissions from 39 countries in all continents. All submissions were evaluated using a double-blind review process: each paper was reviewed by at least two experts belonging to the conference Program Committee. A small number of 23 papers were selected to be presented as full papers at the conference and be published in the conference proceedings as completed research papers. In addition 44 papers, describing work-in-progress, were accepted as short papers and 26 papers were selected for poster presentation. This resulted in a full-paper acceptance ratio of 12%. Then, a short list of excellent and significant papers was selected to appear in this book. We hope that you will find these papers interesting and we hope they represent a helpful reference in the future for all those who need to address any of the research areas mentioned above.

March 2008

Joaquim Filipe
Boris Shishkov
Markus Helfert

# Conference Committee

## Conference Chair

Joaquim Filipe, Polytechnic Institute of Setúbal / INSTICC, Portugal

## Program Co-chairs

Markus Helfert, Dublin City University, Ireland
Boris Shishkov, University of Twente, The Netherlands

## Organizing Committee

Paulo Brito, INSTICC, Portugal
Marina Carvalho, INSTICC, Portugal
Hélder Coelhas, INSTICC, Portugal
Bruno Encarnação, INSTICC, Portugal
Vítor Pedrosa, INSTICC, Portugal
Mónica Saramago, INSTICC, Portugal

## Program Committee

Hamideh Afsarmanesh, The Netherlands
Jacky Akoka, France
Tsanka Angelova, Bulgaria
Keijiro Araki, Japan
Lora Aroyo, The Netherlands
Colin Atkinson, Germany
Juan Carlos Augusto, UK
Elisa Baniassad, China
Mortaza S. Bargh, The Netherlands
Joseph Barjis, USA
Noureddine Belkhatir, France
Fevzi Belli, Germany
Alexandre Bergel, Ireland
Mohamed Bettaz, Jordan
Robert Biddle, Canada
Maarten Boasson, The Netherlands
Wladimir Bodrow, Germany
Marcello Bonsangue, The Netherlands
Jonathan Bowen, UK
Mark van den Brand, The Netherlands
Lisa Brownsword, USA
Barrett Bryant, USA
Cinzia Cappiello, Italy
Antonio Cerone, China
W.K. Chan, China
Kung Chen, Taiwan
Samuel Chong, UK
Chih-Ping Chu, Taiwan
Peter Clarke, USA
Rolland Colette, France
Alfredo Cuzzocrea, Italy
Bogdan Czejdo, USA
David Deharbe, Brazil
Serge Demeyer, Belgium
Steve Demurjian, USA
Nikolay Diakov, The Netherlands
Jan L.G. Dietz, The Netherlands
Jin Song Dong, Singapore

Brian Donnellan, Ireland
Jürgen Ebert, Germany
Paul Ezhilchelvan, UK
Behrouz Far, Canada
Bernd Fischer, UK
Gerald Gannod, USA
Jose M. Garrido, USA
Dragan Gasevic, Canada
Nikolaos Georgantas, France
Paola Giannini, Italy
Paul Gibson, Ireland
Wolfgang Grieskamp, USA
Daniela Grigori, France
Klaus Grimm, Germany
Rajiv Gupta, USA
Tibor Gyimothy, Hungary
Naohiro Hayashibara, Japan
Jang Eui Hong, Korea
Shinichi Honiden, Japan
Ilian Ilkov, The Netherlands
Ivan Ivanov, USA
Tuba Yavuz Kahveci, USA
Krishna Kavi, USA
Khaled Khan, Australia
Roger King, USA
Christoph Kirsch, Austria
Paul Klint, The Netherlands
Alexander Knapp, Germany
Mieczyslaw Kokar, USA
Michael Kölling, UK
Dimitri Konstantas, Switzerland
Jens Krinke, Germany
Tei-Wei Kuo, Taiwan
Rainer Koschke, Germany
Eitel Lauria, USA
Insup Lee, USA
Kuan-Ching Li, Taiwan
Panos Linos, USA
Shaoying Liu, Japan
Zhiming Liu, China
Andrea De Lucia, Italy
Christof Lutteroth, New Zealand
Broy Manfred, Germany
Tiziana Margaria, Germany
Johannes Mayer, Germany
Fergal McCaffery, Ireland
Hamid Mcheick, Canada
Prasenjit Mitra, USA
Dimitris Mitrakos, Greece
Roland Mittermeir, Austria
Birger Møller-Pedersen, Norway
Mattia Monga, Italy
Aldo De Moor, Belgium
Peter Müller, Switzerland
Paolo Nesi, Italy
Elisabetta Di Nitto, Italy
Alan O'Callaghan, UK
Rory O'Connor, Ireland
Claus Pahl, Ireland
Witold Pedrycz, Canada
Massimiliano Di Penta, Italy
Steef Peters, The Netherlands
Mario Piattini, Spain
Arnd Poetzsch-Heffter, Germany
Andreas Polze, Germany
Christoph von Praun, USA
Jolita Ralyte, Switzerland
Juan Fernandez Ramil, UK
Anders P. Ravn, Denmark
Marek Reformat, Canada
Arend Rensink, The Netherlands
Stefano Russo, Italy
Shazia Sadiq, Australia
Kristian Sandahl, Sweden
Bradley Schmerl, USA
Andy Schürr, Germany
Isabel Seruca, Portugal
Marten van Sinderen, The Netherlands
Joao Sousa, USA
George Spanoudakis, UK
Peter Stanchev, USA
Larry Stapleton, Ireland
Stoicho Stoichev, Bulgaria
Kevin Sullivan, USA
Junichi Suzuki, USA
Ramayah Thurasamy, Malaysia
Yasar Tonta, Turkey
Yves Le Traon, France
Enrico Vicario, Italy
Bing Wang, UK
Kun-Lung Wu, USA
Hongwei Xi, USA
Haiping Xu, USA
Hongji Yang, UK

Yunwen Ye, USA
Yun Yang, Australia
Gianluigi Zavattaro, Italy
Xiaokun Zhang, Canada
Jianjun Zhao, China
Hong Zhu, UK
Andrea Zisman, UK

## Auxiliary Reviewers

Alessandro Aldini, Italy
Pete Andras, UK
Xiaoshan Li, China
Shih-Hsi Liu, USA
Michele Pinna, Italy
Riccardo Solmi, Italy
Hongli Yang, China
Chengcui Zhang, USA
Liang Zhao, China
Wei Zhao, USA

## Invited Speakers

Leszek A. Maciaszek, Macquarie University, Australia
Juan Carlos Augusto, University of Ulster at Jordanstown, UK
Tom Gilb, Norway
Dimitris Karagiannis, University of Vienna, Austria
Brian Henderson-Sellers, University of Technology, Australia
Marten J. van Sinderen, University of Twente, The Netherlands

# Table of Contents

## Invited Papers



## Part I: Programming Languages



## Part II: Software Engineering

## Part III: Distributed and Parallel Systems



## Part IV: Information Systems and Data Management

## Part V: Knowledge Engineering

# Invited Papers

# Adaptive Integration of Enterprise and B2B Applications

Leszek A. Maciaszek

Department of Computing, Macquarie University, Sydney, NSW 2109, Australia
leszek@ics.mq.edu.au

**Abstract.** Whether application integration is internal to the enterprise or takes the form of external Business-to-Business (B2B) automation, the main integration challenge is similar – how to ensure that the integration solution has the quality of adaptiveness (i.e. it is understandable, maintainable, and scalable)? This question is hard enough for stand-alone application developments, let alone integration developments in which the developers may have little control over participating applications. This paper identifies main strategic (architectural), tactical (engineering), and operational (managerial) imperatives for buil-ding adaptiveness into solutions resulting from integration projects.

**Keywords:** Application integration, software adaptiveness.

## 1 Introduction

Today's enterprise and e-business systems are rarely developed in-house from scratch. Most systems are the results of evolutionary maintenance of existing systems. Occasionally new systems are developed, but always with the intent to integrate with the existing software. New technologies emerge to facilitate development and integration of enterprise and e-business systems. The current thrust comes from the component technology standards and the related technology of Service Oriented Architecture (SOA).

This paper centers on conditions for developing adaptive complex enterprise and e-business systems. It concentrates on architectural design, engineering principles, and operational imperatives for developing such systems. An *adaptive* system has the ability to change to suit different conditions; the ability to continue into the future by meeting existing functional and nonfunctional requirements and by adjusting to accommodate any new and changing requirements. In some ways, an adaptive system is an antonym of a legacy system. A necessary condition of adaptiveness is the identification and minimization of *dependencies* in software. A software element `A` depends on an element `B`, if a change in `B` may necessitate a change in `A`.

Enterprise and e-business systems are *complex* – their properties and behavior cannot be fully explained by the understanding of their component parts. The software crisis has been looming at our doorsteps for years. Cobol legacy systems and the millennium bug are well known examples on the global scale, and the examples of individual software disasters are countless. Each time when faced with a crisis, we have been engaging thc next technological gear to solve the problem. But also each time we have been introducing an additional level of complexity to the software and with it

J. Filipe, B. Shishkov, and M. Helfert (Eds.): ICSOFT 2006, CCIS 10, pp. 3–15, 2008.

new and more troublesome non-adaptive solutions. The premise of this paper is that, unless we start producing adaptive systems, we are yet to face the first true software crisis.

## 2 Development or Integration?

Business has embraced the Internet-age technology with zeal. Thanks to application integration technologies, organizations can function as loosely connected networks of cooperating units. Development of stand-alone applications is all but history. Accordingly, the term "application development" is being replaced by the more accurate term – "integration development".

There are three *integration levels* [2]. Figure 1 shows how the three levels are related to each other. All integration projects imply exchange of *data* between integrated applications. Typically this means that an application `A` gets access to application's `B` database either directly or by data replication techniques.

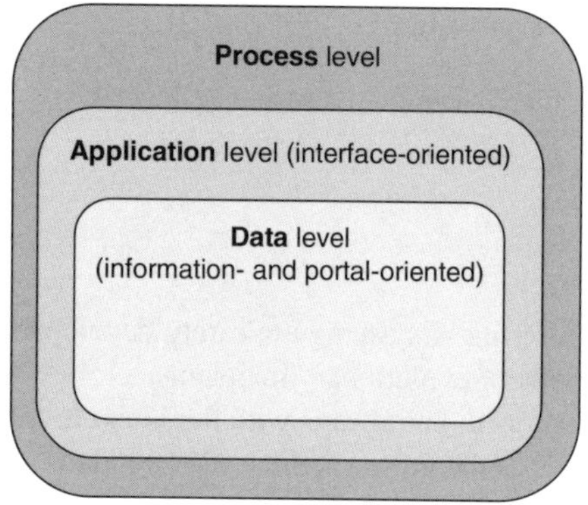


**Fig. 1.** Integration levels

At *application level*, application `A` uses the interfaces (services) of application `B` to request data or to request execution of the services provided by application `B`. A classic example is a *Loan Broker* application [3], in which the integration solution negotiates loan terms with the banks for the customers. Although Loan Broker negotiates with many banks, the negotiations are separate for each bank. Hence, this is dyadic (point-to-point) integration.

Another successful example of dyadic integration is VMI (Vendor-Managed Inventory). In VMI integration, a vendor/supplier is responsible for monitoring and replenishing customer inventory at the appropriate time to maintain predefined levels.

However, the ultimate goal of integration is the much more complex integration of detailed business processes performed by applications (and resulting in services and data production). At the *process level* new workflows of processes are designed that integrate processes already available in existing applications to provide a new value-added functionality.

Clearly, process level integration blurs the line between development and integration. At this level, an integration project is also a new development project. A new umbrella application is produced providing solutions that go beyond the sum of relationships between participating applications/enterprises and that go beyond simple integration effects.

The need for process-level integrations arises when businesses want to enter into collaborative environments to achieve joint outcomes, not just their own outcomes. Electronic marketplaces (*e-markets*) subscribe to that goal, but business factors limit e-market expansion. It is simply the case, that "sharing price or capacity information is often not advantageous to all parties in a supply chain or vertical market" [1].

Where the business conditions are right, process-level integrations can flourish. [1] provides two illustrative examples – transportation optimization and cash netting. *Transportation optimization* is a collaborative logistic application that consolidates various in-transit status messages for the trucks traveling around Europe so that empty trucks can be dynamically hired to take loads. Cash netting is designed to replace point-to-point invoice-payment processes by the "cash-netting" at the end of the day, i.e. once a day payment to/from each account, which is the value of shipments minus the value of receipts.

The main and overriding technology that drives integration development is *SOA* (service-oriented architecture) [2]. SOA uses XML *Web services* as its implementation principle and introduces a new logical layer within the distributed computing platform. This new Integration layer defines a common point of integration across applications and across enterprises. In effect, SOA blurs the distinction between integration and new distributed applications (because the reason for calling a service on the Integration layer is transparent to SOA – and the reason could be an integration or brand new application development).

Moreover, and not out of context, SOA blurs the distinction between a business process and a technology process (one no longer exclusively drives the other).

## 3 Classifying Integration

Integration projects are as much about the strategy as about the technology. As such, they have many dimensions and various mixing of these dimensions is needed to ensure the project's business objectives and to choose the appropriate technology. Tables 1 and 2 provide two different two-dimensional viewpoints on integration projects.

**Table 1.** Two-dimensional view on integration projects

| | Internal integration | External integration → SOA |
|---|---|---|
| Dyadic integration | File sharing<br>Remote procedures | EDI<br>Web services |
| Hub integration | Shared database<br>Workflows | Message brokers<br>Orchestration |

**Table 2.** Another two-dimensional view on integration projects

| | Synchronous integration | Asynchronous integration |
|---|---|---|
| Data integration | Data replication<br>Portal sharing | File transfer<br>Shared database |
| Process integration → SOA | Remote procedures<br>Workflows | Messaging |

The two criteria used in Table 1 are the integration business target and the number of integration participants. The *business target* can be the enterprise itself (internal integration) or another business (external B2B (business-to-business) integration) (cp.[4]). The *number of participants* can be just two parties in a point-to-point supply chain (dyadic integration) or more than two parties in a network structure (hub integration) (cp. [1]).

The two criteria used in Table 2 are the degree of coupling and the integration target (approach). The *degree of coupling* distinguishes between loosely coupled and tight coupled integration (cp. [7]). The *integration target* defines the spectrum of approaches to integration – from integration through data, via integration through interfaces, to direct integration of executing processes.

The cells in both tables serve the purpose of providing examples of integration solutions. The simplest solutions referred to in Table 1 are file sharing and remote procedures. These two solutions are typically used in internal dyadic integration, but they are applicable in other more complicated integration projects as well.

Integration by means of *file sharing* means that files are transferred between applications. The integration effort concentrates merely on re-formatting the files to suit the receiving application.

Integration by means of *remote procedures* is based on an old piece of wisdom that data should be encapsulated by procedures. Accordingly, to access the data in another application, the client application invokes remotely appropriate procedures, which in turn supply the data.

Two other internal integration solutions mentioned in Table 1 are shared database and workflows. Although shown as examples of hub integration, they are often used in simpler dyadic integrations. Also, when the business conditions are right, they can be used in external integration.

Any database is by definition shared, so talking about *shared database* emphasizes only the point that the database is used as an integration solution. Because a database can be shared by any number of users and applications, a shared database is an obvious vehicle for hub integration on the level of data.

The hub integration on the level of processes can be based on *workflows*. A workflow is a distributed business transaction governed by a process management function that ensures the integrated flow of execution of transactional tasks between many systems/applications.

The last column in Table 1 refers to external integration. Modern external integration solutions are based on SOA. A primitive forerunner of SOA as a technology for external integration has been *EDI* (Electronic Data Interchange) – a set of computer interchange standards for business documents. *Web services* are also defined by a collection of protocols and standards used for exchanging data between applications or enterprises. However, they make themselves available over the Internet for integration with applications and systems and they can be part of SOA infrastructure.

Message brokers and orchestration engines are used for hub external integration. *Message broker* is a layer of software between applications integrated within the hub. It is a SOA component that ensures data transformation, merging and enrichment so that applications in the hub can communicate and collaborate.

*Orchestration* is a value-added component to encapsulate and execute new business process logic. It implements workflows that involve collaborators in the hub. Integration solutions obtained via orchestration engines are often classified as *virtual applications* [2].

Integration solutions listed in Table 1 can be analyzed from other angles, including the viewpoints taken in Table 2, namely degree of coupling and integration targets. In general, data integration is more aligned with (better suited for) loosely coupled integration. This is because data can be easily put aside for later use. Conversely, process integration is more aligned with tightly couple integration.

Data replication, portal sharing, and messaging are the three integration solutions listed in Table 2 but missing in Table 1. *Data replication* is classified as synchronous integration because replication servers of databases can be programmed to perform replications continuously and replicate data as the primary data is changing.

Probably slightly controversially, *portal sharing* is classified in Table 2 as synchronous data integration. Portals are web sites or applications that provide access to a number of sources of information and facilities (portlets). They aggregate information from multiple sources into a single display. The display of information is synchronous but no any sophisticated process-level communication between portlets is normally assumed – hence, data integration.

*Messaging* is the primary technology for asynchronous process integration [3]. Based on the Publish/Subscribe model, messaging frameworks guarantee reliable delivery of messages in program-to-program communication while recognizing that synchronous communication with remote applications is difficult to achieve (yet asynchronous communication is frequently acceptable).

## 4 Assuring Adaptive Integration

Building adaptiveness into enterprise and e-business systems engages all three traditional levels of management – strategic, tactical and operational. From the system's development perspective, the strategic level refers to the *architectural* solutions, the tactical level to the *engineering* decisions, and the operational level to the project *controlling* tasks. These three levels of management are used in the conventional top-down fashion when software is developed. We can say that system architecture *defines* adaptiveness, engineering activities *deliver* adaptiveness, and controlling tasks *verify* the existence of adaptiveness in an implemented system.

### 4.1 Defining Adaptiveness

A well known truth, unfortunately frequently forgotten in practice, is that the necessary condition for assuring adaptive integration (and any large software development for that matter) is that the integration adheres to strict and transparent *architectural design*. The architectural design itself must conform to a *meta-architecture* that is known to ensure the quality of adaptiveness in any compliant complex system. Meta-architecture determines the layers of the (necessary) hierarchical structure in a complex system and specifies allowed dependencies between and inside the layers.

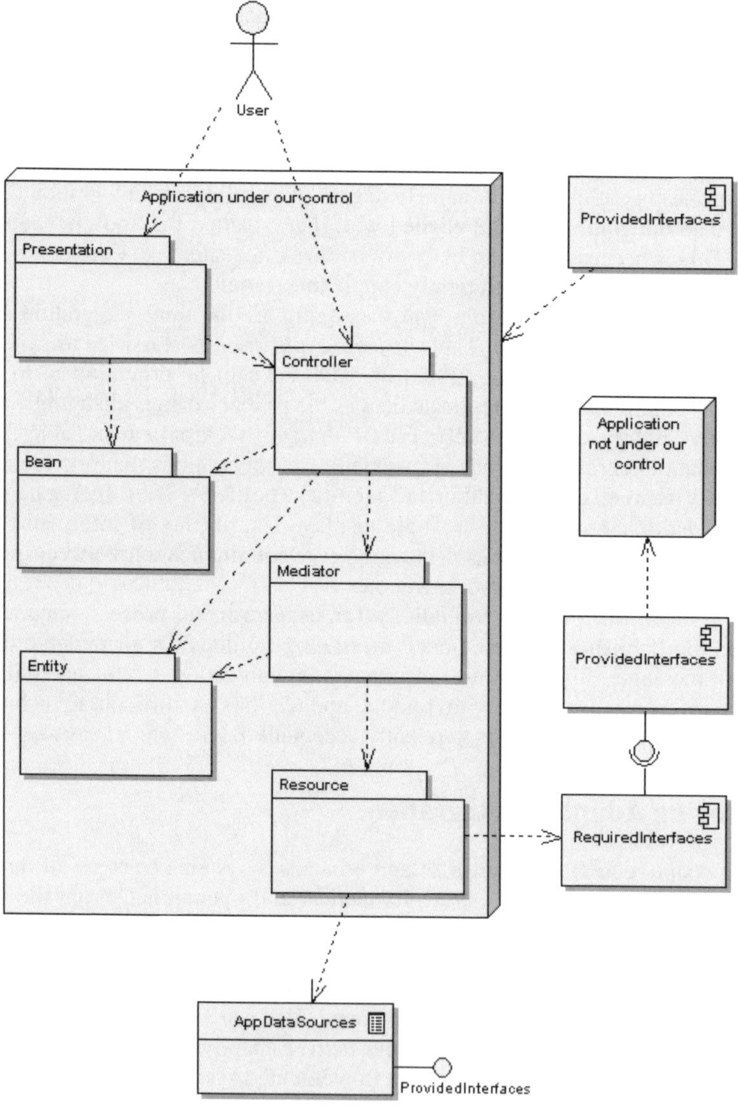


**Fig. 2.** PCBMER-A meta architecture

There are many meta-architectures that in principle can support the quality of adaptiveness. However, most meta-architectures are undefined for that purpose. To be useful, a meta-architecture must classify possible dependencies according to their ripple effect, i.e. adverse chain reactions on client objects once a supplier object is modified in any way [6]. It must determine metrics to compute cumulative dependencies for particular designs in order to be able to select a design that minimizes dependencies

[7] It must then offer guidelines, principles and patterns, which assist system developers in their quest to adhere to the architectural design while not "binding their hands and brains" too much [8].

The pivotal meta-architecture, which we advocate, is called *PCBMER*. The PCBMER framework defines six hierarchical layers of software objects – Presentation, Controller, Bean, Mediator, Entity and Resource.

The PCBMER meta-architecture has evolved from earlier frameworks [8] and has aimed at new development projects. We believe, however, that PCBMER can easily accommodate to integration projects. That belief is consistent with the earlier discussion that the demarcation line between development and integration is blurred and that any more sophisticated process-level integration is really a form of new application development. Nevertheless, PCBMER requires some extensions to account in the meta-architecture for the integration layer.

An important starting point for any extensions of PCBMER is that we can only ensure the quality of adaptiveness in the software that remains under our control. We can then only trust that parties that our software integrates with will also be adaptive. With this understanding in mind, we can distinguish between two meta-architectures that apply in integration projects.

The first integration meta-architecture applies to application-level integrations. This architecture subsumes also data-level integration. We call this architecture PCBMER-A (where A signifies an application-centric integration). Figure 2 shows a high-level view of the PCBMER-A meta-architecture.

The second integration meta-architecture applies to process-level integrations, providing support for inter-application and inter-organization communication. We call this architecture PCBMER-U (where U refers to a utility service that such meta-architectural solutions promise to deliver). Figure 3 is a high-level view of the PCBMER-U meta-architecture.

Figures 2 and 3 illustrate that the integration meta-architectures retain the Core PCBMER framework. Dependencies (dotted arrowed lines) between the core packages remain unchanged in integration projects. Hence, for example, Presentation depends on Controller and on Bean, and Controller depends on Bean. Note that the PCBMER hierarchy is not strictly linear and a higher-layer can have more than one adjacent layer below (and that adjacent layer may be an intra-leaf, i.e. it may have no layers below it).

The *Bean* package represents the data classes and value objects that are destined for rendering on user interface. Unless entered by the user, the bean data is built from the entity objects (the Entity package). The Core PCBMER framework does not specify or endorse if access to Bean objects is via message passing or event processing as long as the Bean package does not depend on other packages.

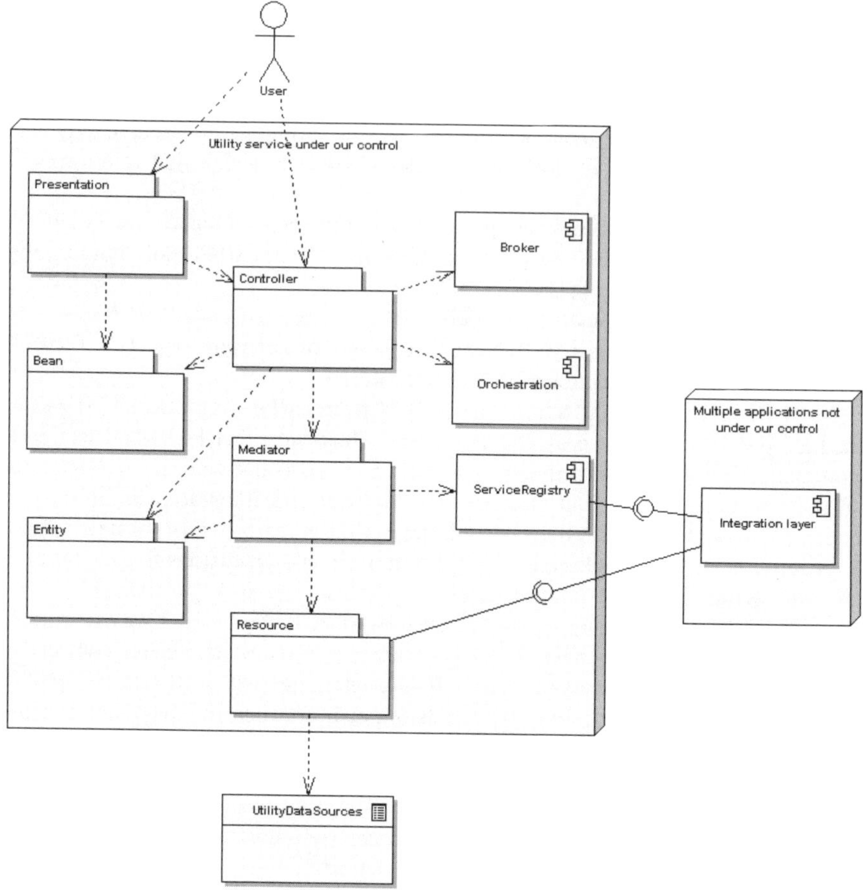


**Fig. 3.** PCBMER-U meta architecture

The *Presentation* package represents the screen and UI objects on which the beans can be rendered. It is responsible for maintaining consistency in its presentation when the beans change. So, it depends on the Bean package. This dependency can be realized in one of two ways – by direct calls to methods (message passing) using the pull model or by event processing followed by message passing using the push model (or rather push-and-pull model)

The *Controller* package represents the application logic. Controller objects respond to the UI requests that originate from Presentation and that are results of user interactions with the system. In a programmable GUI client, UI requests may be menu or button selections. In a web browser client, UI requests appear as HTTP Get or Post requests.

The *Entity* package responds to Controller and Mediator. It contains classes representing "business objects". They store (in the program's memory) objects retrieved from the database or created in order to be stored in the database. Many entity classes are container classes.

The *Mediator* package establishes a channel of communication that mediates between Entity and Resource classes. This layer manages business transactions, enforces business rules, instantiates business objects in the Entity package, and in general manages the memory cache of the application. Architecturally, Mediator serves two main purposes. Firstly, to isolate the Entity and Resource packages so that changes in any one of them can be introduced independently. Secondly, to mediate between the Controller and Entity/Resource packages when Controller requests data but it does not know if the data has been loaded to memory or it is available in the database or it can be obtained from external sources.

The *Resource* package is responsible for all communications with external persistent data sources (databases, web services, etc.). This is where the connections to the database and SOA servers are established, queries to persistent data are constructed, and the database transactions are instigated.

For application-centric integration projects (Figure 2), the PCBMER-A meta-architecture enriches the Resource package with the RequiredInterfaces component. This component provides access to external applications. Although the component is called RequiredInterfaces, the access is not restricted to invoking Java-style interfaces implemented in collaborating applications. Any other integration levels are assumed and allowed, such as direct access to data, access to data encapsulated by accessor methods or by stored procedures, access to data rendered in portals, or access to web services.

For utility-centric integration projects (Figure 3), the PCBMER-U meta-architecture is explicitly extended with new "integration automation" components – Broker, Orchestration, and Service Registry. The first two implement the automation logic and depend on the utlity's application logic in Controller. Service Registry implements the "service discovery" and depends on the utility's business logic in Mediator. All access to the integration layers of participating applications originates from either Mediator or Resource.

## 4.2 Delivering Adaptiveness

Once defined in a meta-architecture, an adaptive solution can be delivered through engineering work. It is the responsibility of engineers to ensure that architectural advantages are retained in the engineered product. The task is not easy because as always "the devil is in the detail". To do the job, the engineers must be equipped with principles, patterns, implementation techniques, etc. that instrument the meta-architectural advantages and that explicitly address the adaptiveness criteria in the solution.

The Core PCBMER framework has a number of advantages resulting in minimization of dependencies. The main advantage is the *separation of concerns* between packages allowing modifications within one package without affecting the other (independent) packages or with a predictable and manageable effect on the other (dependable) packages. For example, the Presentation package that provides a Java application UI could be switched to a mobile phone interface and still use the existing implementation of Controller and Bean packages. That is, the same pair of Controller and Bean packages can support more than one Presentation UI at the same time.

The second important advantage is the *elimination of cycles* between dependency relationships and the resultant six-layer hierarchy with downward only dependencies. Cycles would degenerate a hierarchy into a network structure. Cycles are disallowed both between PCBMER packages and within each PCBMER package.

The third advantage is that the framework ensures a significant degree of *stability*. Higher layers depend on lower layers. Therefore, as long as the lower layers are stable (i.e. do not change significantly, in particular in interfaces), the changes to the higher layers are relatively painless. Recall also that lower layers can be extended with new functionality (as opposed to changes to existing functionality), and such extensions should not impact on the existing functionality of the higher layers.

The Core PCBMER meta-architecture enforces other properties and constraints that are not necessarily directly visible in Figures 2 and 3. Below is the list of the most important PCBMER *engineering principles* (cp. Maciaszek and Liong, 2005):

Downward Dependency Principle (DDP)

The DDP states that the main dependency structure is top-down. Objects in higher layers depend on objects in lower layers. Consequently, lower layers are more *stable* than higher layers. Interfaces, abstract classes, dominant classes and similar devices should encapsulate stable packages so that they can be extended when needed.

Upward Notification Principle (UNP)

The UNP promotes low coupling in a bottom-up communication between layers. This can be achieved by using asynchronous communication based on event processing. Objects in higher layers act as subscribers (observers) to state changes in lower layers. When an object (publisher) in a lower layer changes its state, it sends notifications to its subscribers. In response, subscribers can communicate with the publisher (now in the downward direction) so that their states are synchronized with the state of the publisher.

Neighbor Communication Principle (NCP)

The NCP demands that a package can only communicate directly with its neighbor package as determined by direct dependencies between packages. This principle ensures that the system does not disintegrate to a network of intercommunicating objects. To enforce this principle, the message passing between non-neighboring objects uses delegation or forwarding (the former passes a reference to itself; the latter does not). In more complex scenarios, a special acquaintance package can be used to group interfaces to assist in collaboration that engages distant packages.

Explicit Association Principle (EAP)

The EAP visibly documents permitted message passing between classes. This principle recommends that associations are established on all directly collaborating classes. Provided the design conforms to PCBMER, the downward dependencies between classes (as per DDP) are legitimized by corresponding associations. Associations resulting from DDP are unidirectional (otherwise they would create circular dependencies). It must be remembered, however, that not all associations between classes are due to message passing. For example, both-directional associations may be needed to implement referential integrity between classes in the entity package.

Cycle Elimination Principle (CEP)

The CEP ensures that circular dependencies between layers, packages and classes within packages are resolved. Circular dependencies violate the separation of concerns guideline and are the main obstacle to reusability. Cycles can be resolved by placing offending classes in a new package created specifically for that purpose or by forcing one of the communication paths in the cycle to communicate via an interface.

Class Naming Principle (CNP)

The CNP makes it possible to recognize in the class name the package to which the class belongs. To this aim, each class name is prefixed in PCBMER with the first letter of the package name (e.g. EInvoice is a class in the Entity package). The same principle applies to interfaces. Each interface name is prefixed with two capital letters – the first is the letter "I" (signifying that this is an interface) and the second letter identifies the package (e.g. ICInvoice is an interface in the Controller package).

Acquaintance Package Principle (APP)

The APP is the consequence of the NCP. The acquaintance package consists of interfaces that an object passes, instead of concrete objects, in arguments to method calls. The interfaces can be implemented in any PCBMER package. This effectively allows communication between non-neighboring packages while centralizing dependency management to a single acquaintance package.

### 4.3 Verifying Adaptiveness

The PCBMER meta-architecture together with the seven principles defines a desired model to produce adaptive systems. However, the meta-architecture is a theoretical objective which may or may not be fulfilled in practice. Also, it is possible to have multiple designs (and corresponding implementations), all of which conforming to the meta-architecture, yet exhibiting various levels of "goodness". What we need is to be able to measure how "good" particular software solution is and whether or not it conforms to the meta-architecture. The overall task is called the *roundtrip architectural modelling* in [5].

Therefore, to verify adaptiveness in an integration solution, we need to define *structural complexity* metrics able to compute *cumulative dependencies* between the solution's implementation objects. The dependencies can be defined on messages, events, classes, components, etc. [7].

From the perspective of a system architect and maintainer, the dependencies between classes provide the most valuable metric of system complexity and adaptiveness. It is, therefore, important to make these dependencies explicit and to uncover any hidden dependencies. The *Cumulative Class Dependency* (*CCD*) is a measure of the total number of class dependencies in a system.

DEFINITION: *Cumulative Class Dependency (CCD)* is the total "adaptiveness" cost over all classes $C_{i\{i=1,\ldots,n)}$ in a system of the number of classes $C_{j(j<=1,\ldots,n)}$ to be potentially changed in order to modify each class $C_i$.

The *CCD* definition is intentionally simple. In particular it does not, by itself, judge the quality of the design. Its value is in comparisons between two or more designs for the same system. To this aim, the *CCD* computation strives to validate if a particular design conforms to a chosen meta-architecture (such as PCBMER). Uncovering a class dependency that invalidates the architectural framework leads to the only sensible assumption that the required dependency structure in the system is broken. This in turn means that any dependency is possible and the system adaptiveness has eluded management controls.

The calculation of *CCD* for a particular design starts by assuming the adherence to the architectural framework. If the framework is found to be broken, the *CCD* is calculated as if each class depended on any other class in the system. Such worst-scenario *CCD* can be computed using a probability theory method called *combinations counting rule*. It computes the number of different combinations of pairs of dependent classes which can be formed from the total number of classes in the design.

With the above in mind, the generic *cumulative class dependency* formula for the Core PCBMER is shown in the equation below (this is a generic formula and other formulas may apply to specific PCBMER architectures derived from the Core framework). The formula assumes that access to packages is encapsulated by *hub objects* (Maciaszek, 2006). These could be Java-style interfaces, dominant classes, and similar devices, which force single channels of communication between packages.

$$_{hubPCBMER}CCD = \sum_{i=1}^{root} \frac{o_i(o_i - 1)}{2} + \sum_{j=1}^{root} p_{j+1}$$

where:

$o$ is the number of objects in each package $i$ including any hub objects,

$p_{j+1}$ is the number of objects in each directly adjacent package above any leave package minus any hub object (this computes the number of potential downward paths to all hub objects in the adjacent packages),

$_{hubPCBMER}CCD$ is a cumulative class dependency in a hub hierarchy representing the PCBMER meta-architecture.

Note that the formula accommodates the fact that the PCBMER framework permits a lower-layer package to be communicated from more than one higher-layer package. These higher-layer packages are considered to be "directly adjacent", hence the formula applies as stands. Note that because only downward dependencies are allowed, the communication from higher-layer packages retains the hierarchical properties of the PCBMER framework.

The *CCD* equation ensures *polynomial* growth of dependencies between architectural layers represented as packages, while allowing *exponential* growth of class dependencies within layers. However, the exponential growth can be controlled by grouping classes within a layer into *nested packages* (as packages can contain other packages). The communication between nested packages can then be performed using hubs.

Measuring adaptiveness of designs and programs cannot be done manually. Maciaszek and Liong (2003) describes a tool, called *DQ* (Design Quantifier), which is able to analyse any Java program, establish its conformance with a chosen adaptive meta-architecture, compute complete set of dependency metrics, and visualize the computed values in UML class diagrams.

Although not supported by DQ, tools like DQ should be able to visualize dependencies by producing *call graphs*. Ideally, a call graph could be a variant of a UML sequence diagram. A call graph can be used for the change impact analysis and to answer "what-if" questions such as "which methods are affected if a particular method is modified?"

## 5 Summary

The purpose and all-overriding importance of achieving the quality of adaptiveness in software is in ensuring that the software becomes a long-lasting business asset, not just business cost. This observation is particularly true for integration projects, which by definition tend to deliver software with greater competitive advantages.

In [6] and elsewhere, we explained the interplay between software complexity and adaptiveness, showed that hierarchical structures with hubs minimize complexity, talked about lessons from studying structure and behaviour of living systems, provided classifications of object dependencies, and introduced the PCBMER meta-architecture.

In this paper, we extended earlier work related to new application developments to the software integration projects. We argued that the demarcation line between new development and integration is blurry and that the similar strategies and principles of software production apply. In particular, the Core PCBMER meta-architecture can be successfully adapted to integration projects and we showed necessary architectural extensions. We addressed software engineering practices and technologies that could guarantee the compliance of an implemented software system with the PCBMER meta-architecture and its principles. Finally, we talked about reverse-engineering verification procedures to substantiate in metrics the level of compliance of an integration solution with the adaptivity criteria.

## References

1. Christiaanse, E.: Performance Benefits Through Integration Hubs. Comm. ACM 48(4), 95–100 (2005)
2. Erl, T.: Service-Oriented Architecture. A Field Guide to Integrating XML and Web Services, p. 536. Prentice Hall, Englewood Cliffs (2004)
3. Hohpe, G., Woolf, B.: Enterprise Integration Patterns, p. 650. Addison-Wesley, Reading (2003)
4. Linthicum, D.S.: Next Generation Application Integration. From Simple Information to Web Services, p. 488. Addison-Wesley, Reading (2004)
5. Maciaszek, L.A.: Roundtrip Architectural Modeling. In: Hartmann, S., Stumper, M. (eds.) Australian Computer Science Communications, Newcastle, Australia, January 30 – February 4, 2005, vol. 27(6), pp. 17–23 (2005) (invited paper)
6. Maciaszek, L.A.: From Hubs Via Holons to an Adaptive Meta-Architecture – the AD-HOC Approach. In: Proceedings IFIP Working Conference on Software Engineering Techniques - SET 2006, Warsaw, Poland, October 17-20, 2006, p. 13 ( to appear, 2006)
7. Maciaszek, L.A., Liong, B.L.: Designing Measurably-Supportable Systems. In: Niedzielska, E., Dudycz, H., Dyczkowski, M. (eds.) Advanced Information Technologies for Management, Research Papers No 986. pp. 120–149, Wroclaw University of Economics (2003)
8. Maciaszek, L.A., Liong, B.L.: Practical Software Engineering. A Case-Study Approach, Harlow England, p. 864. Addison-Wesley, Reading (2005)

# Ambient Intelligence: Basic Concepts and Applications

Juan Carlos Augusto

School of Computing and Mathematics, University of Ulster, Jordanstown, U.K.
jc.augusto@ulster.ac.uk

**Abstract.** Ambient Intelligence is a multi-disciplinary approach which aims to enhance the way environments and people interact with each other. The ultimate goal of the area is to make the places we live and work in more beneficial to us. Smart Homes is one example of such systems but the idea can be also used in relation to hospitals, public transport, factories and other environments. The achievement of Ambient Intelligence largely depends on the technology deployed (sensors and devices interconnected through networks) as well as on the intelligence of the software used for decision-making. The aims of this article are to describe the characteristics of systems with Ambient Intelligence, to provide examples of their applications and to highlight the challenges that lie ahead, especially for the Software Engineering and Knowledge Engineering communities.

**Keywords:** Ambient Intelligence, Software Engineering, Knowledge Engineering, Sensor Networks, Smart Homes.

## 1 Introduction

The steady progress in technology have not only produced a plethora of new devices and spread computing power into various levels of our daily lives, it is also driving a transformation on how society relates to computer science. The miniaturization process in electronics has made available a wide range of small computing devices which can now help us when we wash clothes and dishes, cook our meals, and drive our cars. Inspired on those successful applications which are now embedded in our daily lives many new technological developments are spreading little computing devices everywhere possible (see as an example recent developments on RFID technology [1]). Several new devices of that kind are being investigated and produced every year. These developments are being quickly absorbed by the research community (see for example recent reports on [2,3]) and by several leading companies around the world (see for example [4]).

This richness in technology and computer power has been continuously progressing since the very inception of computer science. First a machine was shared by many highly trained programmers. Then it became possible in many countries around the world that many people, not necessarily with a high level of training, will have access to one PC in an individual basis. Now many people can have access to several computing devices like a PC, a laptop and a PDA at work plus a PC at home and various smaller processing units embedded in electro-domestic appliances. All seems to indicate this trend will continue. Slowly systems are being designed in such a way that people do not need to be a computer specialist to benefit from computing power. This technical

J. Filipe, B. Shishkov, and M. Helfert (Eds.): ICSOFT 2006, CCIS 10, pp. 16–26, 2008.

possibility is being explored in an area called *Ambient Intelligence* (AmI) where the idea of making computing available to people in a non-intrusive way is at the core of its values. The benefit of an AmI system is measured by how much can give to people whilst minimizing explicit interaction. The aim is to enrich specific places (a room, a building, a car, a street) with computing facilities which can react to people's needs and provide assistance.

Given the evolution of markets and industry people is now more willing to accept technologies participating and shaping their daily life. At the same time there are important driving forces at political level which create a fertile terrain for this to happen. An important example of this is the decentralization of health care and development of health and social care assistive technologies. For various reasons governments and health professionals are departing away from the hospital-centric health care system enabling this shift of care from the secondary care environment to primary care. Subsequently, there is an effort to move away from the traditional concept of patients being admitted into hospitals rather to enable a more flexible system whereby people are cared for closer to home, within their communities. Smart homes are one such example of a technological development which facilitates this trend of bringing the health and social care system to the patient as opposed to bringing the patient into the health system.

The aim of this paper is to describe more specifically the relationship in between AmI and related areas (Section 2), to describe some possible scenarios of application (Section 6), and finally to highlight the technical difficulties and opportunities laying ahead (Section 7) which, in the view of the author, will shape the course of important areas of computer science (Section 8).

## 2 Ambient Intelligence

"Ambient Intelligence" (AmI) [5,6] is growing fast as a multi-disciplinary approach which can allow many areas of research to have a significant beneficial influence into our society. The basic idea behind AmI is that by enriching an environment with technology (mainly sensors and devices interconnected through a network), a system can be built to take decisions to benefit the users of that environment based on real-time information gathered and historical data accumulated. AmI has a decisive relationship with many areas in computer science. The relevant areas are depicted in Figure 1.

Here we must add that whilst AmI nourishes from all those areas, it should not be confused with any of those in particular. Networks, sensors, interfaces, ubiquitous or pervasive computing and AI are all relevant but none of them conceptually covers AmI. It is AmI which puts together all these resources to provide flexible and intelligent services to users acting in their environments.

AmI is aligned with the concept of the *"disappearing computer"* [7,8]:

> *"The most profound technologies are those that disappear. They weave themselves into the fabric of everyday life until they are indistinguishable from it."*

The notion of a disappearing computer is directly linked to the notion of "Ubiquitous Computing" [9], or "Pervasive Computing" [10] as IBM called it later on. Some authors equate "Ubiquitous Computing" and "Pervasive Computing" with "Ambient Intelligence". Here we argue that Ubiquitous and Pervasive systems are different as they

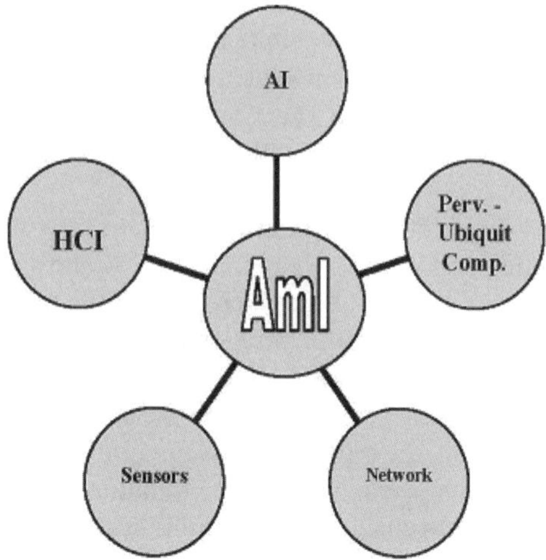


**Fig. 1.** Relationship between AmI and other areas

emphasize the physical presence and availability of resources and miss a key element: the explicit requirement of *"Intelligence"*. This we think, is the ground of Artificial Intelligence (AI) [11] and should not be ignored. Here we refer to AI in a broad sense, encompassing areas like agent-based software and robotics. What matters is that AmI systems provide flexibility, adaptation, anticipation and a sensible interface in the interest of human beings. The same observations can be made about alternatives to *"Ubiquitous"* or *"Pervasive"* like the most recent, and less used, term: *"Everyware"* [12].

This paper will be based in a more suitable definition which expands Raffler's [13] to emphasize Intelligence as a fundamental element of an AmI system:

> *"A digital environment that proactively, but sensibly, supports people in their daily lives."*

In order to be sensible, a system has to be intelligent. That is how a trained assistant, e.g. a nurse, typically behaves. It will help when needed but will restrain to intervene unless is necessary. Being sensible demands recognizing the user, learning or knowing her/his preferences and the capability to exhibit empathy with the user's mood and current overall situation.

Although the term Ambient Intelligence will be used in this article to describe this area of research in Europe, the reader should be aware that similar developments on USA and Canada are usually referred as *"Smart Environments"* or *"Intelligent Environments"*. We keep here the European denomination as it emphasizes the intelligence factor of these systems as opposed to the physical infrastructure.

Important for Ubiquitous/Pervasive computing are the "5Ws" (Who, Where, What, When and Why) principle of design [14] :

**Who:** the identification of a user of the system and the role that user plays within the system in relation to other users. This can be extended to identifying other important elements like pets, robots and objects of interest within the environment.

**Where:** the tracking of the location where a user or an object is geographically located at each moment during the system operation. This can demand a mix of technologies, for example technology that may work well indoors may be useless outdoors and viceversa.

**When:** the association of activities with time is fundamental to build a realistic picture of a system's dynamic. For example, users, pets and robots living in a house will change location very often and knowing when those changes happened and for how long they lasted are fundamental to the understanding of how an environment is evolving.

**What:** the recognition of activities and tasks users are performing is fundamental in order to provide appropriate help if required. The multiplicity of possible scenarios that can follow an action makes this very difficult. Spatial and temporal awareness help to achieve task awareness.

**Why:** the capability to infer and understand intentions and goals behind activities is one of the hardest challenges in the area but with no doubt a fundamental one which allows the system to anticipate needs and serve users in a sensible way.

An important aspect of AmI has to do with interaction. On one side there is a motivation to reduce the human-computer interaction (HCI) [15] as the system is supposed to use its intelligence to infer situations and user needs from the recorded activities, as if a passive human assistant were observing activities unfold with the expectation to help when (and only if) required. On the other hand, a diversity of users may need or voluntarily seek direct interaction with the system to indicate preferences, needs, etc. HCI has been an important area of computer science since the inception of computing as an area of study. Today, with so many gadgets incorporating computing power of some sort, HCI continues to thrive as an important area. An example of an attempt to conciliate both worlds as been reported at: [16] where image processing is done locally inside the context where images are gathered and then a text-based summary is used for diagnosis of the situation. This allows the use of a rich source of information whilst at the same time retaining privacy.

## 3 Smart Homes

An example of an environment enriched with AmI is a "Smart Home". See for example Figure 2 for a depiction of a basic layout and [17,18] for more technical details on how this Smart Homes can operate intelligently.

By Smart Home here we understand a house equipped to bring advanced services to its users. Naturally, how smart a house should be to qualify as a Smart Home is, so far, a subjective matter. For example, a room can have a sensor to decide when its occupant is in or out and on that basis keep lights on or off. However, if sensors only rely on movement and no sensor in, say, the door can detect when the person left, then a person reading and keeping the body in a resting position can confuse the system which will leave the room dark. The system will be confusing absence of movement with absence

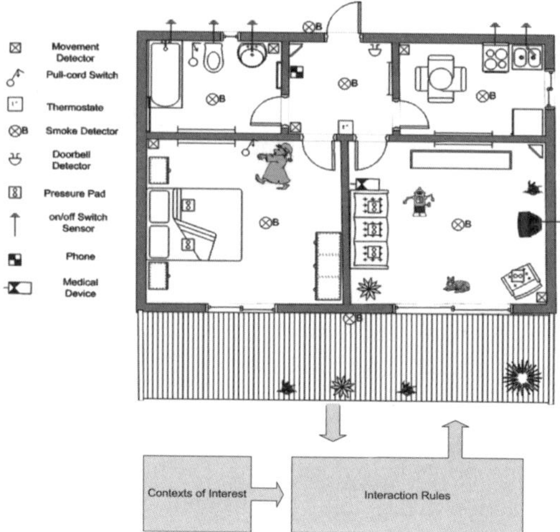


**Fig. 2.** A Smart Home as an AmI instance

of the person, that inference will certainly not be considered as particularly "bright", despite the lights.

Technology available today is rich. Several artifacts and items in a house can be enriched with sensors to gather information about their use and in some cases even to act independently without human intervention. Some examples of such devices are electrodomestics (e.g., cooker and fridge), household items (e.g., taps, bed and sofa) and temperature handling devices (e.g., air conditioning and radiators). Expected benefits of this technology can be: (a) increased safety (e.g., by monitoring lifestyle patterns or the latest activities and providing assistance when a possibly harmful situation is developing), (b) comfort (e.g., by adjusting temperature automatically), and (c) economy (e.g., controlling the use of lights). There is a plethora of sensing/acting technology, ranging from those that stand alone (e.g., smoke or movement detectors), to those fitted within other objects (e.g., a microwave or a bed), to those that can be worn (e.g., shirts that monitor heart beat). For more about sensors and their applications the reader may like to consider [1], and [2].

Recent applications include the use of Smart Homes to provide a safe environment where people with special needs can have a better quality of life. For example, in the case of people at early stages of senile dementia (the most frequent case being elderly people suffering from Alzheimer's disease) the system can be tailored to minimize risks and ensure appropriate care at critical times by monitoring activities, diagnosing interesting situations and advising the carer. There are already many ongoing academic research projects with well established Smart Homes research labs in this area, for example Domus [19], Aware Home [20], MavHome [21], and Gator Tech Smart Home [22].

## 4 Other Environments and Applications for AmI

Other applications are also feasible and relevant and the use of sensors and smart devices can be found in:

- Health-related applications. Hospitals can increase the efficiency of their services by monitoring patients' health and progress by performing automatic analysis of activities in their rooms. They can also increase safety by, for example, only allowing authorized personnel and patients to have access to specific areas and devices.
- Public transportation sector. Public transport can benefit from extra technology including satellite services, GPS-based spatial location, vehicle identification, image processing and other technologies to make transport more fluent and hence more efficient and safe.
- Education services. Education-related institutions may use technology to track students progression on their tasks, frequency of attendance to specific places and health related issues like advising on their diet regarding their habits and the class of intakes they opted for.
- Emergency services. Safety-related services like fire brigades can improve the reaction to a hazard by locating the place more efficiently and also by preparing the way to reach the place in connection with street services. The prison service can also quickly locate a place where a hazard is occurring or is likely to occur and prepare better access to it for security personnel.
- Production-oriented places. Production-centred places like factories can self-organize according to the production/demand ratio of the goods produced. This will demand careful correlation between the collection of data through sensors within the different sections of the production line and the pool of demands via a diagnostic system which can advice the people in charge of the system at a decision-making level.

Well-known leading companies have already invested heavily in the area. For example, Philips [23] has developed Smart Homes for the market including innovative technology on interactive displays. Siemens [24] has invested in Smart Homes and in factory automation. Nokia [25] also has developments in the area of communications where the notion of ambience is not necessarily restricted to a house or a building. VTT [26] has developed systems which advise inhabitants of Smart Homes on how to modify their daily behaviour to improve their health.

In the next section we give one step in the direction of identifying some of the important issues and how to consider them explicitly within a system.

## 5 System Flow

An AmI system can be built in many ways. Mainly they will need sensors and devices to surround occupants of an environment with technology (we can call this an "*e-bubble*") that can provide accurate feedback to the system on the different contexts which are continuously developing. The information collected has to be transmitted by a network and pre-processed by what is called middleware. Finally, in order to make decision-making easier and more beneficial to the occupants of the environment the will have a higher-level layer of reasoning which will accomplish diagnosis and advice or assist other humans which have the final responsibility on the operation of the system. Some elements that may be included are for example an Active Database where the events are collected to record sensors that have been stimulated and a reasoner which will apply spatio-temporal reasoning and other techniques to take decisions [27]. A typical information flow for AmI systems is depicted in Figure 3.

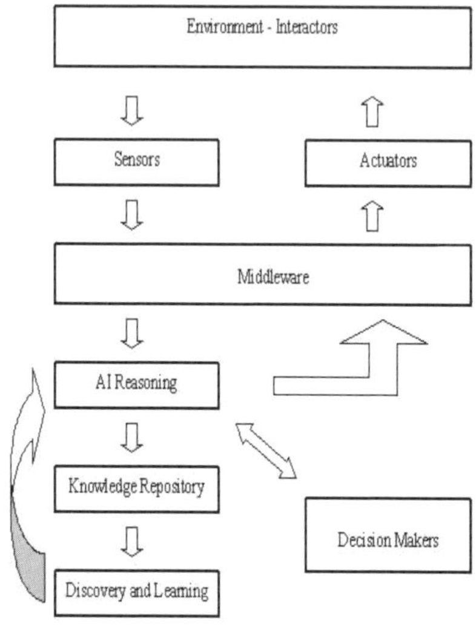


**Fig. 3.** Information flow in AmI systems

As the interactors perform their tasks, some of these tasks will trigger sensors and those in turn will activate the reasoning system. Storing frequency of activities and decisions taken during relevant parts of the system's life time allow the system to learn information which is useful to decision makers, e.g., for doctors and nurses to decide if a change in the medication of a patient suffering Alzheimer's disease may be needed. It also allows learning which can improve the system itself, e.g., to make interaction rules more personalized and useful for a particular person. For example, peoples' habits in winter are different than in summer in terms of what is the usual time to get up or the time they spend watching TV or sleeping.

Lets examine in the following section what the possible intelligent environments can be. The reader more interested in a formal treatment of AmI concepts is referred to [28] where some of the following scenarios are formalized with regards to an abstract AmI architecture.

## 6 AmI Scenarios

AmI systems with the general architecture described in the previous section can be deployed in many possible environments. Below we describe some of these environments in order to better illustrate the scope of the idea.

*Scenario 1:* An instance of the concept of Ambient Intelligence is a Smart Home. Here an AmI specification may include the following details. The meaningful environment is the house, including the backyard and a portion of the front door as these areas also have sensors. Objects are plants, furniture, and so on. Figure 2 have three interactors depicted and therefore $I$ has three elements: a person in the bedroom, a cat, and a floor cleaning robot in the living room. There are also multiple sensors in $S$, movement sensors, pull cord switch, smoke detector, doorbell detector, pressure pad, plus switch sensors for taps, a cooker and a TV. In addition, there is a set of actuators, as the taps, cooker and TV also have the capacity to be turned on and off without human assistance. Medical devices can also exhibit autonomous behaviour by making recommendations before and after their usage. Contexts of interest can be "cooker is left on without human presence in the kitchen for more than 10 minutes", "occupant is still sleeping after 9AM". Interaction rules specified may consider that "if occupant is in bed and is later than 9AM and contact has been attempted unsuccessfully then carer should be notified".

*Scenario 2:* Let us consider a specific room of a hospital as the environment, whit a patient monitored for health and security reasons. Objects in the environment are furniture, medical equipment, specific elements of the room like a toilet and a window. Interactors in this environment will be the patient, relatives and carers (e.g., nurses and doctors). Sensors can be movement sensors and wrist band detectors for identifying who is entering or leaving the room and who is approaching specific areas like a window or the toilet. Actuators can be microphones within the toilet to interact with the patient in an emergency. Contexts of interest can be "the patient has entered the toilet and has not returned after 20 minutes" or "frail patient left the room". Interaction rules specified in $IR$ can consider, for example, that "if patient is leaving the room and status indicates that this is not allowed for this particular patient then nurses should be notified".

*Scenario 3:* Assume a central underground coordination station is equipped with location sensors to track the location of each unit in real-time. Based on the time needed to connect two locations with sensors, the system can also predict the speed of each unit. Examples of objects in this environment are tracks and stations. Interactors are trains, drivers and command centre officers. Sensors are used for identification purposes based on ID signals sent from the train. Other signals can be sent as well, e.g., emergency status. Actuators will be signals coordinating the flow of trains and messages that can be delivered to each unit in order to regulate their speed and the time they have to spend at a stop. Contexts of interest can be "delays" or "stopped train". One interaction rule can be "if line blocked ahead and there are intermediate stops describe the situation to passengers".

*Scenario 4:* Lets assume a school where students are monitored to best advise on balancing their learning experience. The objects within a classroom or play ground are tables and other available elements. The interactors are students and teachers. The sensors will identify who is using what scientific kit and that in turn will allow monitoring of how long students are involved with a particular experiment. Actuators can be recommendations delivered to wristwatch-like personalized displays. Contexts of interest can be "student has been with a single experimentation kit for too long" or "student has not engaged in active experimentation". The first context will trigger a rule "if student has been interacting with one single kit for more than 20 minutes advise the student to try the next experiment available" whilst the second one can send a message to a tutor, such as "if student S has not engaged for more than 5 minutes with an experiment then tutor has to encourage and guide S".

*Scenario 5:* When a fire brigade has to act then the environment can be a city or a neighborhood. Streets can be equipped with sensors to measure passage of traffic within the areas through which the fire brigade truck might go through in order to reach the place where the emergency is located. Objects here will be streets and street junctions. Interactors will be cars. Actuators can be traffic lights as they can help speed the fire brigade through. A context will be a fire occurring at peak time with a number of alternative streets to be used. An interaction rule can be "if all streets are busy, use traffic lights to hold traffic back from the vital passage to be used".

*Scenario 6:* If a production line is the environment then different sensors can track the flow of items at critical bottlenecks in the system and the system can compare the current flow with a desired benchmark. Decision makers can then take decisions on how to proceed and how to react to the arrival of new materials and to upcoming demands. Different parts of the plant can be de/activated accordingly. Similarly, sensors can provide useful information on places where there has been a problem and the section has stopped production, requiring a deviation in flow. Objects here are transportation belts and elements being manufactured whilst actuators are the different mechanisms dis/allowing the flow of elements at particular places. A context can be "a piece of system requiring maintenance" and a related interaction rule can be "if section A becomes unavailable then redirect the flow of objects through alternative paths".

## 7 Are We There Yet...?

A variety of technology that can be deployed and distributed along different environments is being produced. People and organizations are opening to this transformation. Computing, after five decades of unrelenting growth, is in the position to offer systems that will permeate people's daily life as never before.

However, this branch of science has already experienced the pain caused by rushed expectations. Remember AI in the 60s and the AI winter? And then Software Engineering had its ups and downs as well. Despite good success in achieving techniques [29] and tools [30] to increase the reliability of software, major disasters occur from time to time with disastrous consequences for people and companies counted in deaths, injured and multi-millon losses.

Given that in AmI systems people is the main beneficiary (but also mainly affected when the system does not deliver as expected) previous lessons learnt should be considered carefully and enough preparation should be done before widespread use occur.

Looking backwards to how systems have been developed and witnessing the commercial success of faulty systems driven be effective marketing, it is quite likely that systems will be developed unsystematically and deployed prematurely. AmI systems are different to previous one and need different methods and tools to flourish.

## 8 Conclusions

In this chapter, we have reviewed the notion of Ambient Intelligence and associated emerging areas within computer science. We highlighted that an essential component of the area is the distribution of technology intelligently orchestrated to allow an environment to benefit its users. We illustrated the concept by describing briefly a number of different areas of possible application. We expanded in what currently is the driving force of AmI: Smart Homes.

Although the area is very new it has attracted significant attention, sometimes under different names like "intelligent ubiquitous systems" or "intelligent environments". An indication of this is that there is a good number (rapidly increasing) of scientific events, books published, commercial exhibitions and governmental projects being launched every year.

AmI has a strong emphasis on forcing computing to make an effort to reach and serve humans. This may sound the obvious expectation from computing systems but the reality is that so far humans have to do the effort to specialize themselves in order to enjoy the advantages of computing. It is expected that enforcing this requirement at the core of the area will constitute a major driving force and a turning point in the history of computer science. The technological infrastructure seems to be continuously evolving in that direction, and there is a fruitful atmosphere on all sides involved: normal users/consumers of technology, technology generators, technology providers and governmental institutions, that this paradigm shift is needed and feasible.

Still, achieving that capability is far from easy and certainly is not readily available at the moment. The short history of computer science is full of problems which turned to be harder than expected and there is plenty of examples of important systems that crashed. The very fact that makes AmI systems strong can be also their more serious weakness. If humans are put at the centre of the system and made more dependant on the technological environment (we called this an e-bubble), reliability on that e-bubble will be at the level of safety critical systems.

Since these systems are autonomous and proactive, predictability and reliability should not be underestimated if we want the environments where we live and work to be helpful and safe.

## References

1. Want, R.: Rfid – a key to automating everything. Scientific American 290, 46–55 (2004)
2. Nugent, C.D., Augusto, J.C.(eds.): Proceedings of 4th International Conference On Smart homes and health Telematic (ICOST 2006). Smart Homes and Beyond. Assistive Technology Research Series. IOS Press, Belfast, UK (in print, 2006)
3. Augusto, J.C., Shapiro, D. (eds.): Proceedings of the 2nd Workshop on Artificial Intelligence Techniques for Ambient Intelligence (AITAmI 2007). Co-located event of IJCAI 2007. IJCAI, Hyderabad, India (2007)
4. Philips: Homelab (2007), http://www.research.philips.com/technologies/misc/homelab/
5. IST Advisory Group: The european union report, scenarios for ambient intelligence in 2010 (2001), ftp://ftp.cordis.lu/pub/ist/docs/istagscenarios2010.pdf

6. Augusto, J., Cook, D.: Ambient Intelligence: applications in society and opportunities for AI. IJCAI, Hyderabad, India. Lecture Notes for the tutorial given during 20th International Joint Conference on Artificial Intelligence (IJCAI 2007) (2007)
7. Weiser, M.: The computer for the twenty-first century. Scientific American 165, 94–104 (1991)
8. Streitz, N., Nixon, P.: Special issue on 'the disappearing computer. In: Communications of the ACM, vol. 48(3), pp. 32–35. ACM Press, New York (2005)
9. Weiser, M.: Hot topics: Ubiquitous computing. IEEE Computer 26, 71–72 (1993)
10. Saha, D., Mukherjee, A.: Pervasive computing: A paradigm for the $21^{st}$ century. IEEE Computer 36, 25–31 (2003)
11. Russell, S.J., Norvig, P.: Artificial Intelligence: A Modern Approach, 2nd edn. Prentice-Hall, Englewood Cliffs (2003)
12. Greenfield, A.: Everyware: The Dawning Age of Ubiquitous Computing. Peachpit Press (2006)
13. Raffler: Other perspectives on ambient intelligence (2006) www.research.philips.com/password/archive/23/pw23_ambintel_other.html
14. Brooks, K.: The context quintet: narrative elements applied to context awareness. In: Proceedings of the International Conference on Human Computer Interaction (HCI 2003), Erlbaum Associates, Inc, Mahwah (2003)
15. Dix, A., Finlay, J., Abowd, G.D., Beale, R.: Human–Computer Interaction, 3rd edn. Prentice-Hall, Englewood Cliffs (2003)
16. Augusto, J., McCullagh, P., McClelland, V., Walkden, J.-A.: Enhanced Healthcare Provision Through Assisted Decision-Making in a Smart Home Environment. In: Proceedings of the 2nd Workshop on Artificial Intelligence Techniques for Ambient Intelligence (AITAmI 2007), IJCAI, pp. 27–32 (2007)
17. Cook, D.J., Das, S. (eds.): Smart Environments: Technologies, Protocols and Applications. John Wiley and Sons, Chichester (2004)
18. Augusto, J., Nugent, C.: Designing Smart Homes: the role of Artificial Intelligence. Springer, Heidelberg (2006)
19. Pigot, H., Mayers, A., Giroux, S., Lefebvre, B., V. Rialle, N.N.: Smart house for frail and cognitive impaired elders (2002)
20. Abowd, G.A., Bobick, I., Essa, E., Mynatt, W.: The aware home: Developing technologies for successful aging (2002)
21. Cook, D.: Health monitoring and assistance to support aging in place. JUCS 12, 15–29 (2006)
22. Helal, A., Mann, W., Elzabadani, H., King, J., Kaddourah, Y., Jansen, E.: Gator tech smart house: A programmable pervasive space. IEEE Computer magazine, 64–74 (2005)
23. Philips (2006), www.research.philips.com/technologies/syst_softw/ami/background.html
24. Siemens (2006), networks.siemens.de/smarthome/en/index.htm
25. Nokia (2006), research.nokia.com/research/projects/sensorplanet/index.html
26. VTT (2006), www.vtt.fi/uutta/2006/20060602.jsp
27. Augusto, J., Nugent, C.: A new architecture for smart homes based on adb and temporal reasoning. In: Zhang, D., Mokhtari, M. (eds.) Toward a Human Friendly Assistive Environment (Proceedings of 2nd International Conference On Smart homes and health Telematic), September 15-17, 2004. Assistive Technology Research Series, vol. 14, pp. 106–113. IOS Press, Amsterdam (2004)
28. Augusto, J.C.: Ambient intelligence: The confluence of pervasive computing and artificial intelligence. Springer, Heidelberg (2007)
29. ACM-Pnueli (1996), www.acm.org/announcements/turing.html
30. ACM-SPIN (2001), www.acm.org/announcements/ss2001.html

# How to Quantify Quality: Finding Scales of Measure

Tom Gilb

Norway
Tom@Gilb.com

**Abstract.** Quantification is key to controlling system performance attributes. This paper describes how to quantify performance requirements using Planguage – a specification language developed by the author. It discusses in detail how to develop and use *tailored* scales of measure.

## 1 Finding and Developing Scales of Measure and Meters

The basic advice for identifying and developing scales of measure and meters (practical methods for measuring) for scalar attributes is as follows:

1. Try to re-use previously defined Scales and Meters. *Examples [1], [3].*

2. Try to modify previously defined Scales and Meters.

3. If no existing Scale or Meter can be reused or modified, use common sense to develop innovative home-grown quantification ideas.

4. Whatever Scale or Meter you start off with, you must be prepared to learn. Obtain and use early feedback, from colleagues and from field tests, to redefine and improve your Scales and Meters.

### 1.1 Reference Library for Scales of Measure

'Reuse' is an important concept for, sharing experience and saving time when developing Scales. You need to build reference libraries of your 'standard' scales of measure.

Tag: <assign a tag name to this Scale>.

Version: <date of the latest version or change>.

Owner: <role/email of who is responsible for updates/changes>.

Status: <Draft, SQC Exited, Approved>.

Scale: <specify the Scale with defined [qualifiers]>.

Alternative Scales: <reference by tag or define other Scales of interest as alternatives and supplements>.

Qualifier Definitions: <define the scale qualifiers, like 'for defined [Staff]', list their options, like {Nurse, Doctor, Orderly}>.

Meter Options: <suggest Meter(s) appropriate to the Scale>.

Known Usage: <reference projects & specifications where this Scale was actually used in practice with designers' names>.

Known Problems: <list known or perceived problems with this Scale>.

Limitations: <list known or perceived limitations with this Scale>.

J. Filipe, B. Shishkov, and M. Helfert (Eds.): ICSOFT 2006, CCIS 10, pp. 27–36, 2008.

Remember to maintain details supporting each 'standard' Scale, such as Source, Owner, Status and Version (Date). If the name of a Scale's designer is also kept, you can probably contact them for assistance and ideas. Here is a template for keeping reusable scales of measure.

*Example: This is a draft template, with <hints>, for specification of scales of measure in a reference library. Many of the terms used here are defined in Competitive Engineering [2] & [3]. See example below for sample use of this template.*

Tag: Ease of Access.

Version: 11-Aug-2003.

Owner: Rating Model Project (Bill).

Scale: Speed for a defined [Employee Type] with defined [Experience] to get a defined [Client Type] operating successfully from the moment of a decision to use the application.

Alternative Scales: None known yet.

Qualifier Definitions:

Employee Type: {Credit Analyst, Investment Banker, …}.

Experience: {Never, Occasional, Frequent, Recent}.

Client Type: {Major, Frequent, Minor, Infrequent}.

Meter Options:

Test all frequent combinations of qualifiers at least twice. Measure speed for the combinations.

Known Usage: Project Capital Investment Proposals [2001, London].

Known Problems: None recorded yet.

Limitations: None recorded yet.

*Example of a 'Scale' specification for a Scale reference library. This exploits the template in the previous example.*

### 1.2 Reference Library for Meters

Another important standards library to maintain is a library of 'Meters.' Meters support scales of measure by providing practical methods for actually measuring the numeric Scale values. 'Off the shelf' Meters from standard reference libraries can save time and effort since they are already developed and are more or less 'tried and tested' in the field.

It is natural to reference suggested Meters within definitions of specific scales of measure (as in the template and example above). Scales and Meters belong intimately together.

## 2 Managing 'What' You Measure

It is a well-known paradigm that you can manage what you can measure. If you want to achieve something in practice, then quantification, and later measurement, are essential first steps for making sure you get it. If you do not make critical performance attributes measurable, then it is likely to be less motivating for people to find ways to deliver the necessary performance levels. They have no clear targets to work towards, and there are no precise criteria for judgment of failure or success.

## 3 Practical Example: Scale Definition

'User-friendly' is a popular term. Can you specify a scale of measure for it? Here is my advice on how to tackle developing a definition for this.

1. If we assume there is no 'off-the-shelf' definition that could be used (in fact there are, see [1] and [3]):

- Be more specific about the various aspects of the quality. There are many distinct dimensions of quality for user-friendly such as 'user acceptance', 'user training', 'user errors made', 'user customization' and 'environmentally friendly'. List about 5 to 15 aspects of some selected quality that is critical to your project.
- For this example, let's select 'environmentally friendly' as the one of many aspects that we are interested in, and we shall work on this below as an example.

2. Invent and specify a Tag: 'Environmentally Friendly' is sufficiently descriptive. Ideally, it could be shorter, but it is very descriptive left as it is. Let's indicate a 'formally defined concept' by capitalizing the tag.

> *Tag: Environmentally Friendly.*

Note, we usually don't explicitly specify 'Tag:', but this sometimes makes the tag identity clearer.

3. Check there is an Ambition statement, which briefly describes the level of requirement ambition. 'Ambition' is a defined Planguage parameter. More parameters follow, below.

> *Ambition: A high degree of protection, compared to competitors, over the short-term and the long-term, in near and remote environments for health and safety of living things.*

4. Ensure there is general agreement by all the involved parties with the Ambition definition. If not, ask for suggestions for modifications or additions to it. Here is a simple improvement to my initial Ambition statement. It actually introduces a 'constraint'.

> *Ambition: A high degree of protection, compared to competitors, over the short-term and the long-term, in near and remote environments for health and safety of living things, which does not reduce the protection already present in nature.*

5. Using the Ambition description, define an initial 'Scale' (of measure) that is somehow quantifiable (meaning – you can meaningfully attach a number to it). Consider 'what will be sensed by the stakeholders' if the level of quality changes. What would be a 'visible effect' if the quality improved? My initial, unfinished attempt, at finding

a suitable 'Scale' captured the ideas of change occurring, and of things getting 'better or worse':

> *Scale: The % change in positive (good environment) or negative directions for defined [Environmental Changes].*

*My first Scale parameter draft, with a single scalar variable.*

However, I was not happy with it, so I made a second attempt. I refined the Scale by expanding it to include the ideas of specific things being effected in specific places over given times:

> *Scale: % destruction or reduction of defined [Thing] in defined [Place] during a defined [Time Period] as caused by defined [Environmental Changes].*

*This is the second Scalar definition draft with four scalar variables. These will be more-specifically defined whenever the Scale is applied in requirement statements such as 'Goal'.*

This felt better. In practice, I have added more [qualifiers] into the Scale, to indicate the variables that must be defined by specific things, places and time periods whenever the Scale is used.

6. Determine if the term needs to be defined with several different scales of measure, or whether one like this, with general parameters, will do. Has the Ambition been adequately captured? To determine what's best, you should list some of the possible sub-components of the term (that is, what can it be broken down into, in detail?). For example:

> *Thing: {Air, Water, Plant, Animal}.*
>
> *Place: {Personal, Home, Community, Planet}.*
>
> *Thing: = {Air, Water, Plant, Animal}.*
>
> *Place: Consists of {Personal, Home, Community, Planet}.*

*Definition examples of the scale qualifiers used in the examples above. The first example means: 'Thing' is defined as the set of things Air, Water, Plan and Animal (which, since they are all four capitalized, are themselves defined elsewhere). Instead of just the colon after the tag, the more explicit Planguage parameter 'Consists Of' or '=' can be used to make this notation more immediately intelligible to novices in reading Planguage.*

Then consider whether your defined Scale enables the performance *levels* for these sub-components to be expressed. You may have overlooked an opportunity, and may want to add one or more qualifiers to that Scale. For example, we could potentially add the scale qualifiers '*... under defined [Environmental Conditions] in defined [Countries]...*' to make the scale definition even more explicit and more general.

Scale qualifiers (such as ...*'defined [Place]'*...) have the following advantages:

- they add clarity to the specifications
- they make the Scales themselves more reusable in other projects

• they make the Scale more useful in this project: specific benchmarks, targets and constraints can be specified for any interesting combination of scale variables (such as, 'Thing = Air').

7. Start working on a 'Meter' – a specification of how we intend to test or measure the performance of a real system with respect to the defined Scale. Remember, you should first check there is not a standard or company reference library Meter that you could use.

Try to imagine a practical way to measure things along the Scale, or at least sketch one out. My example is only an initial rough sketch.

*Meter: {scientific data where available, opinion surveys, admitted intuitive guesses}.*

*This Meter specification is a sketch defined by a {set} of three rough measurement concepts. These at least suggest something about the quality and costs with such a measuring process. The 'Meter' must always explicitly address a particular 'Scale' specification.*

The Meter will help confirm your choice of Scale as it will provide evidence that practical measurements can feasibly be obtained on a given scale of measure.

8. Now try out the Scale specification by trying to use it for specifying some useful levels on the scale. Define some reference points from the past (Benchmarks) and some future requirements (Targets and Constraints). For example:

*Environmentally Friendly:*
*Ambition: A high degree of protection, compared to competitors, over the short-term and the long-term, in near and remote environments for health and safety of living things, which does not reduce the protection already present in nature.*
*Scale: % destruction or reduction of defined [Thing] in defined [Place] during a defined [Time Period] as caused by defined [Environmental Changes].*
*============= Benchmarks ==================*
*Past [Time Period = Next Two Years, Place = Local House, Thing = <u>Water</u>]: 20% <- intuitive guess.*
*Record [Last Year, Cabin Well, Thing = Water]: 0% <- declared reference point.*
*Trend [Ten to Twenty Years From Now, Local, Thing = Water]: 30% <- intuitive. "Things seem to be getting worse."*
*============ Scalar Constraint ==========*
*Fail [End Next Year, Thing = Water, Place = Eritrea]: 0%. "Not get worse."*
*=============== Targets ====================*
*Wish [Thing = Water, Time = Next Decade, Place = Africa]: <3% <- Pan African Council Policy.*
*Goal [Time = After Five Years, Place = <our local community>, Thing = Water]: <5%.*

If this seems unsatisfactory, then maybe I can find another, more specific, scale of measure? Maybe use a 'set' of different Scales to express the measured concept better? See examples below.

Here is an example of a single more-specific Scale:

*Scale: % change in water pollution degree as defined by UN Standard 1026.*

Here is an example of some other and more-specific set of Scales for the 'Environmentally Friendly' example. They are perhaps a complimentary set for expressing a complex *Environmentally Friendly* idea.

*Environmentally Friendly:*

*Ambition: A high degree of protection, compared to competitors, over the short-term and the long-term, in near and remote environments for health and safety of living things, which does not reduce the protection already present in nature.*

*--Some scales of measure candidates – they can be used as a complimentary set --*

*Air: Scale: % of days annually when <air> is <fit for all humans to breath>.*

*Water: Scale: % change in water pollution degree as defined by UN Standard 1026.*

*Earth: Scale: Grams per kilo of toxic content.*

*Predators: Scale: Average number of <free-roaming predators> per square km, per day.*

*Animals: Scale: % reduction of any defined [Living Creature] who has a defined [Area] as their natural habitat.*

*Many different scales can be candidates to reflect changes in a single critical factor.*

Environmentally Friendly is now defined as a 'Complex Attribute,' because it consists of a number of 'elementary' attributes: {Air, Water, Earth, Predators, Animals}. A different scale of measure now defines each of these elementary attributes. Using these Scales we can add corresponding Meters, benchmarks (such as Past), constraints (such as Fail), and target levels (such as Goal) to describe exactly how Environmentally Friendly we want to be.

**Level of Specification Detail.** How much detail you need to specify, depends on what you want control over, and how much effort it is worth. The basic paradigm of Planguage is you should only elect to do what pays off for you. You should not build a more detailed specification than is meaningful in terms of your project and economic environment. Planguage tries to give you sufficient power of articulation to control both complex and simple problems. You need to scale up, or down, as appropriate. This is done through common sense, intuition, experience and organizational standards (reflecting experience). But, if in doubt, go into more detail. History says we have tended in the past to specify too little detail about requirements. The result consequently has often been to lose control, which costs a lot more than the extra investment in requirement specification.

## 4 Language Core: Scale Definition

This section discusses in more detail the specification of Scales using qualifiers.

**The Central Role of a 'Scale' within Scalar Attribute Definition.** The specified Scale of an elementary scalar attribute is used (re-used!) within all the scalar parameter specifications of the attribute (that is, within all the benchmarks, the constraints and the targets). In other words, a Scale parameter specification is the heart of a specification. A Scale is essential to support all the related scalar parameters: for example Past, Record, Trend, Goal, Budget, Stretch, Wish, Fail and Survival. (As these parameters specify the levels using the Scale.)

Each time a different scalar level parameter is specified, the Scale specification dictates what has to be defined numerically and in terms of Scale Qualifiers (like 'Staff = Nurse'). And then later, each time a scalar parameter definition is read, the Scale specification itself has to be referenced to 'interpret' the meaning of the corresponding scale specification. So the Scale is truly central to a scalar definition. For example '*Goal [Staff = Nurse]* 23%' only has meaning in the context of the corresponding scale: for example '*Scale: % of defined [Staff] attending the operation'*, Well-defined scales of measure are well worth the small investment to define them, to refine them, and to re-use them.

**Specifying Scales using Qualifiers.** The scalar attributes (performance and resource) are best measured in terms of specific times, places and events. If we fail to do this, they lose meaning. People wrongly guess other times, places and events than you intend, and cannot relate their experiences and knowledge to your numbers. If we don't get more specific by using qualifiers, then performance and resource continues to be a vague concept, and there is ambiguity (which times? which places? which events?).

Further, it is important that the set of *different* performance and resource *levels* for different specific time, places and events are identified. It is likely that the levels of the performance and resource requirements will differ across the system depending on such things as time, location, role and system component.

> Decomposing complex performance and resource ideas, and finding market-segmenting qualifiers for differing target levels is a key method for competing for business.

**Embedded Qualifiers within a Scale.** A Scale specification can set up useful qualifiers by declaring embedded scale qualifiers, using the format 'defined [<qualifier>]'. It can also declare default qualifier values that apply by default if not overridden, 'defined [<qualifier>: default: <<u>User-defined Variable</u> or numeric value>]'. For example, [...default: Novice].

**Additional Qualifiers.** However, embedded qualifiers should not stop you adding any other useful additional qualifiers later, as needed, during scale related specification (such as Goal or Meter). But, if you do find you are adding the same type of parameters in almost all related specifications, then you might as well design the Scale to include those qualifiers. A Scale should be built to ensure that it forces the user to define the critical information needed to understand and control a critical performance or resource attribute. This implies that scale qualifiers serve as a checklist of good practice when defining scalar specifications, such as Past and Goal.

Here is an example of how locally defined qualifiers (see the example in the Goal specification) can make a quality specification more specific. In this example we are also going to show how a requirement can be made conditional upon an event. If the event is not true, the requirement does not apply.

First, some *basic definitions* are required:

*Assumption A:* ***Basis*** *[This Financial Year]: Norway is still not a full member of the European Union.*

*EU Trade:* ***Source****: Euro Union Report "EU Trade in Decade 2000-2009".*

*Positive Trade Balance:* ***State*** *[Next Financial Year]: Norwegian Net Foreign Trade Balance has Positive Total to Date.*

*The Planguage parameters {Basis, Source, & State} are in* **bold** *text for readability of this example.*

Now we apply those definitions below:

*Quality A:*

*Type: Quality Requirement.*

*Scale: % by value of Goods delivered that are returned for repair or replacement by consumers.*

*Meter [Development]: Weekly samples of 10,*

*[Acceptance]: 30 day sampling at 10% of representative cases,*

*[Maintenance]: Daily sample of largest cost case.*

*Fail [European Union, Assumption A]: 40% <- European Economic Members.*

*Goal [EU and EEU members, Positive Trade Balance]: 50% <- EU Trade.*

*Some of the user-defined terms used here (like EU Trade) are more fully defined in the example above this one.*

The Fail and the Goal requirements are now defined partly with the help of qualifiers. The Goal to achieve 50% (or more, is implied) is only a valid plan if 'Positive Trade Balance' is true. The Fail level requirement of 40% (or worse, less, is implied) is only valid if 'Assumption A' is true. *All* qualifier conditions must be true for the level to be valid.

## 5 Principles: Scale Specification

Here is a set of principles to help summarise the key points about Scales – see below.

**The Principle of 'Defining a Scale of Measure'**

If you can't define a scale of measure, then the goal is out of control.

*Specifying any critical variable starts with defining its units of measure.*

**The Principle of 'Quantification being Mandatory for Control'**

If you can't quantify it, you can't control it[1].

*If you cannot put numbers on your critical system variables, then you cannot expect to communicate about them, or to control them.*

**The Principle of 'Scales should Control the Stakeholder Requirements'**

Don't choose the easy Scale, choose the powerful Scale.

*Select scales of measure that give you the most direct control over the critical stakeholder requirements. Chose the Scales that lead to useful results.*

**The Principle of 'Copycats Cumulate Wisdom'**

Don't reinvent Scales anew each time – store the wisdom of other Scales for reuse.

*Most scales of measure you will need, will be found somewhere in the literature, or can be adapted from existing literature.*

**The Cartesian Principle**

Divide and conquer said René – put complexity at bay.

*Most high-level performance attributes need decomposition into the list of sub-attributes that we are actually referring to. This makes it much easier to define complex concepts, like 'Usability', or 'Adaptability,' measurably.*

**The Principle of 'Quantification is not Measurement'**

You don't have to measure in order to quantify!

*There is an essential distinction between quantification and measurement.*

*Be clear about one thing. Quantification is not the same as Estimation and Measurement.*

*"I want to take a trip to the moon in nine picoseconds" is a clear requirement specification without measurement."*
*The well-known problems of measuring systems accurately are no excuse for avoiding quantification – Quantification allows us to communicate about how good scalar attributes are or can be – before we have any need to measure them in the new systems.*

[1] Paraphrasing a well-known old saying.

**The Principle of 'Meters Matter'**

Measurement methods give real world feedback about our ideas.

*A 'Meter' definition determines the quality and cost of measurement on a scale; it needs to be sufficient for control and for our purse.*

**The Principle of 'Horses for Courses'**[2]

Different measuring processes will be necessary for different points in time, different events, and different places.[3]

**The Principle of 'The Answer always being '42''**[4]

Exact numbers are ambiguous unless the units of measure are well-defined and agreed.

*Formally-defined scales of measure avoid ambiguity. If you don't define scales of measure well, the requirement level might just as well be an arbitrary number.*

**The Principle of 'Being Sure About Results'**

If you want to be sure of delivering the critical result – then quantify the requirement.

*Critical requirements can hurt you if they go wrong – and you can always find a useful way to quantify the notion of 'going right; to help you avoid doing so.*

## 6 Conclusions

This paper has shown how Planguage specifies performance scales of measure, and how such scales can be used to define benchmarks, targets and constraints. Further discussion and additional information about Planguage can be found in 'Competitive Engineering' [3].

## References

1. Gilb, T.: Principles of Software Engineering Management, p. 442. Addison-Wesley, Reading (1988), ISBN 0-201-19246-2 (See particularly page 150 (Usability) and Chapter 19 Software Engineering Templates)
2. Various free papers, slides, and manuscripts on, `http://www.Gilb.com/`
3. Gilb, T.: Competitive Engineering, p. 474. Elsevier Butterworth-Heinemann, Amsterdam (2005), ISBN 0-7506-6507-6 (An Indian edition is also available)

---

[2] 'Horses' for Courses is UK expression indicating something must be appropriate for use, fit for purpose.

[3] There is no universal static scale of measure. You need to tailor them to make them useful.

[4] Concept made famous in Douglas Adams, The Hitchhiker's Guide to the Galaxy.

# Metamodeling as an Integration Concept

Dimitris Karagiannis and Peter Höfferer

University of Vienna, A-1210 Vienna, Brnner Strae 72, Austria
dk@dke.univie.ac.at, ph@dke.univie.ac.at

**Abstract.** This paper aims to provide an overview of existing applications of the metamodeling concept in the area of computer science. In order to do so, a literature survey has been performed that indicates that metamodeling is basically applied for two main purposes: *design* and *integration*. In the course of describing these two applications we are also going to briefly describe some of the existing work we came across. Furthermore, we provide an insight into the important area of semantic integration and interoperability, whereas we show how metamodels can be brought together with ontologies in this context. The paper is concluded with an outlook on relevant future work in the field of metamodeling.

**Keywords:** Metamodeling, design, semantic integration, semantic interoperability, ontologies.

## 1 Introduction

This paper intends to provide a brief overview on the use of metamodels in different areas of computer science. In order to do so, we chose the empirical method of a literature survey. Alas, current state-of-the-art of search technologies like renown search engines on the Internet are still not able to process requests like "How are metamodels applied in the software engineering community?" producing results with sufficent recall and precision values[1]. Therefore, we had to choose a different approach that shall be briefly described in the following.

We examined the Internet archives of 18 renown journals in different communities of computer science that is to say software engineering, databases, knowledge engineering, and information systems. These high quality journals are published by IEEE, ACM, Springer, Elsevier, or IOSPress – a full list can be found in appendix A in section 5.

When using the provided search facilities of the journals an interesting observation was made. The idea of enhancing computer systems with a better understanding of semantics has been in the spotlight for some years now, just think of the intitiative in the context of the Semantic Web. One of the goals of this initiative is to improve search results in the way to really provide us with the information we need. In order to do so the use of ontology-like constructs (cf. [2], [3]) shall help us to cope with different

[1] Recall is calculated as the ratio of the number of documents retrieved that are relevant compared to the total number of documents that are relevant. Precision on the other hand is the number of documents retrieved that are relevant divided by the total number of documents that are retrieved. [1]

J. Filipe, B. Shishkov, and M. Helfert (Eds.): ICSOFT 2006, CCIS 10, pp. 37–50, 2008.

writings, synonyms, homonyms, and the like. But reality is still different. It is still impossible to use "metamodeling" (American English) as search string also getting hits with the keyword "metamodelling" (British English) and vice versa. Different ways of writing like "metamodeling", "meta-modeling", and "meta modeling" are still resulting in different search results. There is also a long way to go until we are really able to obtain semantically related hits when searching for "metamodeling" that might for example be indexed with "conceptual modeling".

In the end we used "metamodeling, "meta-modeling, "meta modeling" each written once in American and once in British English as well as, "metamodel", "meta-model", and "meta model" as nine different search strings. The actual inquiry was accomplished in July 2006 and provided us with a corpus of 77 articles dealing with our subject of interest whereas no search restrictions concerning the time the papers have been published have been applied. This corpus was used as basis for the further analysis of classification possibilities for metamodel applications.

It has to be stressed here that we concentrated on the tasks that can be handled with metamodels which means investigating their practical *use* to solve real-world problems. We are not talking about a classification of how metamodels can actually be represented. If we would have concentrated on this *design* issue we would have spoken about the application of logical rules or object-orientation and their reproduction in computer systems. We also have not taken into account the actual practical *implementation* of the identified usages which would have implied dealing with service-oriented architectures, databases and the like (see figure 1).

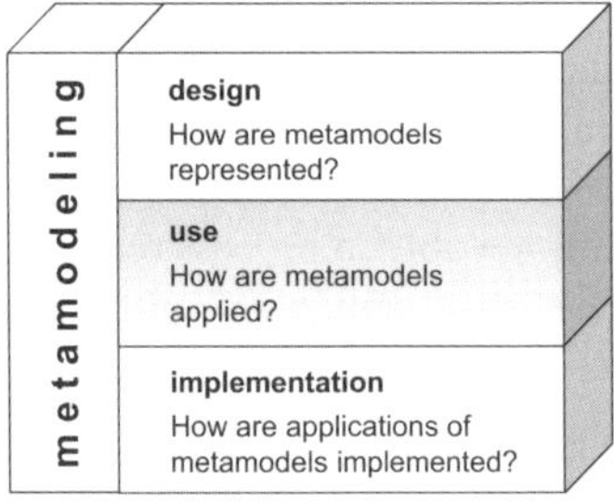


**Fig. 1.** Three different aspects for reasearch work on metamodeling

The remainder of the paper is organized as follows: Section 2 gives a brief overview on modeling and metamodeling in order to ensure a common understanding and to stress the difference between linguistic and ontological metamodeling. Thereafter, in section 3 a simple classification of metamodel applications is presented as well as some of the existing work from literature. Hereby, integration is identified as an important and powerful application of metamodeling and described in more detail. Section 4 continues this topic in showing how semantic aspects of integration and interoperability can be tackled combining metamodels and ontologies. The paper is concluded in section 5 also giving an outlook on important future work.

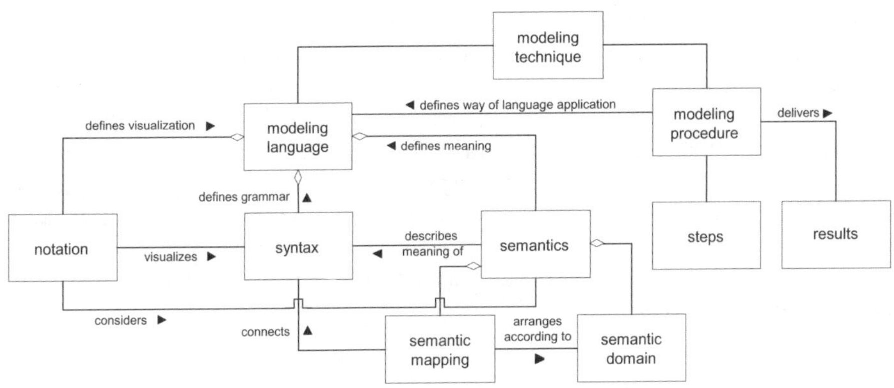


**Fig. 2.** Elements of a modeling technique [8]

## 2 Modeling and Metamodeling

Basically, in the area of computer science models are seen as *"a representation of either reality or vision."* ([4], p. 187) Therefore, they describe things either as they are or as they should be. Of course, this representation is not able to include all aspects of the original but can only focus on some of them (property of reduction) and a model is always intended for a specific purpose (property of pragmatics). [5]

Models can be classified according to the language that is used for their creation. Non-linguistic or iconic models use signs and symbols that have an apparent similarity to the concepts of the real world that are being modeled. Linguistic models on the other hand use basic primitives (i.e. signs, characters, numbers, ...) that do not contain any apparent relationship to the part of reality being modeled except the one that is defined in an explicit way. [6] Nearly all models used in computer science are of the latter linguistic type[2] on which we restrain ourselves hereafter.

The next step before we can talk about metamodels is to clarify how models are actually built. Here the notion of a *modeling technique* comes into play which describes the modeling constructs of a *modeling language* (usually entities, relationships and attributes) and a *modeling procedure* that defines how these constructs have to be combined in order to create a valid model (see figure 2). Following Harel and Rumpe a modeling language now consists of *syntax* which focuses "purely on notational aspects" and *semantics* which defines the meaning [7]. Khn extends this view in that he seperates notation from syntax as he defines *notation* as the *"representation of the elements of the language"* [8]. Syntax then is how the representation elements are allowed to be combined. We consider this distinction as important as it allows for changing only syntax or only notation in the context of method engineering without affecting the other.

A metamodel is most generally defined as being *"a model of models"* [9]. A graphical representation similar to figure 3 is typically used to explain this. On the bottom

[2] Linguistic models can be further distinguished in being realized with textual and graphical/diagrammatic languages [7].

layer 0 there is a subject under consideration that shall be modeled. This is done with the help of a modeling language. For instance, when creating a database for let's say the management of student data we can use the Entity-Relationship modeling technique (ERM, [10]) in order to abstract reality. The available modeling primitives of ERM (i.e. entities, weak entities, relationships with different cardinalities) are described in the metamodel on layer 2 using a meta modeling language. This modeling primitives can be defined by another meta layer, layer 3, which is called meta-meta-layer or meta$^2$-layer containing a meta$^2$-model using a meta$^2$ modeling language. Thus, here the concept of metamodeling is used as a means of language definition. Atkinson and Khne denote this as *linguistic metamodeling* [11].

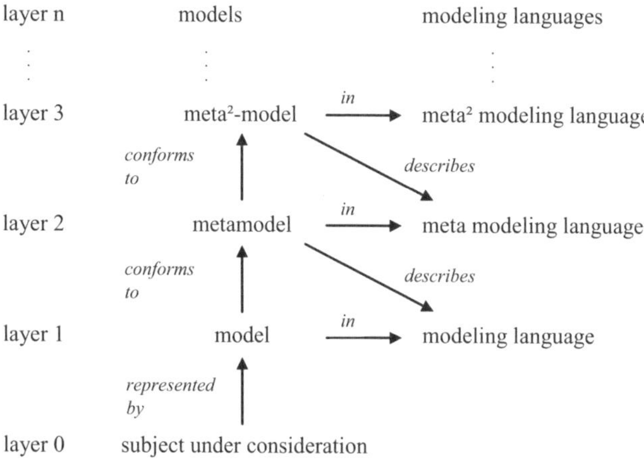


**Fig. 3.** Metamodeling layers (adapted from [6])

But according to these authors this "traditional" point of view on metamodeling covers only one of two important dimensions. It does not explicitely consider that there is not only a linguistic instantiation of concepts like *student* as an instance of *entity*, for example, but also an ontological one as *student* is a *person*, too. In linguistic metamodeling *student* and *person* are situated on the same layer, whereas from an ontological point of view *person* would be on a meta layer. This aspect is called *ontological metamodeling* whereas Atkinson and Khne emphasize that sophisticated metamodeling environments should give equal importance to both indentified metamodeling dimensions. We will make use of these two different metamodeling aspects in section 4.

Now that the basics of models and metamodeling have been recapitulated it is time to proceed to our approach for a classification of metamodeling applications.

## 3 Metamodels in Action: Design and Integration

When we started our literature survey we wondered whether we could find different typical metamodel applications for different areas of computer science. That would be to say that, for instance, the knowledge engineering community is using metamodels in one special way whereas let's say the software engineering community is making a completely different use. After having examined 77 papers (cf. section 1) we saw that this is not the case. In fact, we realized that metamodels are utilized to solve two fundamental types of tasks that we would like to denote as *design* and *integration*.

*Design* involves the creation of metamodels for both the prescriptive definition of not yet existing as well as the descriptive modeling of already existing "subjects" of interest. As will be shown later we distinguish between macro-level and micro-level design.

*Integration* on the other hand denotes the application of metamodeling for bringing together different existing "artefacts" of potentially various kinds that have been generated using different metamodels. This can now, for instance, mean the integration of heterogeneous data sources or the mapping between graphical/diagrammatic models (layer 1 in figure 3) that are all described by different metamodels. As we consider this use of metamodels as a very important and powerful one we are going to describe it in more detail with the help of figure 4 where an example scenario from the area of business process management is depicted.

Here on the meta-level three different metamodels are depicted whereas two of them define modeling languages for creating business process models – i.e. the Business Process Management Systems (BPMS) metamodel [12] and the Event Driven Process Chains (EPC) metamodel [13] – and one more for creating working environment models that describe the organizational structure of enterprises. On the bottom layer three corresponding model instances are shown. In this scenario integration now means to find logical correspondances between instances on the model-layer. There are three different aspects that we would like to define here: *transformation* which allows for the conversion of models that are representing the same aspects of reality (e.g. two business process models that are showing exactly the same chain of activities) from one modeling language (i.e. metamodel) to another. In our case a transformation of the BPMS process model into an EPC process model and vice versa could be performed. *Interoperability* is achieved when semantically identical or related process steps can be identified in different processes that can be even carried out in different organizations. Knowing such tasks allows for seamless inter-organizational cooperation or for the identification of efficiency measures, for instance. Finally, with the help of *references* different elements of models that are conform to different metamodels can be linked. "Acivities" from the business process metamodel can, for example, be related to "roles" or "organizational units" from the working environment model.

Prerequisites for being able to establish these three types of relations on the model-layer are corresponding links on the meta-level. In this context we speak about *mapping* and *integration*. Mapping implies the definition of elements of different metamodels that are related somehow. These relations can be of different types: equivalences, subordinate relationships, complementary relationships, and the like. For instance, the metamodel elements "activity" in the BPMS business process metamodel and "function" in the EPC metamodel are equivalent whereas "role" from the working environment metamodel might be related to "activity" using a "carries out"-relation. The identification of such mappings allows for the creation of integrated metamodels which enables the use of the three aspects that have been defined above for the model-level.[3]

It is important to notice that in order to be able to define mapping relationships a common generic meta$^2$-model is needed to which the elements of the different metamodels correspond to. This common meta$^2$-model ensures the comparability of metamodel elements with one another. Thus, whereas *design* can be realized with only one meta-layer we need a meta$^2$-layer for *integration*. Of course, this also implies that we cannot handle integration tasks without having properly defined all involved metamodels on the design level.

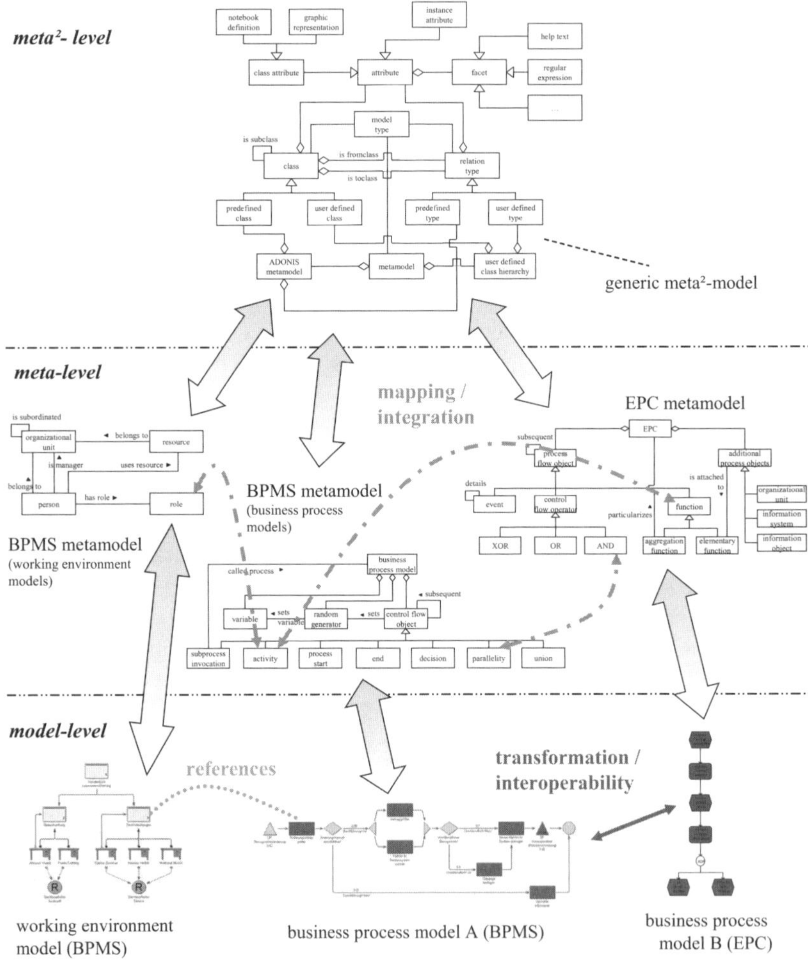


**Fig. 4.** Integration and interoperability using metamodels

[3] Related work concerning the mapping and integration of metamodels can be found in [14] or [15], for instance.

We were able to find examples for these two major use cases of metamodeling in various domains like “data processing”, “knowledge representation”, “requirements engineering”, “information systems”, “business process- & workflow management”, “decision support”, and “business”. In the following, we will describe some existing work we came across during our literature survey.

## 3.1 Design

Considering *design* two different aspects have to be distinguished: On the one hand the metamodeling concept with its abstraction layers (see figure 3) can be used to realize a kind of inheritance mechanism. We call this “micro-level design” as the inner structure of data models, representation languages, or the like is defined here. “Macro-level design” on the other hand generates concrete metamodels (layer 2 in figure 3) that act as templates or reference structures that can be used to deal with a variety of continuative tasks. The macro-perspective metamodels contain a lot of application specific semantics and their implementation is therefore restricted to particular domains while the micro-perspective use is generic.

**Macro-Level Design.** One widespread application of metamodels is top-down as design templates or reference frameworks for certain tasks within a specified domain. The advantage of this use is quite apparent: a commonly accepted understanding of relevant “real-life”-concepts is guaranteed and new model instances can be created in a structured way whereas it is ensured that all relevant aspects are taken into account and nothing of importance is forgotten.

In our literature study we found a wide variety of applications of macro-level design metamodels. In the area of data processing, formal metamodels are, for instance, used to describe basic ETL (extraction-transformation-loading) tasks in the context of the extraction, processing and insertion of operational data in “cleaned” databases of data warehouses. The identified metamodel constructs are hereby used to generate templates (e.g. “domain mismatch”, “fact table”, ...) that can be lined up to compose complex ETL processes ([16], [17]).

Design templates are especially used in the information systems community. Some concrete applications include templates for the implementation of web-based systems [18], an agent-oriented metamodel for organizations and information systems [19], and federated information systems [20].

[21] provide a general metamodel for business processes and [22] define a metamodel for adaptive workflow management systems. As workflow management systems use strict predefined process control structures, they are not that suited for supporting knowledge intensive processes that usually need much more flexibility. For the support of knowledge workers that have to act in such a flexible environment a new type of systems have been proposed: case handling systems that assist rather than strictly guide a user. [23] introduce a metamodel for the cases that are provided by such systems.

Another macro-level design application can be found in [24] where a metamodel of the Bunge-Wand-Weber model (BWWM) is generated which is basically a model for defining requirements when designing and realizing information systems. The original BWWM defines five fundamental and about 30 other constructs (see [25]) and it is argued to be hard to understand because of this complexity. Rosemann and Green see a variety of advantages in their metamodel of the BWWM: it clarifies the understanding, simplifies the communication, is a means for structuring and analyzing, and can finally be used to derive new modeling techniques.

We also came across two metamodels that are directly intended for business use. [26] gives a metamodel for the definition of business rules, and [27] for the creation of eContracts.

**Micro-Level Design.** To recapitulate, micro-level design is concerned with the definition of the inner structure of data models or representation languages. This ability is often used in the context of data processing for enhancing the reflection mechanisms[4] of database management systems. To be more specific, the TIGUKAT object model [28] uses the concept of meta- and meta$^2$-objects in order to define types with specific behavior. This behavior is then passed to the objects of the instance level. In this context metamodeling is used to realize an inheritance mechanism that is known from the area of object-oriented programming, for instance. The FORM data model [29] for representing heterogeneous types of entities and relationships in an organization does a quite similar thing as it uses metamodeling *"to express meta-knowledge that allows a system to enforce generic patterns of object behavior rather than a number of specific actions"*. A recent work proposes a framework for uniform representation of and access to data models, schema, and data. This "uni-level description" (ULD) also makes use of the classic metamodeling layers with one conceptual difference: schema data and instances are each stored on the same meta-layer as compared to earlier approaches that put these on two different meta layers [30].

Another example for micro-level design using metamodeling concepts can be found in the knowledge engineering discipline. Here a knowledge representation language called Telos is introduced that uses meta-layers as classification dimension in order to express "instance of" and "abstraction" relationships [31], [32].

## 3.2 Integration

The basic problem that is dealt with here is how to bring together models (layer 1 of figure 3) that have been realized using different metamodels (layer 2). As Zaniolo has put it *"a direct mapping between different models is a formidable problem"* ([33], p. 33) and is not feasible in practice. Therefore the mapping should rather be realized on the meta-layer or to be more specific at a meta$^2$-layer that acts as "translator" between metamodels that have been instantiated from the same meta$^2$-model. It has to be stressed here that the notion of a meta$^2$-model is sometimes not found literally in work dealing with integration but, nevertheless, the concept of a meta$^2$-layer is used in an implicit way.

This "translator capability" can now be used to integrate heterogeneous data sources. In our document corpus we found three different applications for doing this. [33] realize a schema mapping between multiple data formats. [34] propose a global information resources dictionary (GIRD) that abstracts all classes of enterprise metadata which

[4] Reflection is the ability of a system to manage information about itself and to reason about this information. [28]

is structured in four different categories, that is to say functional models, structural models, software and hardware resources, and enterprise/application families. GIRD is then used to answer global database queries that integrate data of all categories. In the context of enhancing the use of data warehouses by management, [35] integrate data models with enterprise models in order to provide analyses that really correspond to the information needs of the users.

In the areas of information systems and software engineering integration is often used in the context of method engineering. [36], for instance, build one integrated metamodel of object-oriented methodologies concerning all development phases (i.e. analysis, design, and programming). The advantage of this approach is that this single metamodel can also be used to generate object-oriented program code (OOPC) from analysis/design model instances (OOADM) as a mapping between OOADM and OOPC can be established on the metamodel-layer.

Situational method engineering which is *"concerned with the tuning of methods and techniques to the specific characteristics of a certain project"* [37] is an integration application that can be found quite often in literature including the following articles: [38], [39], [37], [40] and [41].

Finally, we came across an application of integration in the requirements engineering domain. [42] explicitely mention the use of a meta$^2$-model to create an integrated view that also enables the computer-based support of team- and goal-oriented analysis methods like the informal method JAD. They also make use of the model transformation capabilities of the meta$^2$-approach that "come for free" in that different requirements models that have been realized in unequal modeling languages (therefore using various metamodels) can be mapped.

It can be seen that metamodels are an adequate means for integration. But so far the described approaches are not able to realize *semantic integration and interoperability* which *"is concerned with the use of explicit semantic descriptions"* [43] most often provided in the form of ontologies [44]. The next section will show how metamodels and ontologies can be combined.

## 4 Semantic Integration and Interoperability Using Metamodels and Ontologies

The topic of integrating data and ensuring the interoperability of information systems is of great practical importance which can already be seen by the fact that according to Gartner up to 40% of the companies' information technology budgets are spent on integration issues [45]. The heterogeneities that have to be dealt with in this context are usually classified to be of syntactical, structural or semantic nature [3])whereas resolving the latter seems to be most laborious as 60-80% of the resources of integration projects are spent on reconciling semantic heterogeneities [46].

In order to be able to overcome the heterogeneities of resources – regardless if they are data, information systems, or anything else – they have to be represented in an adequate way. For this task linguistic, diagrammatic languages (cf. section 2) are often well suited like demonstrated by UML or ERM. These languages together with the concept of metamodeling are able to express syntactical and structural aspects as well as what we would like to denote as *type semantics*. This type semantics is defined through the process of linguistic metamodeling and allows reasoning such as, for instance, *student* is derived from the metamodel construct *entity* and therefore is a kind of real world object and not a relationship. But in this context we are not able to state anything about the semantics of *student* itself. It can by no means be reasoned that this term denotes a human person that can be male or female and who is attending an university-like institution. We would like to call this information *inherent semantics* as it describes a kind of "inner meaning" of modeled resources that is exceeding the type semantics that is being inherited by the elements of the metamodel-layer.

This inherent semantics can now be made explicit by linking model elements representing resources with concepts of ontologies, a process that is called *lifting* [47] or also *ontology anchoring* [38] which is the quintessence of semantic integration. Lifting reflects in our opinion what [11] denote as ontological metamodeling and is of course not limited to the model-layer but can be applied to the meta- or meta$^2$-layer as well. This shall be illustrated in figure 5 using the following example.

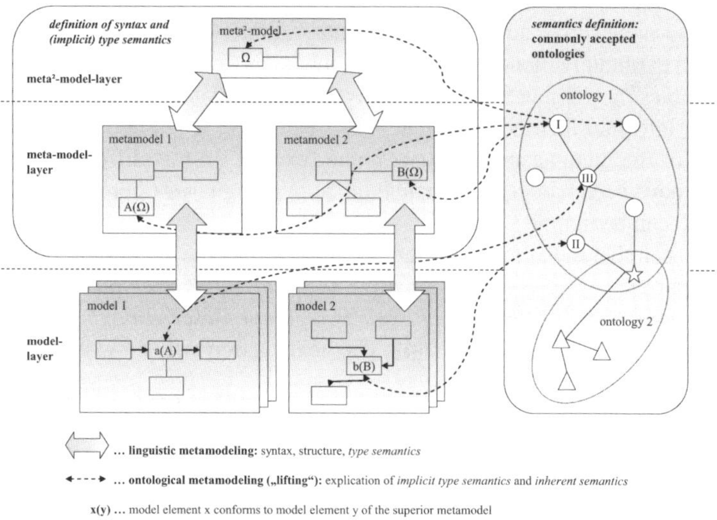


**Fig. 5.** Architecture for semantic interoperability using both metamodels and ontologies

Imagine two models that have been created according to different metamodels and are now to be integrated which means that semantically related model elements have to be found. The "classical" metamodel-based approach would follow the path of linguistic instantiation which means that, for instance, it would be reasoned that meta-classes *A* and *B* are related because they are derived from the same meta$^2$-class $\Omega$ (meta-classes in figure 5 are given in brackets). Then the next step would be to say that model elements *a* and *b* belong together as they are instances of *A* and *B*. To be more specific we could assume *A* and *B* to be meta-classes of performance figures (ontology construct I) and the task would be to sum up all monetary figures (ontology construct II) of models 1 and 2. Ontology construct III now could stand for quantity figures. We see that

$a$ is a quantitiy figure then and should therefore not be added up to $b$. This conclusion can not be drawn with the information provided by the linguistic metamodeling process but only because of the additional information originating from the lifting of the model constructs.

Of course, one could argue now that the distinction between monetary and quantity figures could have been realized on the metamodel-layer as well. Basically, this is true but not really preferable because of a very specific feature of ontologies: They are by definition commonly accepted within communities as they reflect a shared and sometimes even standardized conceptualization[5] compared to metamodels that are often only valid for specific tools or organizations. The advange of lifting (meta-)model concepts to ontology concepts is therefore founded in a reduced mapping complexity. If n metamodels are to be mapped with each other the complexity of $n * (n - 1)/2 \approx O(n^2)$ in the case of bidirectional point-to-point mappings can be reduced to $O(n)$ with one intermediate ontology.

Applying ontologies for semantic markup also allows for making use of all research results in the field of ontology mapping – see [49] or [50] for introducing surveys.

Recapitulatory, we believe that the combination of metamodeling and ontologies provides excellent means to solve the task of extensive integration and interoperability handling all syntactical, structural and semantic heterogeneity. Some related work in this context can be found in [47] who deal with model transformations in the area of software engineering. [51] also utilizes the idea of ontology-based transformation but like in the aforementioned papers the use of lifting is restricted to the metamodel layer. An approach that makes use of lifting on all (meta-)model layers can be found in [52].

## 5 Conclusions and Outlook

In this paper we described the results of a literature survey that aimed for identifying basic applications of the metamodeling concept. In the course of doing so, we found that metamodels are basically used for two main purposes, that is to say *design* and *integration*, and we have described some of the existing work we came across. Furthermore, we delivered an insight into the important field of semantic integration and interoperability showing how metamodels can be enriched with ontology concepts. We are convinced that this approach will greatly enhance *integration and interoperability* both on the conceptual (EMI) as well as on the technical level (EAI).

*Method engineering* for the combination of modeling paradigms is another important metamodel application scenario which will bring together descriptive-, decision support-, and predicative models. *Model-driven business engineering* will help for managing the interdependencies of corporations' elements.

Further research work in the area of metamodels can focus on one of these identified applications. Another option is to elaborate on the two other aspects of metamodeling identified in figure 1: How metamodels are actually realized or how the identified tasks are implemented in practice.

[5] [48] denotes this type of ontologies as *reference ontologies*.

## References

1. Witten, I.H., Frank, E.: Data mining: practical machine learning tools and techniques. Morgan Kaufmann series in data management systems, 2nd edn. Morgan Kaufman, Amsterdam (2005)
2. Garshol, L.M.: Metadata? Thesauri? Taxonomies? Topic Maps! Making sense of it all. Journal of Information Science 30, 378–391 (2004)
3. Obrst, L.: Ontologies for Semantically Interoperable Systems. In: CIKM 2003: Proceedings of the Twelfth International Conference on Information and Knowledge Management, pp. 366–369. ACM Press, New York (2003)
4. Whitten, J.L., Bentley, L.D., Dittman, K.C.: Systems analysis and design methods, 6th edn. McGraw-Hill Irwin, Boston (2004)
5. Stachowiak, H.: Allgemeine Modelltheorie. Springer, Wien (1973)
6. Strahringer, S.: Metamodellierung als Instrument des Methodenvergleichs: eine Evaluierung am Beispiel objektorientierter Analysemethoden. Berichte aus der Betriebswirtschaft. Shaker, Aachen (1996)
7. Harel, D., Rumpe, B.: Modeling Languages: Syntax, Semantics and All That Stuff - Part I: The Basic Stuff (2000)
8. Kühn, H.: Methodenintegration im Business Engineering. PhD thesis, Universität Wien, Wien (2004)
9. Object Management Group: MDA Guide Version 1.0.1 (2003)
10. Chen, P.P.S.: The entity-relationship model - toward a unified view of data. ACM Transactions on Database Systems 1, 9–36 (1976)
11. Atkinson, C., Kühne, T.: Model-Driven Development: A Metamodeling Foundation. IEEE Software 20, 36–41 (2003)
12. Karagiannis, D., Junginger, S., Strobl, R.: Introduction to Business Process Management Systems Concepts. In: Business Process Modelling, pp. 81–106. Springer, Berlin Heidelberg (1996)
13. Keller, G., Nüttgens, M., Scheer, A.W.: Semantische Prozemodellierung auf der Grundlage. Ereignisgesteuerter Prozeketten (EPK) (1992)
14. Kühn, H., Bayer, F., Junginger, S., Karagiannis, D.: Enterprise Model Integration. In: Bauknecht, K., Tjoa, A.M., Quirchmayr, G. (eds.) E-Commerce and Web Technologies, 4th International Conference, EC-Web, Prague, Czech Republic, September 2-5, 2003, pp. 379–392. Springer, Berlin Heidelberg (2003)
15. Kühn, H., Murzek, M.: Interoperability Issues in Metamodelling Platforms. In: Konstantas, D., Bourriéres, J.P., Léonrad, M., Boudjlida, N. (eds.) Interoperability of Enterprise Software and Applications, pp. 216–226. Springer, Heidelberg (2006)
16. Vassiliadis, P., Vagena, Z., Skiadopoulos, S., Karayannidis, N., Sellis, T.K.: ARKTOS: towards the modeling, design, control and execution of ETL processes. Information Systems 26, 537–561 (2001)
17. Vassiliadis, P., Simitsis, A., Georgantas, P., Terrovitis, M., Skiadopoulos, S.: A generic and customizable framework for the design of ETL scenarios. Information Systems 30, 492–525 (2005)
18. Nikolaidou, M., Anagnostopoulos, D.: A Systematic Approach for Configuring Web-Based Information Systems. Distributed and Parallel Databases 17, 267–290 (2005)
19. Wagner, G.: The Agent-Object-Relationship metamodel: towards a unified view of state and behavior. Information Systems 28, 475–504 (2003)
20. Jarke, M., Jeusfeld, M.A., Peters, P., Pohl, K.: Coordinating Distributed Organizational Knowledge. Data & Knowledge Engineering 23, 247–268 (1997)
21. Rolland, C., Souveyet, C., Moreno, M.: An Approach for Defining Ways-of-Working. Information Systems 20, 337–359 (1995)

22. Chiu, D.K.W., Li, Q., Karlapalem, K.: A Meta Modeling Approach to Workflow Management Systems Supporting Exception Handling. Information Systems 24, 159–184 (1999)
23. van der Aalst, W.M.P., Weske, M., Grünbauer, D.: Case handling: a new paradigm for business process support. Data & Knowledge Engineering 53, 129–162 (2005)
24. Rosemann, M., Green, P.: Developing a meta model for the Bunge-Wand-Weber ontological constructs. Information Systems 27, 75–91 (2002)
25. Kayed, A., Colomb, R.M.: Using BWW model to evaluate building ontologies in CGs formalism. Information Systems 30, 379–398 (2005)
26. Herbst, H.: Business Rules in Systems Analysis: a Meta-Model and Repository System. Information Systems 21, 147–166 (1996)
27. Krishna, P.R., Karlapalem, K., Chiu, D.K.W.: An $ER^{EC}$ framework for e-contract modeling, enactment and monitoring. Data & Knowledge Engineering 51, 31–58 (2004)
28. Peters, R.J., Ozsu, M.T.: Reflection in a Uniform Behavioral Object Model. In: Proceedings of the 12th International Conference on Entity-Relationship Approach, pp. 34–45 (1993)
29. Kim, D.H., Park, S.J.: FORM: A Flexible Data Model for Integrated CASE Environments. Data & Knowledge Engineering 22, 133–158 (1997)
30. Bowers, S., Delcambre, L.: Using the uni-level description (ULD) to support data-model interoperability. Data & Knowledge Engineering 59, 511–533 (2006)
31. Mylopoulos, J., Borgida, A., Jarke, M., Koubarakis, M.: Telos: representing knowledge about information systems. ACM Trans. Inf. Syst. 8, 325–362 (1990)
32. Mylopoulos, J.: Conceptual Modeling and Telos. In: Loucopoulos, P., Zicari, R. (eds.) Conceptual Modeling, Databases and Case: an integrated view of information systems development, pp. 49–68. Wiley, New York (1992)
33. Zaniolo, C., Melkanoff, M.A.: A Formal Approach to the Definition and the Design of Conceptual Schemata for Database Systems. ACM Trans. Database Syst. 7, 24–59 (1982)
34. Cheung, W., Hsu, C.: The model-assisted global query system for multiple databases in distributed enterprises. ACM Trans. Inf. Syst. 14, 421–470 (1996)
35. Jarke, M., Jeusfeld, M.A., Quix, C., Vassiliadis, P.: Architecture and Quality in Data Warehouses: An Extended Repository Approach. Information Systems 24, 229–253 (1999)
36. Hillegersberg, J.V., Kumar, K.: Using Metamodeling to Integrate Object-Oriented Analysis, Design and Programming Concepts. Information Systems 24, 113–129 (1999)
37. ter Hofstede, A.H.M., Verhoef, T.F.: On the Feasibility of Situational Method Engineering. Information Systems 22, 401–422 (1997)
38. Brinkkemper, S., Saeki, M., Harmsen, F.: Meta-Modelling Based Assembly Techniques for Situational Method Engineering. Information Systems 24, 209–228 (1999)
39. Dominguez, E., Zapata, M.A.: Noesis: Towards a situational method engineering technique. Information Systems (in press)
40. Beydoun, G., Gonzalez-Perez, C., Low, G., Henderson-Sellers, B.: Synthesis of a generic MAS metamodel. In: ACM SIGSOFT Software Engineering Notes: SELMAS 2005: Proceedings of the fourth international workshop on Software engineering for large-scale multi-agent systems, pp. 1–5. ACM Press, New York (2005)
41. Prakash, N.: On Methods Statics and Dynamics. Information Systems 24, 613–637 (1999)
42. Nissen, H.W., Jarke, M.: Repository Support for Multi-Perspective Requirements Engineering. Information Systems 24, 131–158 (1999)
43. Kalfoglou, Y., Schorlemmer, M., Uschold, M., Sheth, A., Staab, S.: Semantic Interoperability and Integration. In: Kalfoglou, Y., Schorlemmer, M., Sheth, A., Staab, S., Uschold, M. (eds.) Semantic Interoperability and Integration. Internationales Begegnungs- und Forschungszentrum fuer Informatik (IBFI), Dagstuhl, Germany (2005)
44. Alexiev, V., Breu, M., de Bruijn, J., Fensel, D. (eds.): Information Integration with Ontologies: Experiences from an Industrial Showcase. John Wiley & Sons, Chichester (2005)

45. Haller, A., Cimpian, E., Mocan, A., Oren, E., Bussler, C.: WSMX - A Semantic Service-Oriented Architecture. In: IEEE Computer Society(ed.). IEEE International Conference on Web Services (ICWS 2005), pp. 321–328 (2005)
46. Doan, A., Noy, N.F., Halevy, A.Y.: Introduction to the Special Issue on Semantic Integration. SIGMOD Record 33, 11–13 (2004)
47. Kappel, G., Kapsammer, E., Kargl, H., Kramler, G., Reiter, T., Retschitzegger, W., Schwinger, W., Wimmer, M.: On Models and Ontologies - A Layered Approach for Model-based Tool Integration. In: Mayr, H.C., Breu, R. (eds.) Modellierung 2006, pp. 11–27 (2006)
48. Guarino, N.: Formal Ontology and Information Systems. In: Proceedings of FOIS 1998, pp. 3–15. IOS Press, Amsterdam (1998)
49. Kalfoglou, Y., Schorlemmer, M.: Ontology Mapping: The State of the Art. In: Kalfoglou, Y., Schorlemmer, M., Sheth, A., Staab, S., Uschold, M. (eds.) Semantic Interoperability and Integration. Internationales Begegnungs- und Forschungszentrum fuer Informatik (IBFI), Dagstuhl, Germany (2005)
50. Noy, N.F.: Semantic Integration: A Survey Of Ontology-Based Approaches. SIGMOD Record 33, 65–70 (2004)
51. Roser, S., Bauer, B.: Ontology-Based Model Transformation. In: Bruel, J.-M. (ed.) MoDELS 2005. LNCS, vol. 3844, pp. 355–356. Springer, Heidelberg (2005)
52. Terrasse, M.N., Savonnet, M., Leclercq, E., Grison, T., Becker, G.: Do we need metamodels AND ontologies for engineering platforms?. In: GaMMa 2006: Proceedings of the 2006 international workshop on Global integrated model management, pp. 21–28. ACM Press, New York (2006)

## Appendix A

A total of 18 journals has been surveyed whereas 11 delivered search results according to our defined nine search strings (see section 1). The surveyed journals are as follows [number of relevant retrieved documents is given in brackets]:

- Data & Knowledge Engineering [6]
- Expert Systems: The International Journal of Knowledge Engineering and Neural Networks [3]
- IEEE Transactions on Knowledge and Data Engineering (T-KDE) [2]
- ACM Transactions on Information Systems (TOIS; Formerly: ACM Transactions on Office Information Systems) [3]
- Communications of the Association for Information Systems (CAIS) [0]
- Electronic Journal of Information Systems Evaluation (EJISE) [0]
- Information Systems [21]
- Information Systems Research [4]
- Journal of Intelligent Information Systems [0]
- Journal of the Association for Information Systems (JAIS) [0]
- ACM SIGSOFT Software Engineering Notes [10 selected out of 87]
- ACM Transactions on Software Engineering and Methodology (TOSEM) [15]
- Empirical Software Engineering [0]
- IEE Proceedings (sic!) - Software Engineering [0]
- IEEE Transactions on Software Engineering (T-SE) [7]
- VLDB Journal, The - The International Journal on Very Large Databases [0]
- Distributed and Parallel Databases [1]
- ACM Transactions on Database Systems (TODS) [5]

# Engineering Object and Agent Methodologies

B. Henderson-Sellers

University of Technology, Sydney, Australia
brian@it.uts.edu.au

**Abstract.** Method engineering provides an excellent base for constructing situation-specific software engineering methodologies for both object (OO) and agent (AO) software development. Both the OPEN Process Framework (OPF) and the Framework for Agent-oriented Method Engineering (FAME) use an existing repository coupled to an appropriate metamodel (which in the near future will be the new ISO standard metamodel ISO/IEC 24744, itself based on the concept of powertypes). This flexible, yet standardized, repository supplies method fragments that are then configured to support specific projects. In addition, all existing, and new, OO and AO methodologies can be recreated, thus providing an industry strength resource for object-oriented and agent-oriented software development.

## 1 Methodologies and Method Engineering

Methodologies (a.k.a. methods) for software and systems development need to support developers in their endeavours. They need to contain a large number of descriptive/prescriptive elements, including information on appropriate tools and techniques, descriptions of organizational roles, project management advice and an underlying process model [14]. They should support (possibly creating) organizational standards, offering guidance and support, monitoring and control advice such that successful development can be repeated and failures learned from and duplicate mistakes avoided in the future. However, when created in vacuo, they are often not well suited to practice [21] which has led in some authors' views to a backlash [4] or them even being seen as a "waste of time" [5]. Nevertheless, it is worth pursuing the notion that a methodology can offer useful practical support for industry developers but only if the methodology is attuned to a specific organization and/or its projects. Tuning an existing comprehensive or heavyweight methodology can be time-consuming, the alternative being a bottom-up construction of a methodology from small methodological pieces (known as "fragments"). This is called situational method engineering (SME) [20], [23] and will be explored here.

In SME, method fragments [8] are stored in a repository or methodbase [25], [9] and are retrieved on a project-specific basis. These selected fragments are then "glued together" in an appropriate manner to create the situationally-specific methodology.

One such repository of method fragments belongs to the OPEN approach [13], [18], (http://www.open.org.au, http://www.opfro.org). In this paper, we briefly outline this approach, in terms of object-oriented method fragments (Section 2) and then illustrate how this has been more recently extended [15] to accommodate agent-oriented method fragments. Such an extension forms part of the FAME (Framework

J. Filipe, B. Shishkov, and M. Helfert (Eds.): ICSOFT 2006, CCIS 10, pp. 51–58, 2008.

for Agent-oriented Method Engineering) project (Section 3), which not only aims to complete the suite of agent-oriented fragments but also to replace OPEN's metamodel (see Appendix G of [11]) by the new ISO/IEC 24744 SEMDM (Software Engineering Metamodel for Development Methodologies) standard (see e.g. descriptions of early drafts in [17]).

## 2 Existing OO Method Fragments in the Open Repository

The OPEN Process Framework (OPF) [13], [11] is defined by a metamodel that supports the concepts of method engineering (Figure 1). It provides a rich repository of over a thousand method fragments, which can be used in different software projects, together with a set of guidelines offering advice on the fragment selection based on the notion of possibility matrices linking each pair of method fragments [19] – see Figure 2.

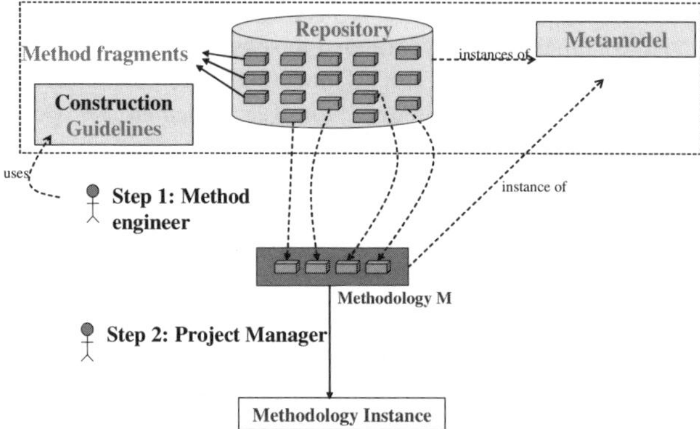


**Fig. 1.** Method engineering using the OPF

| Techniques \ Tasks | | | | |
|---|---|---|---|---|
| | M | D | F | F | F |
| | D | D | F | F | D |
| | D | D | O | O | D |
| | F | O | O | O | F |
| | F | M | O | D | F |
| | R | R | M | R | O |
| | D | R | F | M | O |
| | D | F | M | D | D |
| | R | R | D | R | R |
| | O | D | O | O | R |
| | F | M | O | F | D |

**Fig. 2.** Method fragments are linked using a deontic matrix, here for Tasks and Techniques (showing five levels of possibility from F= Forbidden, through D=Discouraged, O=Optional, R=Recommended to M=Mandatory)

In OPF's metamodel, there are elements to describe process fragments such as Activities, Tasks and Techniques; people components such as Producers and Roles; organizational components, such as Enterprise, Programme and Project, and product fragments in the form of a whole range of Work Products including diagrams and documents, supported by various kinds of languages (natural language, modelling language and coding language). Method construction may be top-down or bottom-up. Using the former as an example, the method engineer would select appropriate Activities from the OPF fragment repository and then, using the possibility (or deontic) matrix approach, choose appropriate Tasks, Techniques, Producers, Work products etc.

# 3 The Fame Project: New AO Method Fragments and a New Metamodel

## 3.1 AO Fragments

In order to add AO fragments to the OPF repository, each of the extant AO methodologies (such as Gaia, Tropos or Prometheus) was analysed and fragments extracted. These fragments were those encapsulating either brand new (AO) concepts (for example, the BDI model [24]) or extensions to existing OO fragments (for example, the Tasks "Determine MAS infrastructure facilities" extends the existing OO Task of "Create a system architecture"). These have been documented in a series of papers, summarized here in Table 1 – a total of 1 new Activity ("Early requirements analysis" – not shown in Table 1), 29 new Tasks, 14 new Subtasks, 23 new Techniques and 29 new Work Products (although these numbers may be reduced when/if overlaps are identified).

When reused (by extraction from the repository), fragments can be selected to recreate any specific AO methodology (e.g. Gaia: [27], Tropos: [7], Prometheus: [22]) or a new hybrid (for example Prometheus enhanced by Tropos - Figure 3, which also shows linkages between OPF OO and AO Task fragments and Technique fragments in the form of a deontic matrix, as shown earlier in Figure 2).

**Table 1.** Summary of Tasks, Techniques and Work Products so far added to OPEN in order to support agent-oriented software developments. [Note that there is no meaning in any horizontal alignments].

| New Tasks | New Techniques | New Work Products |
|---|---|---|
| Assign goals to responsibilities | Activity scheduling | Agent acquaintance diagram |
| Assign and compose roles | Agent delegation strategies | Agent class card |
| Construct agent conversations | Agent internal design | Agent design model |
| Construct the agent model | AND/OR decomposition | Agent overview diagram |
| Define ontologies | Belief revision of agents | Agent protocol diagram |
| Design agent internal structure | Capabilities identification and analysis | Agent structure diagram |
| Determine agent communication protocol | Commitment management | CAMLE behaviour diagram |
| Determine agent interaction protocol | Contract nets | CAMLE scenario diagram |
| Determine delegation strategy | Deliberative reasoning: Plans | Coupling Graph |
| Determine control architecture | Control architecture | Caste diagram |

**Table 1.** (*Continued*)

| **New Tasks** | **New Techniques** | **New Work Products** |
|---|---|---|
| Determine reasoning strategies for agents | Environmental evaluation | Domain knowledge ontology |
| Determine security policy for agents | Environmental resources modelling | Functionality descriptor |
| Determine system operation | FIPA KIF compliant language | Goal hierarchy diagram |
| Gather performance knowledge | Learning strategies for agents | Inference diagram |
| Identify emergent behaviour | Market mechanisms | Network design model |
| Identify system behaviours | Means-end analysis | Platform design model |
| Identify system organization | Organizational rules specification | Protocol schema |
| Model actors | Organizational structure specification | PSM specification |
| Model agent knowledge | Performance evaluation | Role diagram |
| Model agent relationships | Reactive reasoning: ECA rules | Role schema |
| Model agents' roles | Task selection by agents | Service table |
| Model capabilities for actors | 3-layer BDI model | Task hierarchy diagram |
| Model dependencies for actors and goals | | Task knowledge specification |
| Model goals | | Task textual description |
| Model plans | | (Tropos) Actor Diagram |
| Model the agent's environment | | (Tropos) Capability Diagram |
| Specify shared data objects | | (Tropos) Goal Diagram |
| Undertake agent personalization | | (Tropos) Plan Diagram |
| **New Subtasks** | **to the main Task of:** | |
| Define perceptor module<br>Define actuator module Determine agent architecture | Design agent internal structure | |
| Determine agents' organizational behaviours<br>Determine agents' organizational roles<br>Identify sub-organizations<br>Define organizational rules<br>Define organizational structures | Identify System Organization | |
| Model responsibilities<br>Model permissions | Model agents' roles | |
| Model environmental resources<br>Model percepts<br>Model events | Model the agent's environment | |
| Determine MAS infrastructure facilities | Create a system architecture | |

| Technique | Tasks 1 | 2 | 3 | 4 | 5 | 6 |
|---|---|---|---|---|---|---|
| Abstract class identification | | | | | | |
| Agent internal design | | | Y | | | |
| AND/OR decomposition | Y | | | | | |
| Class naming | Y | Y | | | | |
| Control architecture | | Y | | | | |
| Context modelling | Y | | | Y | | |
| Delegation analysis | Y | Y | | | | |
| Event modelling | | | | Y | | |
| Intelligent agent identification | | Y | | | | |
| Means-end analysis | Y | | | | | |
| Role modelling | Y | Y | | | Y | Y |
| State modelling | | Y | | | | |
| Textual analysis | Y | Y | | | | |
| 3-layer BDI model | | Y | Y | | | |

**Key:**
1. Model dependencies for actors and goals; 2. Construct the agent model;
3. Design agent internal structure; 4. Model the agent's environment;
5. Model responsibilities; 6. Model permissions

**Fig. 3.** Deontic matrix showing OPF Techniques-Tasks linkages for Prometheus enhanced by two Techniques from Tropos (viz. AND/OR decomposition and Means-end analysis) (significantly modified from [15])

## 3.2 A New Metamodel

At a higher abstraction level, creating a methodology metamodel to support all current AO methodologies has been attempted as part of a FIPA (Foundation for Intelligent Physical Agents) project, in which the newly created metamodel is essentially an aggregation of a number of single-methodology metamodels [6]. In contrast, in the FAME project, we seek to identify high level commonalities, leaving details for later, methodology-specific elaboration. With that aim in mind, we identified as a basis for the FAME metamodel first the Australian methodology metamodel standard AS4651 [26] and then the emerging ISO/IEC 24744 standard, due to be published in late 2006/early 2007 and with strong similarities to AS4651. This metamodel replaces the strict metamodelling approach (based on instance-of relationships) of the OMG, in which it is not possible to model both the method and endeavour domain at the same time, transmitting information to each of them from the metamodel domain, with a new conceptualization based upon powertype patterns [12]. Although not fitting into a strict metamodelling mindset, powertypes (and their associated set representation – Figure 4) provide a solution more aligned to people and their endeavours. With these three domains (Endeavour, Method and (formal) Metamodel (Figure 5) – in which all conceptual, powertype models exist), attributes can be assigned either to (i) the xxx element in the metamodel from which a method level entity[1] can inherit or (ii) the xxxKind element in the metamodel from which a slot value can be instantiated in the method level entity. In the first case, the inherited attribute is then given a slot value at the next level i.e. that of the Endeavour. In the second case, the value given at the method level acts a little like a Class attribute does in OO programming.

[1] Actually this is a clabject (Atkinson, 1998; Atkinson and Kühne, 2000), which is defined as having both an object facet and a class facet.

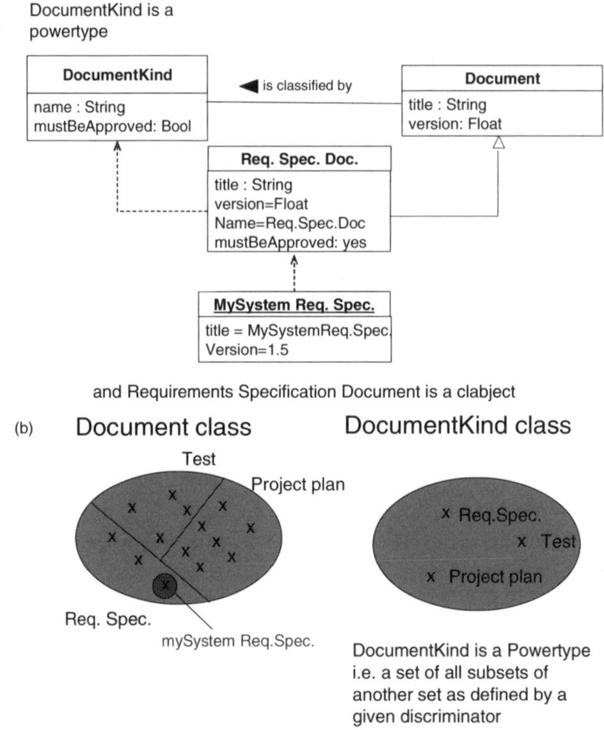


**Fig. 4.** Powertypes expressed in (a) a UML-style diagram and (b) Venn diagrams (representation as sets)

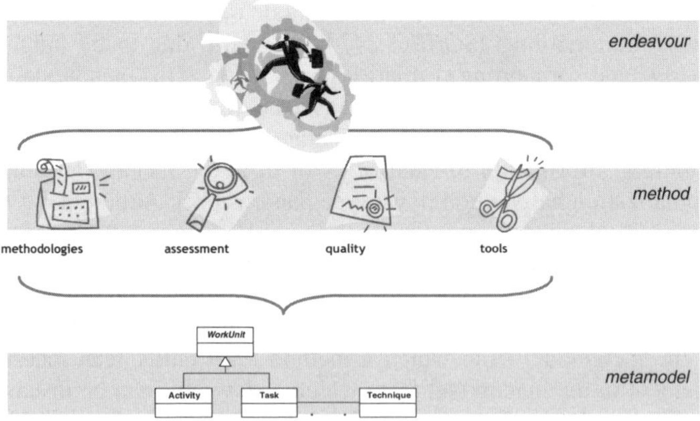


**Fig. 5.** Revised metalevel hierarchy based on practice as well as theory (after [16])

## 4 Summary

Since it is evident that it is not possible to create a "one-size-fits-all" methodology (e.g. [10]), other solutions are sought for the challenge of delivering to an industrial software applications development team an appropriate and site-specific methodology. Based on a "standard" set of method fragments in a repository or methodbase, method engineering provides an excellent paradigm for just such construction of situation-specific software engineering methodologies for both object and agent software development. We have illustrated this approach here both in terms of the OPEN Process Framework (OPF) and the Framework for Agent-oriented Method Engineering (FAME) and indicated future alignment with the new ISO standard metamodel ISO/IEC 24744.

Such a flexible, yet standardized, repository supplies method fragments that are then configured to support specific projects. In addition, all existing, and new, OO and AO methodologies can be recreated, thus providing an industry strength resource for object-oriented and agent-oriented software development.

## Acknowledgements

The work described here was supported by several grants from the Australian Research Council, whom we thank for financial support. Thanks also to Dr Cesar Gonzalez-Perez for his insightful comments on an earlier draft of this manuscript. This is Contribution number 06/07 of the Centre for Object Technology Applications and Research at the University of Technology, Sydney.

## References

1. Atkinson, C.: Supporting and applying the UML conceptual framework. In: Bézivin, J., Muller, P.-A. (eds.) UML 1998. LNCS, vol. 1618, pp. 21–36. Springer, Berlin (1999)
2. Atkinson, C., Kühne, T.: Meta-level independent modelling. In: International Workshop on Model Engineering at 14th European Conference on Object-Oriented Programming (2000)
3. Atkinson, C., Kühne, T.: Processes and Products in a Multi-level Metamodeling Architecture. Int. J. Software Eng. and Knowledge Eng. 11(6), 761–783 (2001)
4. Avison, D.E.: Information systems development methodologies: a broader perspective, in Method Engineering. In: Brinkkemper, S., Lyytinen, K., Welke, R.J. (eds.) Principles of Method Construction and Too Support. Procs. IFIP TC8, WG8.1/8.2 Working Conference on Method Engineering, Atlanta, USA, August 26-28, 1996, pp. 263–277. Chapman & Hall, London (1996)
5. Baddoo, N., Hall, T.: De-motivators for software process improvement: an analysis of practitioners views. Journal of Systems and Software 66, 23–33 (2003)
6. Bernon, C., Cossentino, M., Gleizes, M., Turci, P., Zambonelli, F.: A Study of Some Multi-agent Meta-models. In: Odell, J.J., Giorgini, P., Müller, J.P. (eds.) AOSE 2004. LNCS, vol. 3382, pp. 62–77. Springer, Heidelberg (2005)
7. Bresciani, P., Giorgini, P., Giunchiglia, F., Mylopolous, J., Perini, A.: Tropos: an agent-oriented software development methodology. Autonomous Agents and Multi-Agent Systems 8(3), 203–236 (2004)

8. Brinkkemper, S.: Method Engineering: Engineering of Information Systems Development Methods and Tools. Inf. Software Technol. 38(4), 275–280 (1996)
9. Brinkkemper, S., Saeki, M., Harmsen, F.: Assembly Techniques for Method Engineering. In: Pernici, B., Thanos, C. (eds.) CAiSE 1998. LNCS, vol. 1413, pp. 381–400. Springer, Heidelberg (1998)
10. Cockburn, A.S.: Selecting a project's methodology. IEEE Software 17(4), 64–71 (2000)
11. Firesmith, D.G., Henderson-Sellers, B.: The OPEN Process Framework. An Introduction, p. 330. Addison-Wesley, Reading (2002)
12. Gonzalez-Perez, C., Henderson-Sellers, B.: A Powertype-Based Metamodelling Framework. Software and Systems Modelling 4(4) (2005), DOI 10.1007/210270-005-0099-9
13. Graham, I., Henderson-Sellers, B., Younessi, H.: The OPEN Process Specification, p. 314. Addison-Wesley, Reading (1997)
14. Henderson-Sellers, B.: Who needs an OO methodology anyway? J. Obj.-Oriented Programming 8(6), 6–8 (1995)
15. Henderson-Sellers, B.: Creating a comprehensive agent-oriented methodology - using method engineering and the OPEN metamodel. In: Henderson-Sellers, B., Giorgini, P. (eds.) Agent-Oriented Methodologies, Idea Group, ch. 13, pp. 368–397 (2005)
16. Henderson-Sellers, B.: Method engineering: theory and practice. In: Karagiannis, D., Mayr, H.C. (eds.) 5th International Conference ISTA 2006. Klagenfurt, Austria (eds, Klagenfurt, Austria, May 30-31, 2006. Lecture Notes in Informatics (LNI) – Proceedings, vol. P-84, pp. 13–23. Gesellschaft für Informatik, Bonn (2006)
17. Henderson-Sellers, B., Gonzalez-Perez, C.: A comparison of four process metamodels and the creation of a new generic standard. Inf. Software Technol. 47(1), 49–65 (2005)
18. Henderson-Sellers, B., Simons, A.J.H., Younessi, H.: The OPEN Toolbox of Techniques, p. 426. Addison-Wesley, UK (1998)
19. Henderson-Sellers, B., Haire, B., Lowe, D.: Using OPEN's deontic matrices for e-business. In: Rolland, C., Brinkkemper, S., Saeki), M. (eds.) Engineering Information Systems in the Internet Context, pp. 9–30. Kluwer Academic Publishers, Boston (2002)
20. Kumar, K., Welke, R.J.: Methodology Engineering: a Proposal for Situation-Specific Methodology Construction. In: Cotterman, W.W., Senn, J.A. (eds.) Challenges and Strategies for Research in Systems Development, pp. 257–269. John Wiley & Sons, Chichester (1992)
21. Lyytinen, K.: Different perspectives on information systems: problems and solutions. ACM Computer Surveys 19(1), 5–46 (1987)
22. Pagdham, L., Winikoff, M.: Developing Intelligent Agent Systems: A Practical Guide. Wiley, Chichester (2005)
23. Ralyté, J., Deneckère, R., Rolland, C.: Towards a generic method for situational method engineering, Advanced Information Systems Engineering. In: Eder, J., Missikoff, M. (eds.) CAiSE 2003. LNCS, vol. 2681, pp. 95–110. Springer, Heidelberg (2003)
24. Rao, A.S., Georgeff, M.P.: BDI agents: from theory to practice. In: Procs. First International Conference on Multi Agent Systems, San Francisco, CA, USA, pp. 312–319 (1995)
25. Saeki, M., Iguchi, K., Wen-yin, K., Shinohara, M.: A meta-model for representing software specification & design methods. In: Procs. IFIP WG8.1 Conf. on Information Systems Development Process, Come, pp. 149–166 (1993)
26. Standards Australia, Standard Metamodel for Software Development Methodologies, AS 4651-2004, Standards Australia, Sydney (2004)
27. Zambonelli, F., Jennings, N., Wooldridge, M.: Developing multiagent systems: the Gaia methodology. ACM Transaction on Software Engineering and Methodology 12(3), 317–370 (2003)

# PART I
# Programming Languages

# From Static to Dynamic Process Types

Franz Puntigam

Technische Universität Wien, Argentinierstr. 8, 1040 Vienna, Austria
franz@complang.tuwien.ac.at

**Abstract.** Process types – a kind of behavioral types – specify constraints on message acceptance for the purpose of synchronization and to determine object usage and component behavior in object-oriented languages. So far process types have been regarded as a purely static concept for Actor languages incompatible with inherently dynamic programming techniques. We propose solutions of related problems causing the approach to become useable in more conventional dynamic and concurrent languagues. The proposed approach can ensure message acceptability and support local and static checking of race-free programs.

**Keywords:** Process types, synchronization, type systems, race-free programs.

## 1 Introduction

Process types [1] represent a behavioral counterpart to conventional object types: They support subtyping, genericity, and separate compilation as conventional types. Additionally they specify abstractions of object behavior. Abstract behavior specifications are especially desirable for software components, and they can be used for synchronization. Both concurrent and component-based programming are quickly becoming mainstream programming practices, and we expect concepts like process types to be important in the near future. However, so far process types are not usable in mainstream languages:

1. Their basis are active objects communicating by message passing [2]. Variables are accessible only within single threads. In mainstream languages like Java, threads communicate through shared (instance) variables; one thread reads values written by another. To support such languages we must extend process types with support of shared variables.
2. Process types are static. Object state changes must be anticipated at compilation time. We must adapt process types to support dynamic languages like Smalltalk (using dynamic process type checking).

Support of dynamic languages turns out to be a good basis for supporting communication through shared variables. Hence, we address mainly the second issue and show how dynamic type checking can deal with the first issue.

We introduce the basic static concept of process types for a conventional (Java-like) object model in Section 2. Then, we add support of dynamic synchronization in Section 3 and of shared variables with late type checking in Section 4. Local and static checking of race-free programs is rather easy in our setting as discussed in Section 5.

J. Filipe, B. Shishkov, and M. Helfert (Eds.): ICSOFT 2006, CCIS 10, pp. 61–73, 2008.

$$
\begin{array}{rcl}
P & ::= & \mathit{unit}^* \\
\mathit{unit} & ::= & \texttt{class}\ c\ [\texttt{<}\ c^+]_{\mathrm{opt}}\ \texttt{is}\ [\texttt{token}\ x^+]_{\mathrm{opt}}\ \mathit{def}^+ \mid \\
 & & \texttt{type}\ c\ [\texttt{<}\ c^+]_{\mathrm{opt}}\ \texttt{is}\ [\texttt{token}\ x^+]_{\mathrm{opt}}\ \mathit{decl}^+ \\
\mathit{decl} & ::= & m\ \texttt{(}\ \mathit{par}^*\ [\texttt{with}\ \mathit{ctok}]_{\mathrm{opt}}\ \texttt{)}\ [\texttt{:}\ t]_{\mathrm{opt}} \\
\mathit{def} & ::= & v\ \texttt{:}\ c \mid \mathit{decl}\ \texttt{do}\ s^+ \mid \texttt{new}\ \texttt{(}\ \mathit{par}^*\ \texttt{)}\ \texttt{:}\ t\ \texttt{do}\ s^+ \\
\mathit{par} & ::= & v\ \texttt{:}\ c\ [\ \texttt{[}\ \mathit{ctok}\ \texttt{]}\ ]_{\mathrm{opt}} \\
\mathit{ctok} & ::= & \mathit{tok}^+\ \texttt{->}\ \mathit{tok}^* \mid \texttt{->}\ \mathit{tok}^+ \\
\mathit{tok} & ::= & x\ [\ \texttt{.}\ n]_{\mathrm{opt}} \\
t & ::= & c\ [\ \texttt{[}\ \mathit{tok}^+\ \texttt{]}\ ]_{\mathrm{opt}} \\
s & ::= & v\ \texttt{:}\ t\ \texttt{=}\ e \mid v\ \texttt{=}\ e \mid e \mid \texttt{return}\ [e]_{\mathrm{opt}} \mid \texttt{fork}\ e \\
e & ::= & \texttt{this} \mid v \mid c \mid n \mid e\texttt{.}m\texttt{(}\ e^*\ \texttt{)} \mid \texttt{null}
\end{array}
$$

$c \in$ class and type names
$x \in$ token names
$m \in$ message selectors
$v \in$ variable names
$n \in$ natural number literals

**Fig. 1.** Syntax of TL1

## 2 Static Process Types

Figure 1 shows the grammar of TL1 (Token Language 1) – a simple Java-like language we use as showcase. We differentiate between classes and types without implementations. To create a new object we invoke a creator `new` in a class. Type annotations follow after ":". Token declarations (names following the keyword `token`), tokens occurring within square brackets in types, and `with`-clauses together determine the statically specified object behavior.

The first example shows how tokens allow us to specify constraints on the acceptability of messages:

```
type Buffer is
   token empty filled
   put(e:E with empty->filled)
   get(with filled->empty): E
```

According to the `with`-clause in `put` we can invoke `put` only if we have an `empty`; this token is removed on invocation, and `filled` is added on return. For `x` of type `Buffer[empty]` – a buffer with a single token `empty` – we invoke `x.put(..)`. This invocation changes the type of `x` to `Buffer[filled]`. Next we invoke `x.get()`, then `x.put(..)`, and so on. Static type checking enforces `put` and `get` to be invoked in instances of `Buffer[empty]` in alternation. Type checking is simple because we need only compare available tokens with tokens required by `with`-clauses and change tokens as specified by `with`-clauses [1].

The type `Buffer[empty.8 filled.7]` denotes a buffer with at least 8 filled and 7 empty slots. An instance accepts `put` and `get` in all sequences such that the buffer never contains more that 15 or less than zero elements as far as the client knows.

In the next example we show how to handle tokens in parameter types similarly as in `with`-clauses:

```
class Test is
  play(b:Buffer[filled->filled])
  do e:E = b.get() -- b:Buffer[empty]
     e = e.subst() -- another e
     b.put(e)      -- b:Buffer[filled]
  copy(b:Buffer[empty filled->filled.2])
  do e:E = b.get() -- b:Buffer[empty.2]
     b.put(e) -- b:Buffer[empty filled]
     b.put(e) -- b:Buffer[filled.2]
```

Let `y` be of type `Buffer[empty.2 filled.2]` and `x` of type `Test`. We can invoke `x.play(y)` since `y` has the required token `filled`. This routine gets an element from the buffer, assigns it to the local variable `e` (declared in the first statement), assigns a different element to `e`, and puts this element into the buffer. Within `play` the buffer is known to have a single filled slot on invocation and on return. For the type of `b` specified in the formal parameter list it does not matter that the buffer has been empty meanwhile and the buffer contents changed. After return from `play` variable `y` is still of type `Buffer[empty.2 filled.2]`.

Invocations of `copy` change argument types: On return from `x.copy(y)` variable `y` will be of type `Buffer[empty filled.3]`. Removing tokens to the left of `->` on invocation causes the type to become `Buffer[empty filled]`, and adding the tokens to the right on return causes it to become `Buffer[empty filled.3]`.

Parameter passing does not produce or consume tokens. Tokens just move from the argument type to the parameter type on invocation and vice versa on return. Only `with`-clauses can actually add tokens to and remove them from an object system. This is a basic principle behind the idea of tokens: Each object can produce and consume only its own tokens.

A statement '`fork x.copy(y)`' spawns a new thread executing `x.copy(y)`. Since the execution continues without waiting for the new threads, invoked routines cannot return tokens. The type of `y` changes from `Buffer [empty.2 filled.2]` to `Buffer[empty filled]`. The type of `y` is split into two types – the new type of `y` and the type of `b`. Both threads invoke routines in the same buffer without affecting each other concerning type information.

Assignment resembles parameter passing in the case of spawning threads: We split the type of an assigned value into two types such that one of the split types equals the current static type of the variable, and the remaining type becomes the new type of the assigned value. Thereby, tokens move from the value's to the variable's type. If the statically evaluated type of `v` is `Buffer[empty.2]` and `y` is of type `Buffer[empty.2 filled.2]`, then `v=y` causes `y`'s type to become `Buffer[filled.2]`.

Local variables are visible in just a single thread of control. This property is important because it allows us to perform efficient type checking by a single walk through the code although variable types can change with each invocation. Because of explicit formal parameter types we can check each class separately. If variables with tokens in their types were accessible in several threads, then we must consider myriads of possible interleavings causing static type checking to become practically impossible. Instance variables can be shared by several threads. To support instance variables and still keep

$$def ::= v : c \mid decl\ [\texttt{when}\ ctok]_{\text{opt}}\ \texttt{do}\ s^+ \mid \texttt{new}(\ par^*\ ) : t\ [\texttt{->}\ tok^+]_{\text{opt}}\ \texttt{do}\ s^+$$

**Fig. 2.** Syntax of TL2 (Differences to TL1)

the efficiency of type checking we require their types to carry no token information. We address this restriction in Section 4.

Explicit result types of creators play a quite important role for introducing tokens into the system:

```
class Buffer1 < Buffer is
  s:E     -- single buffer slot
  put(e:E with empty->filled) do s=e
  get(with filled->empty):E do return s
  new(): Buffer1[empty] do null
```

Class `Buffer1` inherits `empty` and `filled` from `Buffer`. An invocation of `Buffer1.new()` returns a new instance with a single token `empty`. No other token is initially available. Since invocations of `put` and `get` consume a token before they issue another one, there is always at most one token for this object. No empty buffer slot can be read and no filled one overwritten, and we need no further synchronization even if several threads access the buffer. The use of tokens greatly simplifies the implementation. However, this solution is inherently static and does not work in more dynamic environments.

## 3 Dynamic Tokens

The language TL2 (see Figure 2) slightly extends TL1 with dynamic tokens for synchronization. This concept resembles more conventional synchronization like that in Java. There is no need to anticipate such synchronization at compilation time.

We associate each object with a multi-set of tokens (token set for short) to be manipulated dynamically. TL2 differs from TL1 by optional `when`-clauses in routines and optional initial dynamic tokens (following `->`) in creators. Tokens to the left of `->` in `when`-clauses must be available and are removed before executing the body, and tokens to the right are added on return. Different from `with`-clauses, `when`-clauses require dynamic tokens to be in the object's token set and change this token set. If required dynamic tokens are not available, then the execution is blocked until they become available. Checks for token availability occur only at run time. The following variant of the buffer example uses static tokens to avoid buffer overflow and underflow, and dynamic tokens to ensure mutual exclusion:

```
class Buffer50 < Buffer is
  token sync
  lst: List
  new(): Buffer50[empty..50] ->sync do
      lst = List.new()
  put(e:E with empty->filled)
```

```
    when sync->sync do lst.addLast(e)
  get(with filled->empty): E
    when sync->sync do
      return lst.getAndDeleteFirst()
```

The creator introduces just a single token `sync`. Both `put` and `get` remove this token at the begin and issue a new one on return. Clients need not know about the mutual exclusion of all buffer operations. Of course we could use only dynamic tokens which is more common and provides easier handling of buffers.

Static and dynamic tokens live in mostly independent worlds. Nonetheless we have possibilities to move tokens from the static to the dynamic world and vice versa as shown in the following example:

```
class StaticAndDynamic is
  token t
  beDynamic(with t->) when ->t do null
  beStatic(with ->t) when t-> do null
  new(): StaticAndDynamic[t] do null
```

There always exists only a single token `t` for each instance, no matter how often and from how many threads we invoke `beDynamic` and `beStatic`.

The major advantage of our approach compared to concepts like semaphores and monitors is the higher level of abstraction. It is not so easy to "forget" to release a lock as often occurs with semaphores, and it is not necessary to handle wait queues using `wait` and `notify` commands as with monitors. For static tokens we need not execute any specific synchronization code at all. This synchronization is implicit in the control flow.

## 4 Dynamic Typing

In TL1 and TL2 we constrained the flexibility of the language to get efficient static type checking: Types of instance variables cannot carry tokens. In this section we take the position that static type checking is no precondition for the token concept to be useful. We want to increase the language's flexibility (by supporting tokens on instance variables) and nonetheless ensure that synchronization conditions expressed in `with`-clauses are always satisfied. An error shall be reported before invocations if required tokens are not available.

Figure 3 shows the grammar of TL3 that differs form TL2 just by missing type annotations on formal parameters and declarations. However, without type annotations there is no explicit information about available tokens. We handle this information dynamically. One kind of type annotation is left in TL3: Types of new instances returned by

$$
\begin{aligned}
\mathit{decl} &::= m\ (\ v^*\ [\texttt{with}\ \mathit{ctok}]_{\text{opt}}\ ) \\
\mathit{def} &::= v\texttt{:}\ \mid \mathit{decl}\ [\texttt{when}\ \mathit{ctok}]_{\text{opt}}\ \texttt{do}\ s^+ \mid \\
&\qquad \texttt{new}\ (\ v^*\ )\ \texttt{:}\ t\ [\texttt{->}\ \mathit{tok}^+]_{\text{opt}}\ \texttt{do}\ s^+ \\
s &::= v\ \texttt{:=}\ e \mid v = e \mid e \mid \texttt{return}\ [e]_{\text{opt}} \mid \texttt{fork}\ e
\end{aligned}
$$

**Fig. 3.** Syntax of TL3 (Differences to TL1–TL2)

creators must be specified explicitly because tokens in this type together with `with`-clauses determine which routines can be invoked. Such types are part of behavior specifications. Except of type annotations the following example in TL3 equals `Buffer50`:

```
type BufferDyn is
  token empty filled
  put(e with empty->filled)
  get(with filled->empty)

class Buffer50Dyn < BufferDyn is
  token sync
  lst:
  new():Buffer50Dyn[empty.50]->sync do
      lst = List.new()
  put(e with empty->filled)
    when sync->sync do lst.addLast(e)
  get(with filled->empty)
    when sync->sync do
      return lst.getAndDeleteFirst()
```

The following example gives an intuition about the use of static tokens in a dynamic language. An open window is displayed on a screen or shown as icon:

```
type Window is
  token displ icon closed
  setup(with closed->displ)
  iconify(with displ->icon)
  display(with icon->displ)
  close(with displ->closed)

class WindowImpl < Window is
  new(): WindowImpl[closed] do ...
  ...

class WManager is
  win:
  new(w):WManager do win=w win.setup()
  onButton1() do win.iconify()
  onButton2() do win.close()
  onButton3() do win.display()
```

Some state changes (directly from an icon to closed, etc.) are not supported. Class `WManager` specifies actions to be performed when users press buttons. Under the assumption that a displayed window has only Button 1 and 2 and an icon only Button 3 the constraints on state changes are obviously satisfied. Since the assumption corresponds to the existence of at most one token for each window we need nothing else to ensure a race-free program. We express the assumption by `with`-clauses and dynamically ensure them to be satisfied. The variable `win` must be associated with a (static) token specifying the window's state. In TL1 and TL2 we cannot express such type information that is implicit in TL3.

TL3 deals with dynamic tokens in the same way as TL2. To dynamically handle information about available static tokens we consider two approaches – TL3flex as a simple and flexible approach, and TL3strict as a more restrictive and safer approach.

*TL3flex.* In TL3flex we tread static tokens in a similar way as dynamic tokens: Each objects contains a pool of static tokens. On invocations tokens to the left of `->` in `with`-clauses are taken from the pool, and on return those to the right are added to the pool. An error is reported if the pool does not contain all required tokens.

This approach is very flexible. Each thread can use all previously issued static tokens no matter which thread caused the tokens to be issued. A disadvantage is a low quality of error messages because there is no information about the control flow causing tokens not to be available. Furthermore, there is a high probability for program runs not to uncover synchronization problems. Thus, program testing is an issue.

*TL3strict.* To improve error messages and the probability of detecting problems we dynamically simulate static type checking: Instead of storing static tokens centralized in the object we distribute them among all references to the object. On invocation we check and update only tokens associated with the corresponding reference. We must find an appropriate distribution of tokens among references. In TL1 and TL2 the programmer had to determine the distribution by giving type annotations. In TL3strict we distribute tokens lazily as needed in the computation.

Instead of splitting a token set on parameter passing or assignment we associate the two references with pointers to the (unsplit) token set as well as with a new empty token set for each of the two references. Whenever required tokens are not available in the (after assignment or parameter passing empty) token set of a reference we follow the pointers and take the tokens where we find them. New tokens are stored in the references' own token sets. This way all references get the tokens they need (if available) and we need not foresee how to split token sets. Repeated application leads to a tree of token sets with pointers from the leaves (= active references) toward the root (= token set returned by creator). We report an error only if tokens required at a leaf cannot be collected from all token sets on the path to the root. On return of invocations we let actual parameters point to token sets of corresponding formal parameters.

Figure 4 shows an example: Immediately after creating a window there is only one reference `n` to it (a). The box contains the single token in the corresponding token set. When invoking `new` in `WManager` using `n` as actual parameter we construct new token sets for `n` and for the formal parameter `w` (b). When the creator assigns `w` to `win` we add new token sets for `w` and `win` (c). An invocation of `setup` on `win` removes the token `closed` and adds `displ`. On return from the creator we let the token set of `n` point to that of `w` (d). Now only `win` carries the single token. We cannot change the window's state through `n`. Therefore, TL3strict is safer and less flexible than TL3flex.

We can build large parts of the structures shown in Figure 4 already at compilation time by means of abstract interpretation. Most checks for the availability of tokens can be performed statically. In fact we need dynamic checks of token availability only for tokens associated with instance variables.

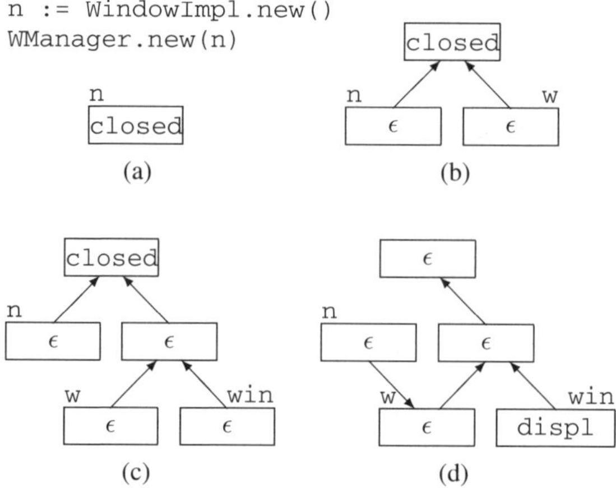


**Fig. 4.** Token Sets per Reference

## 5 Race-Free Programs

It is possible to ensure race-free programs just by analyzing the tokens in classes. We use only a single sufficient (but not always necessary) criterion: No two preconditions in `with`-clauses and `when`-clauses of routines accessing the same variable (where an access is a write) can be satisfied at the same time. To check this criterion we compute upper bounds on the token sets that can be constructed from the tokens of new instances. We analyze each class separately.

In the following description of the algorithm to determine upper bounds of token sets we first consider only static tokens as in TL1. We start with the set of token sets declared in the result types of the analyzed class' creators (one token set per creator). For each `with`-clause in the class we repeatedly construct new token sets by removing tokens to the left of `->` and adding those to the right from/to each token set constructed so far containing all required tokens. If a token set contains all tokens occurring in another token set, then we remove the smaller token set. And if a token set differs from the token set from which it was constructed just by containing more tokens, then we increment the token numbers that differ to the special value $\infty$ indicating infinite grow. Because of this treatment the algorithm always reaches a fixed point. The algorithm is accurate in the sense that

- the token set produced for an instance of the class is always a subset of a token set returned by the algorithm,
- if a token set returned by the algorithm does not contain $\infty$, then there exists a sequence of invocations producing exactly this token set,
- and if a token set returned by the algorithm contains $\infty$, then there exist invocation sequences producing corresponding tokens without upper bounds.

In TL2 and TL3 we must consider static and dynamic tokens together to get most accurate results. Since the static and the dynamic world are clearly separated, static and dynamic tokens must not be intermixed. We have to clearly mark each token as either static or dynamic (for example, by an index) and regard differently marked tokens as different. The algorithm starts with one token set for each creator containing both static and dynamic tokens. A new token set is constructed by simultaneously removing and adding tokens as specified in the `with`- and `when`-clause of a routine. The result shows which dynamic tokens can exist together with static tokens. For example, applied to `StaticAndDynamic` (see Section 3) the algorithm returns two token sets, one containing only a static token `t` and the other only a dynamic token `t`; in this case no dynamic token can exist at the same time as a static one.

Once we know the upper bounds it is easy to perform our check of race-free programs as shown in the following pseudo-code:

let $U$ be the upper-bound set of token sets of class $c$;
for each instance variable $v$ of $c$
  for each routine $r$ write-accessing $v$
    for each routine $s$ (read or write) accessing $v$
      let $p$ be the union of the token sets
                to the left of -> in $r$ and $s$;
      if there is a $u \in U$ containing all tokens in $p$
        then issue a warning about a potential race;
otherwise $c$ is race-free

As an example we apply this check to `Buffer1` (see Section 2). As upper-bound set of token sets $S$ we have {{`empty`}, {`filled`}}; there is always at most one token `empty` or `filled`. The only instance variable `s` is written in `put` and read in `get`. Hence, $r$ ranges just over `put`, $s$ over `put` and `get`, and $p$ over {`empty.2`} and {`empty, filled`}. The class is race-free because no token set in $S$ contains two `empty` or an `empty` and a `filled`.

The set $S$ can become quite large because of combinatorial explosion. For example, $S$ constructed for `Buffer50Dyn` contains 51 different token sets – all possibilities of summing up tokens of two names to 50 tokens. Fortunately, a simple change in the algorithm to compute upper bounds can reduce the size of $S$ considerably: When computing the fixed point we replace all token numbers larger than $2 \cdot n^2 \cdot i$ by $\infty$, where $n$ is the largest total number of tokens to the left of -> in the `with`- and `when`-clause of the same routine, and $i$ is the number of different token names in the class. For `Buffer50Dyn` we have $n = 2$, $i = 3$, $2 \cdot n^2 \cdot i = 24$, and $S$ contains just a single token set {`sync`, `empty.`$\infty$, `filled.`$\infty$}. This optimization does not change the output of the race-freeness check: Soundness is not affected because the multi-set of supposedly reachable tokens in a system can just get larger. No token set $p$ (as in the algorithm) can contain more than $2 \cdot n$ tokens, and a single token of some name can be generated from no more than $n \cdot i$ tokens of another name. Therefore, more than $2 \cdot n^2 \cdot i$ tokens of one name can be ignored for our purpose. Probably there are more accurate estimations, but we expect this simple one to be sufficient because token numbers to the left of -> are usually small.

All information needed to check race-free classes is explicit in TL1, TL2, and TL3. We need no information about formal parameter types and no aliasing information. No global program analysis is necessary.

## 6 Discussion, Related Work

The idea of integrating process types into dynamic languages is new and at a first glance unexpected because such types were developed to move dynamic aspects like synchronization to the static language level whenever possible [3,1,4]. In some sense the integration of more advanced static concepts into dynamic languages is a consistent further development allowing us to use the appropriate (static or dynamic) concept for each task. Such integration helps us to deepen our understanding of related concepts.

We usually regard synchronization of concurrent threads as a purely dynamic concept: If there is a dependence between two control flows, then one of the corresponding threads must wait until the other thread has caught up to meet the synchronization point. Since threads usually run asynchronously and at statically unpredictable speed, it is only possible to decide at run time whether a thread must wait at a synchronization point. However, these considerations are valid only at a very low level (close to the hardware) point of view. From the programmers' higher level point of view it is quite often not clear whether there exist dependences between threads or not. Using explicit synchronization as with monitors, semaphores, rendezvous communication, etc. programmers must add much more synchronization points than are actually necessary. There are optimization techniques that can statically eliminate up to about 90% (about 60% in average) of all locks from Java programs and thereby considerably improve program performance [5]. Probably even more synchronization points are actually not necessary.

Current programming languages allow programmers to write programs with races although there are many proposals to ensure race-free programs [6,7,8,9]. Applications of such techniques may lead to further increase of unnecessary synchronization because no approach can accurately decide between necessary and unnecessary locks. Nonetheless, these techniques are very useful because races are an important practical problem in concurrent programming.

Process types were developed as abstractions over expressions in process calculi [3]. These abstractions specify acceptable messages of active objects and allow the acceptability to change over time (thereby specifying synchronization constraints). Static type checking ensures that only acceptable messages can be sent and enforces all synchronization constraints to be satisfied. In this sense type checking in process types has a similar purpose as ensuring race-free programs. However, process types allow us to specify arbitrary constraints on message acceptability, not just synchronization necessary to avoid races. In fact, the underlying calculi do not support shared data that could suffer from races.

There is a clear tendency toward more and more complex interface specifications going far beyond simple signatures of available routines [10,11,12,13,14,15,16,17,18]. We consider such interfaces to be partial specifications of object behavior [19]. They are especially valuable to specify the behavior of software components as far as needed for component composition. Process types are useful as partial behavior specifications [20,21]. We regard behavior specifications as the major reason for using process types.

Pre- and postconditions in `with`-clauses allow us to specify a kind of contracts between components [22,23]. Such contracts clearly specify responsibilities of software and help us to move responsibilities from one component to another. For example, we move the responsibility of proper synchronization from the server to the client if we use `with`-clauses instead of `when`-clauses.

Behavioral types and synchronization of concurrent threads are related topics: Specifications of object behavior cannot ignore necessary synchronization if we expect components composed according to their behavioral types to work together in concurrent environments, and constraints on message acceptability specify a kind of synchronization. The present work allows programmers to decide between synchronization globally visible through the interface (`with`-clauses) and local synchronization regarded as an implementation detail (`when`-clauses). While `with`-clauses just ensure that clients coordinate themselves (for example, through the control flow allowing `m2()` to be invoked only after `m1()`) `when`-clauses ensure proper synchronization using more conventional techniques. Locking does not get visible in interfaces, just synchronization requirements are visible.

There are good reasons for using locking only for local synchronization: Uncoordinated locking easily leads to deadlocks and other undesirable behavior, and it is much easier to coordinate locking within a single unit. The monitor concept supports just local locking for similar reasons. Furthermore, it is very difficult to deal with globally visible locking at the presence of subtyping and inheritance [24]. Process types express just synchronization conditions in interfaces, they do not provide for locking. Another approach directly expresses locking conditions in interfaces [25,26]. As experience shows, that approach easily leads to undesirable locking where it would be more appropriate to raise exceptions.

There are several approaches similar to process types. Nierstrasz [16] and Nielson and Nielson [27] define behavioral types where subtypes show the same deadlock behavior as supertypes, but message acceptability is not ensured. Many further approaches consider dynamic changes of message acceptability, but do not guarantee message acceptability in all cases [25,28,29,26,30]. Well known in the area of typed $\pi$-calculi [31] is the work of Kobayashi, Pierce and Turner on linearity [32] which ensures all sent messages to be acceptable. Work of Najm and Nimour [33] is very similar to process types except that in their approach at each time only one user can interact with an object through an interface (no type splitting). These approaches specify constraints on the acceptability of messages in a rather direct way and do not make use of a token concept. The use of tokens in behavior specifications gives us high expressiveness and flexibility, allows us to express synchronization in a way similar to well-known concepts like monitors and semaphores, and is easily understandable.

## 7 Conclusions

Behavioral types like process types gain more and more importance especially together with component composition. By partially specifying object behavior these types express synchronization in the form of software contracts clearly determining who is responsible for proper synchronization. Process types use simple token sets as abstractions over object states.

In this paper we explored how to add process types to rather conventional object-oriented programming languages. As a showcase we developed the languages TL1 to TL3. Static type checking in TL1 ensures that all conditions in `with`-clauses are satisfied, this is, all required tokens are available. We can synchronize concurrent threads just by waiting for messages. To overcome the restriction, TL2 adds a new dynamic concept of synchronization based on token sets. Neither TL1 nor TL2 can deal with static token sets associated with instance variables because of possible simultaneous accesses by concurrent threads. In TL3 we dispense with static types and apply one of two methods to dynamically ensure the availability of required tokens – a flexible method and one with better error messages and partial support of static type checking. All variables in TL3 have only dynamic types that can implicitly carry tokens. In the three languages we can ensure race-free programs by checking each class separately, without any need of global aliasing information.

Our approach uses token sets for several related purposes – synchronization of concurrent threads and statically and dynamically checked abstract behavior specifications. It is a major achievement to integrate these concepts because of complicated interrelations. The integration is valuable because it gives software developers much freedom and at the same time clear contracts and type safety.

Much work on this topic remains to be done. For example, currently our algorithm can issue warnings about potential races even in purely sequential program parts. Many other approaches to ensure race-free programs put much effort into detecting sequential program parts. By integrating such approaches into our algorithm we expect to considerably improve the accuracy. Most approaches to remove unnecessary locking from concurrent programs also work on sequential program parts [34,5,35]. We expect a combination of the techniques to improve run time efficiency.

## References


1. Puntigam, F.: Coordination Requirements Expressed in Types for Active Objects. In: Aksit, M., Matsuoka, S. (eds.) ECOOP 1997. LNCS, vol. 1241, pp. 367–388. Springer, Heidelberg (1997)
2. Agha, G., Mason, I.A., Smith, S., Talcott, C.: Towards a theory of actor computation. In: Cleaveland, W.R. (ed.) CONCUR 1992. LNCS, vol. 630, pp. 565–579. Springer, Heidelberg (1992)
3. Puntigam, F.: Flexible types for a concurrent model. In: Proceedings of the Workshop on Object-Oriented Programming and Models of Concurrency, Torino (1995)
4. Puntigam, F.: Concurrent Object-Oriented Programming with Process Types. Der Andere Verlag, Osnabrück (2000)
5. von Praun, C., Gross, T.R.: Static conflict analysis for multi-threaded object-oriented programs. In: PLDI 2003, pp. 115–128. ACM Press, New York (2003)
6. Bacon, D.F., Strom, R.E., Tarafdar, A.: Guava: A dialect of Java without data races. In: OOPSLA (2000)
7. Boyapati, C., Rinard, M.: A parameterized type system for race-free Java programs. In: OOPSLA 2001. ACM, New York (2001)
8. Brinch-Hansen, P.: The programming language Concurrent Pascal. IEEE Transactions on Software Engineering 1, 199–207 (1975)
9. Flanagan, F., Abadi, M.: Types for Safe Locking. In: Swierstra, S.D. (ed.) ESOP 1999. LNCS, vol. 1576. Springer, Heidelberg (1999)
10. Arbab, F.: Abstract behavior types: A foundation model for components and their composition. Science of Computer Programming 55, 3–52 (2005)
11. de Alfaro, L., Henzinger, T.A.: Interface automata. In: Proceedings of the Ninth Annual Symposium on Foundations of Software Engineering (FSE), pp. 109–120. ACM Press, New York (2001)
12. Heuzeroth, D., Reussner, R.: Meta-protocol and type system for the dynamic coupling of binary components. In: OOSPLA 1999 Workshop on Reflection and Software Engineering, Bicocca, Italy (1999)

13. Jacobsen, H.-A., Krämer, B.J.: A design pattern based approach to generating synchronization adaptors from annotated IDL. In: IEEE International Conference on Automated Software Engineering (ASE 1998), Honolulu, Hawaii, USA, pp. 63–72 (1998)
14. Lee, E.A., Xiong, Y.: A behavioral type system and its application in Ptolemy II. Formal Aspects of Computing 16, 210–237 (2004)
15. Mezini, M., Ostermann, K.: Integrating independent components with on-demand remodularization. In: OOPSLA 2002 Conference Proceedings, Seattle, Washington, nov 2002, pp. 52–67. ACM Press, New York (2002)
16. Nierstrasz, O.: Regular types for active objects. ACM SIGPLAN Notices 28, 1–15 (1993); Proceedings OOPSLA 1993
17. Plasil, F., Visnovsky, S.: Behavioral protocols for software components. IEEE Transactions on Software Engineering 28, 1056–1076 (2002)
18. Yellin, D.M., Strom, R.E.: Protocol specifications and component adaptors. ACM Transactions on Programming Languages and Systems 19, 292–333 (1997)
19. Liskov, B., Wing, J.M.: Specifications and their use in defining subtypes. ACM SIGPLAN Notices 28, 16–28 (1993); Proceedings OOPSLA 1993
20. Puntigam, F.: State information in statically checked interfaces. In: Eighth International Workshop on Component-Oriented Programming, Darmstadt, Germany (2003)
21. Südholt, M.: A Model of Components with Non-regular Protocols. In: Gschwind, T., Aßmann, U., Nierstrasz, O. (eds.) SC 2005. LNCS, vol. 3628, pp. 99–113. Springer, Heidelberg (2005)
22. Meyer, B.: Object-Oriented Software Construction, 2nd edn. Prentice-Hall, Englewood Cliffs (1997)
23. Meyer, B.: The grand challenge of trusted components. In: ICSE-25 (International Conference on Software Engineering), Portland, Oregon. IEEE Computer Society Press, Los Alamitos (2003)
24. Matsuoka, S., Yonezawa, A.: Analysis of inheritance anomaly in object-oriented concurrent programming languages. In: Agha, G. (ed.) Research Directions in Concurrent Object-Oriented Programming, MIT Press, Cambridge (1993)
25. Caromel, D.: Toward a method of object-oriented concurrent programming. Communications of the ACM 36, 90–101 (1993)
26. Meyer, B.: Systematic concurrent object-oriented programming. Communications of the ACM 36, 56–80 (1993)
27. Nielson, F., Nielson, H.R.: From CML to process algebras. In: Best, E. (ed.) CONCUR 1993. LNCS, vol. 715, pp. 493–508. Springer, Heidelberg (1993)
28. Colaco, J.-L., Pantel, M., Salle, P.: A set-constraint-based analysis of actors. In: Proceedings FMOODS 1997, Canterbury, United Kingdom. Chapman and Hall, Boca Raton (1997)
29. Kobayashi, N., Yonezawa, A.: Type-theoretic foundations for concurrent object-oriented programming. ACM SIGPLAN Notices 29, 31–45 (1994); Proceedings OOPSLA 1994
30. Ravara, A., Vasconcelos, V.T.: Behavioural types for a calculus of concurrent objects. In: Lengauer, C., Griebl, M., Gorlatch, S. (eds.) Euro-Par 1997. LNCS, vol. 1300, pp. 554–561. Springer, Heidelberg (1997)
31. Milner, R., Parrow, J., Walker, D.: A calculus of mobile processes (parts I and II). Information and Computation 100, 1–77 (1992)
32. Kobayashi, N., Pierce, B., Turner, D.: Linearity and the pi-calculus. ACM Transactions on Programming Languages and Systems 21, 914–947 (1999)
33. Najm, E., Nimour, A.: A calculus of object bindings. In: Proceedings FMOODS 1997, Canterbury, United Kingdom. Chapman and Hall, Boca Raton (1997)
34. Choi, J.-D., Gupta, M., Serrano, M., Sreedhar, V.C., Midkiff, S.: Escape analysis for Java. In: OOPSLA 1999, Denver, Colorado (1999)
35. Vivien, F., Rinard, M.: Incrementalized pointer and escape analysis. In: PLDI 2001. ACM, New York (2001)

# Aspectboxes: Controlling the Visibility of Aspects

Alexandre Bergel[1], Robert Hirschfeld[2], Siobhán Clarke[1], and Pascal Costanza[3]

[1] Distributed Systems Group, Trinity College Dublin, Ireland
{Alexandre.Bergel,Siobhan.Clarke}@cs.tcd.ie
[2] Hasso-Plattner-Institut,Universität Potsdam, Germany
hirschfeld@hpi.uni-potsdam.de
[3] Programming Technology Lab, Vrije Universiteit Brussel, Belgium
pascal.costanza@vub.ac.be

**Abstract.** Aspect composition is still a hot research topic where there is no consensus on how to express where and when aspects have to be composed into a base system. In this paper we present a modular construct for aspects, called *aspectboxes*, that enables aspects application to be limited to a well defined scope. An aspectbox encapsulates class and aspect definitions. Classes can be imported into an aspectbox defining a base system to which aspects may then be applied. Refinements and instrumentation defined by an aspect are visible *only* within this particular aspectbox leaving other parts of the system unaffected.

**Keywords:** Aspect-oriented programming, aspect composition, scoping change, aspects, classboxes, squeak.

## 1 Introduction

Aspect-oriented programming (AOP) promises to improve the modularity of programs by providing a modularity construct called aspect to clearly and concisely capture the implementation of crosscutting behavior. An aspect instruments a base software system by inserting pieces of code called advices at locations designed by a set of pointcuts.

An important focus of current research in AOP is on aspect composition [7,11,13,5]. Ordering and nesting are commonly used when composing aspects and advices [10, 16]. Whereas most aspect languages provide means to compose aspects at a very fine grained level, experience has shown that ensuring a sound combination of aspects is a challenging and difficult task [12]. First steps are already taken by AspectJ [10] by restricting pointcuts to a Java package or a class through the use of dedicated pointcuts primitives such as within and withincode primitive pointcuts.

If we regard an aspect as an extension to a base system, multiple extensions are difficult to manage and control, even if they are not interacting with each other. We believe that the reason for this is the lack of a proper scoping mechanism.

In this paper we define a new modular construct for an aspect language called an aspectbox. An *aspectbox* is a modular unit that may contain class and aspect definitions. Classes can be imported into an aspectbox and the aspect is then applied to the imported classes. Refinements originated from such aspects are visible *only* within the aspectbox that defines this aspect. Outside this aspectbox the base system behaves as if there were no aspect. Other parts outside a particular aspectbox remain unaffected.

J. Filipe, B. Shishkov, and M. Helfert (Eds.): ICSOFT 2006, CCIS 10, pp. 74–83, 2008.

In Section 2 we provide an example illustrating the issues when composing aspects. In Section 4 we describe the aspectboxes module system and its properties. In Section 5 we present our Squeak-based implementation of aspectboxes. Related work is discussed in Section 6. We conclude by summarizing the presented work in Section 7.

## 2 Motivation

To motivate the need for limiting the scope of aspects, we use an example based on the design of a small four-wheel electric car, and its implementation based on a mainstream aspect language, AspectJ [10].

The CyCab [2] is an electric four wheel car designed to transport up to two people. The mechanics is taken from a small electrical golf car frame. Functionalities implemented in a CyCab range from an autonomous driving facility (like a coach in a train) to ultrasonic sensors for collision avoidance. A CyCab is composed of three different units (*driving control*, *position control* and *safety control*). Each unit is composed of one or more modules. Figure 1 illustrates the architecture of a CyCab.

**Driving Control.** A CyCab is steered with a joystick emitting electrical pulses used by the motion engine to activate the four motored wheels. This feature is provided by three modules within the *driving control* unit. The *joystick* module emits signals that are captured by the *motion engine* module. This module controls the *wheels*.

**Position Control.** The *position control* unit computes the velocity and the location of the CyCab based on the acceleration given by the *motion engine* to the *wheels* and their angle between the car head.

**Safety Control.** The *safety control* unit verifies the interactions between the three modules of the *driving control* unit. For example, it asserts that pulses emitted by the joystick

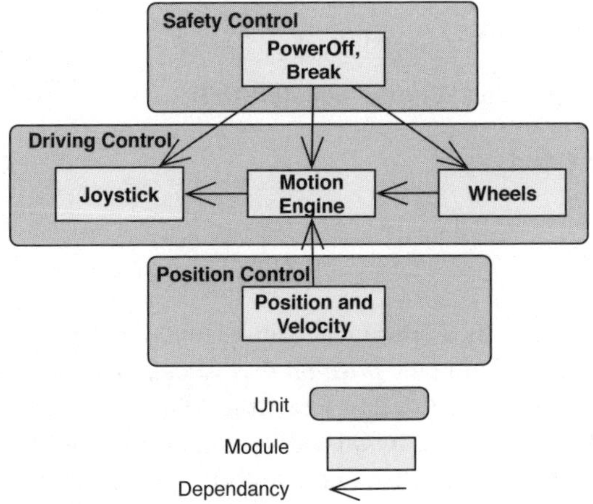


**Fig. 1.** The three units and their modules that compose the CyCab electrical car

trigger the correct reaction in the engine and the wheels reflect the heading dictated by the joystick. In addition, in the event of failure the power is shut down and communication between the three modules is cut off.

## 3 Example Analysis

Behavior defined by the *safety control* unit crosscuts the whole *driving control* unit. For example, the impact of a power shut-down is that the *joystick*, the *motion engine* and *wheels* are disconnected. This can be easily captured in an aspect that adds behaviour to check the power status into each affected module, as implemented by the following AspectJ aspect:

```
aspect PowerOff {
  private boolean hasPower = ...;
  pointcut drivingControl():
    target(Joystick) && call(public * *(..)) ||
    target(Engine) && call(public * *(..)) ||
    target(Wheels) && call(public * *(..));
  void around(): drivingControl() {
    if (hasPower == true)
      proceed();
  }
  ...
}
```

The PowerOffAndBreak aspect inserts a check before all public methods of the classes Joystick, Engine and Wheels to proceed only if power is equal to true. This aspect is applied to the *driving control* unit and has to be composed with the PositionAndVelocity aspect defined by the *position control* unit:

```
aspect PositionAndVelocity {
  double speed;
  pointcut speedUp() : call (* Engine.accelerate());
  after(): speedUp() {
    //... Speed calculation
  }
  ...
}
```

PositionAndVelocity inserts a speed calculation functionality after the execution of the accelerate method. Defining the *position and velocity module* as an aspect has the benefit to leave the *driving control* unit free from referring to the speed and position computation. The two aspects PowerOffAndBreak and PositionAndVelocity are woven into the base system, the *driving control* unit, to form a deployable system. With current aspect languages such as AspectJ, *extensions defined by all aspects are automatically applied to all the modules in the system* (*i.e.*, the physical display screen, the electronic control unit in charge of the safety).

This facility is particularly dangerous regarding the implicit sharing of the control flow of the application. A failure raised by the PositionAndVelocity aspect may easily impact the PowerOffAndBreak aspect affecting the electronic control unit in charge of the safety.

Whereas most of current aspect languages offer sophisticated pointcut primitives to express location of join points, they do not provide a means to limit the impact of an aspect into a well-defined system area. In the up coming section we define a module system for an aspect-oriented programming environment in which one or more aspect compositions are effective only in the context of a well-defined subset of the base system.

## 4 Scoping Aspects with Aspectboxes

Most of today's aspect languages do not provide a way to limit the impact of an aspect within a delimited scope. In this section, we describe a module system for an aspect-oriented programming language that allows for controlling the visibility of a set of aspects relative to a well-defined system area.

### 4.1 Aspectboxes in a Nutshell

*Aspectboxes* is a namespace mechanism for aspects. An aspect lives in an aspectbox and the effects of this aspect is limited to the aspectbox in which it is defined and to other aspectboxes that rely on the base system extended by this aspect. An aspectbox can (i) define classes, (ii) import classes from another aspectbox and (iii) define aspects.

The import relationship is transitive: If an aspectbox AB2 imports a class C from another aspectbox AB1, then a third aspectbox AB3 can import C from AB2. From the point of view of the importing aspectbox AB3, there is no difference if the class is defined or imported in the provider aspectbox AB2. Because aspects cannot be reused across multiple base systems, aspects cannot be imported.

A pointcut definition contained in an aspect refers only to classes that are imported (*i.e.*, visible within the aspectbox that defines this aspect). An aspect in an aspectbox refines the behavior of the classes that are imported or defined, for instance by adding some code before and after some methods. The classes augmented with the aspect can also be imported from another aspectbox. From the point of view of an importing aspectbox, there is no distinction between classes defined within the aspectbox and those imported.

### 4.2 Namespace for Classes and Aspects

An aspectbox defines a namespace for class definitions, aspect definitions and aspect compositions.

**Aspectbox as Namespace for Classes.** The class Engine contained in the aspectbox DrivingControlAB[1] as illustrated in Figure 1 is defined as the following[2]:

[1] We end the name of aspectboxes by AB to clearly make a distinction between them and regular class names.

[2] Since our aspectboxes prototype is implemented in Squeak, we therefore use the Squeak syntax to describe them.

```
(Aspectbox named: #DrivingControlAB)
  createClassNamed: #Engine
  instanceVariableNames: ''
```

The class Engine does not have any instance variables and two methods accelerateWheels: anAcceleration and setAnglewithHeading: anAngle are defined on it.

```
DrivingControlAB.Engine>>
      accelerateWheels: anAcceleration
  "accelerate the wheels with a given acceleration"
  ...

DrivingControlAB.Engine>>
      setAnglewithHeading: anAngle
  "set the heading of the car by setting
  appropriately the wheel angle"
  ...
```

An aspectbox acts as a code packaging mechanism and constrains aspect visibility. A class is *visible* within an aspectbox if this class is defined in or imported to this aspectbox. Any class visible within an aspectbox AB1 can be imported from AB1 by other aspectboxes. The aspectbox PositionControlAB imports the class Engine from DrivingControlAB

```
(Aspectbox named: #PositionControlAB)
  import: #Engine from: #DrivingControlAB
```

An instantiation of a class can occurs in any aspectbox as long as this class is visible in the aspectbox that contains the code performing the instantiation. Class instances (*i.e.,* objects) do not belong to an aspectbox.

**Aspectbox as Namespace for Aspect Definitions.** The module *position and velocity* is implemented by the PositionAndVelocity aspect:

```
(Aspectbox named: #PositionControlAB)
  createAspectNamed: #PositionAndVelocity
  instanceVariableNames: 'heading velocity'
```

Because the aspect PositionAndVelocity has to be applied to the class Engine, this class has to be imported from the DrivingControlAB aspectbox. This aspect also defines advices to be applied to the methods accelerateWheels: anAcceleration and setAnglewithHeading: anAngle that compute the velocity and the heading, respectively, as illustrated in Figure 2.

**Aspectbox as Namespace for Aspect Compositions.** An aspect, which is defined in an aspectbox, is applied to classes that are visible in this aspectbox (*i.e.,* classes that are imported or defined). The effect of this aspect is limited to the aspectbox in which this aspect is defined. Outside this aspectbox, it is as if no aspect would have been applied to the base system.

```
PositionControlAB. PositionAndVelocity>> adviceComputeVelocity
  ^AfterAdvice
    pointcut: (JoinPointDescriptor
                  targetClass: Engine targetSelector: #accelerateWheels:)
    afterBlock: [:receiver :arguments :aspect |
                  "computation of the velocity according to the speed of the wheels"
                  velocity := ...]

PositionControlAB. PositionAndVelocity>> adviceComputeHeading
  ^AfterAdvice
    pointcut: (JoinPointDescriptor
                  targetClass: Engine targetSelector: #setAnglewithHeading:)
    afterBlock: [:receiver :arguments :aspect |
                  "computation of the heading according to the speed of the wheels"
                  heading := ...]
```

**Fig. 2.** The velocity and the heading are computed by two advices adviceComputeVelocity and adviceComputeHeading, respectively

The aspectbox SafetyControlAB defines the aspect PowerOff. This aspect has one advice, adviceDrivingControl that proceed a method call if the hasPower is true.

```
(Aspectbox named: #SafetyControlAB)
  createAspectNamed: #PowerOff
  instanceVariableNames: 'hasPower'.

SafetyControlAB.PowerOff>>
      adviceDrivingControl
  | joinpoints |
  joinpoints := JointPointDescriptor
        targetClasses: {Joystick . Engine . Wheels}.
  ^AroundAdvice
    pointcut: joinpoints
    aroundBlock: [:receiver :arguments :aspect |
        hasPower ifTrue: [ aspect proceed ]
```

Aspects PowerOff and PositionAndVelocity described above have a common pointcut: public method of the class Engine. Because these two aspects belongs to different aspectboxes (SafetyControlAB and PositionControlAB, respectively), they do not conflict with each other.

### 4.3 Executing Code in an Aspectbox

Triggering a program execution in an aspectbox is achieved by the method eval:.

```
(Aspectbox named: #SafetyControlAB) eval: [
  | app |
  app := SafetyApplication new.
  app run].
```

The code above instantiates the class SafetyApplication and invokes the method run. The code invoked by this method run will benefit from aspects defined in SafetyControlAB (*i.e.,* PowerOff). Similarly, an application invoked in the aspectbox PositionControlAB will benefit from PositionAndVelocity without being affected by SafetyControlAB.

### 4.4 Absolute Isolation of Aspects

It is widely accepted that encapsulating different functionalities of a system in distinct modular units aids their comprehensibility and maintainability [15].

Figure 1 illustrates a modular architecture. Because it is closely linked to the physical and external physical mechanic events, the *driving control* unit needs special care and should not be altered by other units that are not necessary for its execution. Also, for safety reasons, the *position control* unit has to be built on top of the motion engine without affecting its execution. *Different concerns composed into a system have to be well modularized and isolated from the base system.*

The aspectboxes module system has the following properties:

- *Conflicts between aspects are avoided.* By living in different scopes, aspects are kept separated. Even if aspects defined in different aspectboxes have the same join points, there is no need to define precedence rules for composition ordering.
- *Minimal extension of the aspect language.* Combining the aspectboxes module system with AspectS [8] did not require any modification of the aspect language syntax. Static references contained in the definition of pointcuts are resolved using the classes visible in the aspectbox in which these pointcuts are defined in.

## 5 Implementation

A prototype of aspectboxes is implemented in Squeak. Figure 3 describes how the *safety control* and the *position control* are hooked into the *driving control* module.

**AspectS.** AspectS [8] is an approach to general-purpose aspect-oriented programming in the Squeak[3] Smalltalk environment [9]. It extends the Squeak metaobject protocol to accommodate the aspect modularity mechanism. In contrast to systems like AspectJ, weaving and unweaving in AspectS happens dynamically at runtime, on-demand, employing metaobject composition. Instead of introducing new language constructs, AspectS utilizes Squeak itself as its pointcut language. AspectS benefits from the expressiveness and uniformity of Squeak.

**Activation Blocks.** AspectS uses Method Wrappers [4] to instrument both message sends and receptions. Such wrappers support execution of additional code before, after, around, or instead of an existing method. The core of the aspect activation mechanism is implemented in the isActive method of the class MethodWrapper. All additional code provided by a wrapper is to be activated only if all activation blocks associated with it evaluate to true. Activation blocks are treated as predicate methods, returning either true or false as the outcome of their execution.

[3] Squeak is an open-source Smalltalk available from http://www.squeak.org

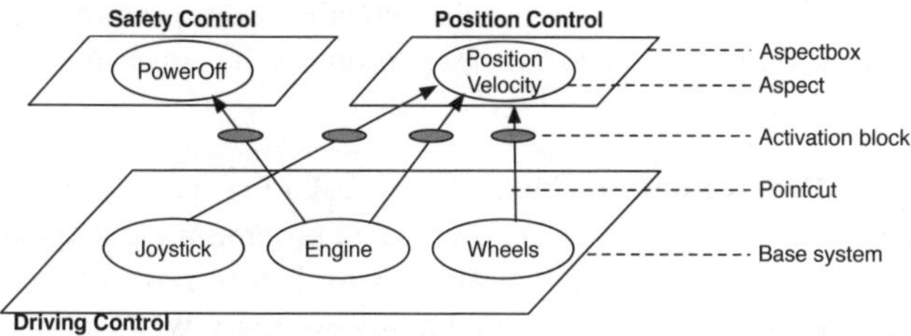


**Fig. 3.** The PowerOff. and PositionAndVelocity aspects hooked into the driving control module.

**Aspectboxes.** The aspectboxes module system is fully integrated in the Squeak environment. When an aspect is woven, activation blocks are created and placed at join points shadows. When the control flow of the application reaches a join point, the isActive methods is executed in order to determine if this potential join point is within the scope of an aspectbox defining this aspect to yield activation or not (*i.e.,* if it is associated with the current control flow).

## 6 Related Work

**AspectJ.** The pointcut language offered by AspectJ provides a mechanism to restrict a pointcut definition to a package or a class (*i.e.,* within and withincode pointcut primitives). The purpose of these constructs is to restrict the location of join points between a base system and an aspect, however advices hooked at those join points remain globally visible. Therefore, the restricting pointcut primitives of AspectJ do not help in scoping an aspect application.

**CaesarJ.** Aspects, packages and classes are unified in CaesarJ [1] under a single notion, a cclass. Aspect deployment can either be global or thread local.

Aspectboxes promotes a syntactic scoping of aspects: an aspect is scoped to the aspectbox that defines it. In CaesarJ, an aspect is scoped to the thread it was installed in.

**Classboxes.** The Classbox module system allows a class to be extended by means of class member additions and redefinitions. These extensions are visible in a locally and well-delimited scope. Several versions of a same class can coexist at the same time in the same system. Each class version corresponds to a particular view of this class [3].

Classboxes and aspectboxes have a common root which is the scoping mechanism for refinement. Whereas classboxes support structural refinement (*i.e.,* class members addition and redefinition), aspectboxes offer a scoping mechanism for behavioral refinement.

**Context-aware Aspects.** Context awareness promotes software program behaviour to depend on "context". Context-aware aspects [17] offers language constructs to handle contexts. A context is defined by the programmer as a plain standard object. The pointcut language is extended with primitives such as inContext(c) and createdInContext(c) that restrict a pointcut expression to a particular context c and to objects that were created in a context c, respectively.

Whereas context-aware aspects trigger the activation of aspects based on some arbitrary context activation function, aspectboxes promote the concurrent applications of aspects by restricting them to different scope.

**Context-oriented Programming.** ContextL [6], a CLOS-based implementation for Context-Oriented Programming, provides dedicated programming language constructs to associate partial class and method definitions with layers. Layers activation and deactivation is driven by the control flow of a running program. When a layer is activated, the partial definitions become part of the program until this layer is deactivated.

Whereas scoping software system refinement is the common problem for context-oriented programming and aspectboxes, the approaches are different. A layer in ContextL encapsulate structural definitions, whereas aspectboxes encapsulate behavioral definitions.

**AWED.** Aspects with Explicit Distribution (AWED) [14] is an approach for defining crosscutting behaviour on remote locations (*i.e.,* distributed applications). AWED is an aspect language supporting remote pointcuts, distributed advices and distributed aspects. A distributed aspect allows for state sharing and aspect instance to be distributed across multiple hosts.

## 7 Conclusions

Aspectboxes provide a new aspect modularity construct limiting the scope of aspect composition with a base software system. Modifications to the base system are visible only in the aspectbox the aspect is defined in. This allows one to deploy multiple concurrent modifications in the same base system, avoiding conflicting situations across aspectboxes.

In the work presented in this paper, an aspect cannot be imported from an aspectbox. The reason for this is that aspects are not generic (*i.e.,* cannot be applied to other base systems). As future work, we plan to refine the notion of import to enable reuse of aspects within multiple aspectboxes.

Our prototypical implementation is based on AspectS. It integrates the composition mechanisms of AspectS and Classboxes to achieve the desired composition and scoping behavior.

## Acknowledgements

We gratefully acknowledge the financial support of the Science Foundation Ireland and Lero — the Irish Software Engineering Research Centre. We also like to thank Parinaz Davari and Daniel Rostrup for their valuable comments.

## References

1. Aracic, I., Gasiunas, V., Mezini, M., Ostermann, K.: An overview of caesarj. Transactions on Aspect-Oriented Software Development 3880, 135–173 (2006)
2. Baille, G., Garnier, P., Mathieu, H., Pissard-Gibollet, R.: Le cycab de l'inria rhône-alpes. Technical Report RT-0229, INRIA (1999)

3. Bergel, A., Ducasse, S., Nierstrasz, O.: Classbox/J: Controlling the scope of change in Java. In: Proceedings of Object-Oriented Programming, Systems, Languages, and Applications (OOPSLA 2005), pp. 177–189. ACM Press, New York (2005)
4. Brant, J., Foote, B., Johnson, R., Roberts, D.: Wrappers to the Rescue. In: Jul, E. (ed.) ECOOP 1998. LNCS, vol. 1445, pp. 396–417. Springer, Heidelberg (1998)
5. Brichau, J., Mens, K., Volder, K.D.: Building Composable Aspect-Specific Languages with Logic Metaprogramming. In: Batory, D., Consel, C., Taha, W. (eds.) GPCE 2002. LNCS, vol. 2487. Springer, Heidelberg (2002)
6. Costanza, P., Hirschfeld, R.: Language constructs for context-oriented programming. In: Proceedings of the Dynamic Languages Symposium 2005 (2005)
7. Douence, R., Fradet, P., Südholt, M.: Composition, reuse and interaction analysis of stateful aspects. In: AOSD 2004: Proceedings of the 3rd international conference on Aspect-oriented software development, pp. 141–150. ACM Press, New York (2004)
8. Hirschfeld, R.: AspectJ(tm): Aspect-Oriented Programming in Java. In: Aksit, M., Mezini, M., Unland, R. (eds.) NODe 2002. LNCS, vol. 2591, pp. 1–1. Springer, Heidelberg (2003)
9. Ingalls, D., Kaehler, T., Maloney, J., Wallace, S., Kay, A.: Back to the future: The story of Squeak, A practical Smalltalk written in itself. In: Proceedings OOPSLA 1997, ACM SIGPLAN Notices, pp. 318–326. ACM Press, New York (1997)
10. Kiczales, G., Hilsdale, E., Hugunin, J., Kersten, M., Palm, J., Griswold, W.G.: An Overview of AspectJ. In: Knudsen, J.L. (ed.) ECOOP 2001. LNCS, vol. 2072, pp. 327–353. Springer, Heidelberg (2001)
11. Klaeren, H., Pulvermüller, E., Raschid, A., Speck, A.: Aspect Composition Applying the Design by Contract Principle. In: Butler, G., Jarzabek, S. (eds.) GCSE 2000. LNCS, vol. 2177, pp. 57–69. Springer, Heidelberg (2001)
12. Lopez-Herrejon, R., Batory, D., Lengauer, C.: A disciplined approach to aspect composition. In: PEPM 2006: Proceedings of the 2006 ACM SIGPLAN symposium on Partial evaluation and semantics-based program manipulation, pp. 68–77. ACM Press, New York (2006)
13. Nagy, I., Bergmans, L., Aksit, M.: Composing aspects at shared join points. In: Hirschfeld, R., Ryszard Kowalczyk, A.P., Weske, M. (eds.) Proceedings of International Conference NetObjectDays, NODe2005, Erfurt, Germany. Lecture Notes in Informatics, vol. P-69, Gesellschaft für Informatik (GI) (2005)
14. Navarro, L.D.B., Südholt, M., Vanderperren, W., Fraine, B.D., Suvée, D.: Explicitly distributed AOP using AWED. In: Proceedings of the 5th Int. ACM Conf. on Aspect-Oriented Software Development (AOSD 2006), ACM Press, New York (2006)
15. Parnas, D.L.: On the criteria to be used in decomposing systems into modules. CACM 15(12), 1053–1058 (1972)
16. Tanter, É.: Aspects of Composition in the Reflex AOP Kernel. In: Löwe, W., Südholt, M. (eds.) SC 2006. LNCS, vol. 4089, pp. 99–114. Springer, Heidelberg (2006)
17. Tanter, É., Gybels, K., Denker, M., Bergel, A.: Context-aware aspects. In: Löwe, W., Südholt, M. (eds.) SC 2006. LNCS, vol. 4089, pp. 229–244. Springer, Heidelberg (2006)

# On State Classes and Their Dynamic Semantics

Ferruccio Damiani[1], Elena Giachino[1], Paola Giannini[2], and Emanuele Cazzola[2]

[1] Dipartimento di Informatica, Università degli Studi di Torino
Corso Svizzera 185, 10149 Torino, Italy
{damiani,giachino}@di.unito.it

[2] Dipartimento di Informatica, Università del Piemonte Orientale
Via Bellini 25/G, 15100 Alessandria, Italy
giannini@mfn.unipm.it

**Abstract.** We introduce *state classes*, a construct to program objects that can be safely concurrently accessed. State classes model the notion of object's *state* (intended as some abstraction over the value of fields) that plays a key role in concurrent object-oriented programming (as the *state* of an object changes, so does its coordination behavior). We show how state classes can be added to Java-like languages by presenting STATEJ, an extension of JAVA with state classes. The operational semantics of the state class construct is illustrated both at an abstract level, by means of a core calculus for STATEJ, and at a concrete level, by defining a translation from STATEJ into JAVA.

**Keywords:** Java, concurrent object-oriented language, small-step semantics, core calculus, implementation by translation.

## 1 Introduction

The notion of object's state, intended as some abstraction on the values of fields, plays a key role in concurrent object-oriented programming. Various language constructs for expressing object's state abstractions have been proposed in the literature (see, e.g., [1] for a survey). We propose *state classes*, a programming feature that could be added to JAVA-like programming languages. The main novelties in our proposal are: (1) The ability of states to carry values, thanks to the fact that states may be parameterized by special fields, that we call *attributes*; and (2) The presence of a static type and effect system guaranteeing that, even though the state of the objects may vary through states with different attributes, no attempt will be made to access non-existing attributes (this is, for state attributes, the standard requirement that well typed programs cannot cause a *field not found error*).

This paper focuses on the dynamic semantics of state classes. Typing issues are addressed in [2]. The paper is organized as follows: Section 2 introduces STATEJ, an extension of JAVA with state classes, through an example. Section 3 gives the FSJ calculus (a core calculus for STATEJ). Section 4 outlines how STATEJ can be implemented by translation into plain JAVA. Sections 5 and 6 conclude by discussing related and further work, respectively.

J. Filipe, B. Shishkov, and M. Helfert (Eds.): ICSOFT 2006, CCIS 10, pp. 84–96, 2008.

```
public state class ReaderWriter {
 state FREE {
  public void shared() {this!!SHARED(1);}
  public void exclusive() {this!!EXCLUSIVE;}
 }
 state SHARED(int n) {
  public void shared() {n++;}
  public void releaseShared()
   {n--; if (n==0) this!!FREE;}
 }
 state EXCLUSIVE {
  public void releaseExclusive()
   {this!!FREE;}
 }
}
```

**Fig. 1.** A multiple-reader, single-writer lock

## 2 An Example

In this section we motivate STATEJ through an example. The *state class construct* is designed to program objects that can be safely concurrently accessed. Therefore, in a state class, all the fields are private and all the methods are synchronized (that is, they are executed in mutual exclusion on the receiver object). A state class may extend an *ordinary* (i.e., non-state) class, but only state classes may extend state classes. Each state class specifies a collection of states. Each state is parameterized by some special fields, called *attributes*, and declares some methods. The state of an object o can be changed only inside methods of o, by means of a state *transition statement*, this!!S($e_1, \ldots, e_n$), where "S" is the name of the target state and "$e_1, \ldots, e_n$" ($n \geq 0$) supply the values for all the attributes of S. An object belonging to a state class is always in one of the states specified in its class. Each state class constructor must set the state of the created object. The default constructor of the root of a hierarchy of state classes sets the state to the first state defined in the class.

The class ReaderWriter (in Fig. 1) implements a *multiple reader, single-writer lock* — see [3], for an implementation using traditional concurrency primitives in a dialect of MODULA 2, and [4], for an implementation using chords in POLYPHONIC C♯.

When a thread $e$ invokes a method m on an object o belonging to a state class (e.g., to the class ReaderWriter in Fig. 1), if either o is in a state that does not support the invoked method (e.g., shared invoked on an EXCLUSIVE ReaderWriter) or some other thread is executing a method on o, then the execution of $e$ is blocked until o reaches (because of the action of some other thread) a state where the invoked method is available and no other thread is executing a method on o.

The policy implemented by the ReaderWriter class above is prone to writers' starvation. The class ReaderWriterFair (in Fig. 2) extends the class ReaderWriter to implement a writer starvation free policy.

An extending class inherits all the states of the extended class, and may add/override methods and introduce new states. Thus, class ReaderWriteFair has states FREE,

```
public state class ReaderWriterFair
             extends ReaderWriter {
 state SHARED(int n) {
  public void exclusive()
   {this!!PENDING_WRITER(n);
    pre_exclusive();
    this!!EXCLUSIVE;}
 }
 state PENDING_WRITER(int n) {
  public void releaseShared()
   {n--; if (n==0) this!!PRE_EXCLUSIVE;}
 }
 state PRE_EXCLUSIVE {
  private void pre_exclusive() { }
 }
}
```

**Fig. 2.** A fair multiple-reader, single-writer lock

SHARED, EXCLUSIVE, PENDING_WRITER and PRE_EXCLUSIVE. When the request exclusive is received by an object o in state SHARED($n$), then the state of o is set to PENDING_WRITER($n$) and the method body suspends; in this state o can only execute up to $n$ requests of releaseShared; after the $n$-th such request, the state of o is set to PRE_EXCLUSIVE; in state PRE_EXCLUSIVE the method body for exclusive can continue, and will set the state of o to EXCLUSIVE.

The ReaderWriterFair class illustrates a common pattern in state class programming: the private method pre_exclusive has an empty body, and acts as a test that the receiver has reached the state PRE_EXCLUSIVE.

## 3 A Calculus for StateJ

This section gives syntax and operational semantics of FSJ, a minimal imperative core calculus for StateJ. FSJ models the innovative features of the state construct (namely state classes, state attributes and methods, and state transitions) and multi-threaded computations.

A FSJ program consists of a set of class definitions plus an expression to be evaluated, that we will call the *main expression* of the program.

### 3.1 Syntax

The abstract syntax of FSJ class declarations (L), class constructor declarations (K), state declarations (N), method declarations (M), and expressions (e) is given in Fig. 3. The metavariables A, B, C, and D range over class names; S ranges over state names; f and g range over attribute names; m ranges over method names; x ranges over method parameter names; and a, b, c, d, and e range over expressions.

We write "$\bar{\mathtt{e}}$" as a shorthand for a possibly empty sequence "$\mathtt{e}_1, \cdots, \mathtt{e}_n$" (and similarly for C, f, S, x) and write "$\bar{\mathtt{N}}$" as a shorthand for "$\mathtt{N}_1 \cdots \mathtt{N}_n$" with no commas (and

**Syntax:**

$$
\begin{array}{lll}
\texttt{L} & ::= & \texttt{state class C extends C } \{\texttt{K } \bar{\texttt{N}}\} \\
\texttt{K} & ::= & \texttt{C}(\bar{\texttt{C}}\ \bar{\texttt{f}})\{\texttt{this!!S}(\bar{\texttt{f}})\} \\
\texttt{N} & ::= & \texttt{state S } (\bar{\texttt{C}}\ \bar{\texttt{f}})\{\bar{\texttt{M}}\} \\
\texttt{M} & ::= & \texttt{C m } (\bar{\texttt{C}}\ \bar{\texttt{x}})\ \{\texttt{e}\} \\
\texttt{e} & ::= & \texttt{x} \mid \texttt{this} \mid \texttt{this.f} \mid \texttt{e; e} \mid \texttt{new C}(\bar{\texttt{e}}) \\
 & & \mid \texttt{this!!S}(\bar{\texttt{e}}) \mid \texttt{spawn(e)} \mid \texttt{e.m}(\bar{\texttt{e}})
\end{array}
$$

**Subtyping:**

$$
\texttt{C} <: \texttt{C} \qquad \frac{\texttt{C}_1 <: \texttt{C}_2 \qquad \texttt{C}_2 <: \texttt{C}_3}{\texttt{C}_1 <: \texttt{C}_3}
$$

$$
\frac{\texttt{state class C}_1 \texttt{ extends C}_2 \cdots \{\cdots\}}{\texttt{C}_1 <: \texttt{C}_2}
$$

**State attributes lookup:**

$$
\frac{\texttt{state class C} \cdots \{\cdots \texttt{state S}(\bar{\texttt{C}}\ \bar{\texttt{f}})\{\cdots\}\cdots\}}{attributes(\texttt{C},\texttt{S}) = \bar{\texttt{C}}\ \bar{\texttt{f}}}
$$

$$
\frac{\texttt{state class C extends D } \{\texttt{K } \bar{\texttt{N}}\} \qquad \texttt{S} \notin \bar{\texttt{N}}}{attributes(\texttt{C},\texttt{S}) = attributes(\texttt{D},\texttt{S})}
$$

**Method definition lookup:**

$$
\frac{\texttt{state class C} \cdots \{\texttt{K } \bar{\texttt{N}}\} \qquad \texttt{state S}\{\bar{\texttt{M}}\} \in \bar{\texttt{N}} \qquad \texttt{A m}(\bar{\texttt{A}}\ \bar{\texttt{x}})\ \{\texttt{e}\} \in \bar{\texttt{M}}}{mDef(\texttt{m},\texttt{C},\texttt{S}) = \texttt{A m}(\bar{\texttt{A}}\ \bar{\texttt{x}})\ \{\texttt{e}\}}
$$

$$
\frac{\texttt{state class C extends D } \{\texttt{K } \bar{\texttt{N}}\} \qquad (\texttt{S} \notin \bar{\texttt{N}} \text{ or } (\texttt{state S}\{\bar{\texttt{M}}\} \in \bar{\texttt{N}} \text{ and } \texttt{m} \notin \bar{\texttt{M}}))}{mDef(\texttt{m},\texttt{C},\texttt{S}) = mDef(\texttt{m},\texttt{D},\texttt{S})}
$$

**Fig. 3.** FSJ syntax, subtyping rules, and lookup functions

similarly for $\bar{\texttt{M}}$). We write the empty sequence as "•" and denote the concatenation of sequences using either comma or juxtaposition, as appropriate. We abbreviate operations on pair of sequences by writing "$\bar{\texttt{C}}\ \bar{\texttt{f}}$" for "$\texttt{C}_1\ \texttt{f}_1, \ldots, \texttt{C}_n\ \texttt{f}_n$", where $n$ is the length of $\bar{\texttt{C}}$ and $\bar{\texttt{f}}$. We assume that sequences of state declarations or names, attribute declarations or names, method parameter declarations or names, method declarations do not contain duplicate names.

The class declaration

$$\texttt{state class C extends D } \{\texttt{K } \bar{\texttt{N}}\}$$

defines a state class of name C with superclass D. The new class has a single constructor K and a set of states $\bar{\texttt{N}}$. The state declarations $\bar{\texttt{N}}$ may either refine (by adding/overrinding methods) states that are already present in D or add new states.

The constructor declaration $\texttt{C}(\bar{\texttt{C}}\ \bar{\texttt{f}})\ \{\texttt{this!!S}(\bar{\texttt{f}})\}$ specifies how to initialize the state and the state attributes of an instance of C. It takes exactly as many parameters as there are attributes of the state S and its body consists of a state transition statement.

The state declaration $\texttt{state S}(\bar{\texttt{C}}\ \bar{\texttt{f}})\ \{\bar{\texttt{M}}\}$ introduces a state with name S and attributes of names $\bar{\texttt{f}}$ and types $\bar{\texttt{C}}$. The declaration provides a suite of methods $\bar{\texttt{M}}$ that are available in the state S of the class C containing the state declaration. A state S declared in a class C inherits all the (not overridden) methods that are defined in the (possible) declarations of S contained in the superclasses of C.

The method declaration $\texttt{C m}\ (\bar{\texttt{C}}\ \bar{\texttt{x}})\ \{\texttt{e}\}$ introduces a method named m with result type C, parameters $\bar{\texttt{x}}$ of types $\bar{\texttt{C}}$, and body e. The variables $\bar{\texttt{x}}$ and the pseudo-variable this are bound in e.

The class declarations in a program must satisfy the following conditions: (1) Object is a distinguished class name whose declaration does *not* appear in the program; (2) For every class name C (except Object) appearing anywhere in the program, one and only one class with name C is declared in the program; and (3) The subtype relation induced by the class declarations in the program (denoted by <: and formally defined in the middle of Fig. 3) is acyclic. To simplify the notation in what follows (as in [5]), we always assume a *fixed* program.

The lookup functions are given at the bottom of Fig. 3. We write $\mathtt{S} \notin \bar{\mathtt{N}}$ to mean that no declaration of the state S is included in $\bar{\mathtt{N}}$, and $\mathtt{m} \notin \bar{\mathtt{M}}$ to mean that no declaration of the method m is included in $\bar{\mathtt{M}}$. Lookup of the attributes of a state S of a class C, written $attributes(\mathtt{C}, \mathtt{S})$, returns a sequence $\bar{\mathtt{C}}\,\bar{\mathtt{f}}$ pairing the type of each attribute declared in the state with its name. Lookup of the definition of the method m in the state S of a state class C is denoted by $mDef(\mathtt{m}, \mathtt{C}, \mathtt{S})$.[1] Note that $attributes(\mathtt{C}, \mathtt{S})$ and $mDef(\mathtt{m}, \mathtt{C}, \mathtt{S})$ are undefined when $\mathtt{C} = \mathtt{Object}$.[2]

### 3.2 Operational Semantics

In this section we introduce the operational semantics of FSJ, by defining the reduction rules that transform *configurations* representing multi-threaded computation. A configuration is a pair "$\bar{e}, \mathcal{H}$", where $\bar{e}$ is a sequence of $n \geq 1$ *runtime expressions* and $\mathcal{H}$ is a *heap* mapping *addresses* to *objects*. *Addresses*, ranged over by the metavariable $\iota$, are the elements of the denumerable set **I**. *Objects* are finite mappings associating: (1) the distinguished name "class" to a class name indicating the class of the object; (2) the distinguished name "state" to a state name indicating the state of the object; and (3) a mapping associating a finite number (possibly zero) of state attribute names to addresses. Objects will be denoted by $[\![\mathsf{class} : \mathtt{C},\ \mathtt{state} : \mathtt{S},\ \bar{\mathtt{f}} : \bar{\iota}]\!]$.

The first component of a configuration, $\bar{e}$, will be called "sequence of threads". A thread of computation is represented by the evaluation of a runtime expression $e_i$ in the heap $\mathcal{H}$. The different threads share the same heap $\mathcal{H}$. Threads do not have, as in full STATEJ and JAVA, an associated stack, keeping the association between parameters and values. In fact, since FSJ does not include assignment, method calls are evaluated by directly substituting the formal parameters and the metavariable this with the corresponding values (in FSJ the only values are addresses). We call the result of this substitution, which is no longer an expression of the source language, a *simple runtime expression*. Simple runtime expressions, ranged over by $s$, are obtained from the pseudo grammar defining expressions (in Fig. 3) by replacing the clauses "$\mathtt{x} \mid \mathtt{this} \mid \mathtt{this.f} \mid$" with the clauses "$\iota \mid \iota.\mathtt{f} \mid$" (see the top of Fig. 4).

*Runtime expressions*, ranged over by $e$, are defined by the grammar at top of Fig. 4. In FSJ every method is synchronized, therefore on method call the lock of the object receiving the call must be acquired, unless the call is inside a method of the object itself, in which case the call can proceed (the lock is *reentrant*). Moreover, when the method call is on a method not defined in the current state, the lock of the object must be released. This gives to other threads a chance to change the state of the object to a state in which the method is defined. Both these situations are modelled by particular *runtime expressions*: (1) $\mathtt{ret}(\iota, \mathtt{m}, e)$, where $e$ does not contain occurrences of $\mathtt{unlock}(\iota.\cdots(\cdots))$,

[1] In full STATEJ, like in JAVA, the lookup functions take into account method overloading, that (for simplicity) is not included in FSJ.

[2] In full STATEJ the class Object has several methods.

**Simple runtime expressions, runtime expressions, evaluation contexts, redexes, and auxiliary functions:**

$$
\begin{array}{lcl}
s & ::= & \iota \mid \iota.\mathtt{f} \mid s;s \mid \mathtt{new}\ \mathtt{C}(\bar{s}) \mid \iota\mathtt{!!S}(\bar{s}) \mid \mathtt{spawn}(s) \mid s.\mathtt{m}(\bar{s}) \\
e & ::= & \iota \mid \iota.\mathtt{f} \mid e;s \mid \mathtt{new}\ \mathtt{C}(\bar{\iota}, \dot{e}, \bar{s}) \mid \iota\mathtt{!!S}(\bar{\iota}, \dot{e}, \bar{s}) \mid \mathtt{spawn}(e) \mid e.\mathtt{m}(\bar{s}) \mid \iota.\mathtt{m}(\bar{\iota}, \dot{e}, \bar{s}) \\
 & & \mid \mathtt{ret}(\iota, \mathtt{m}, e) \mid \mathtt{unlock}(\iota.\mathtt{m}(\bar{\iota})) \\
\mathcal{E} & ::= & [\,] \mid \mathcal{E};s \mid \mathtt{new}\ \mathtt{C}(\bar{\iota}, \mathcal{E}, \bar{s}) \mid \iota\mathtt{!!S}(\bar{\iota}, \mathcal{E}, \bar{s}) \mid \mathtt{spawn}(\mathcal{E}) \mid \mathcal{E}.\mathtt{m}(\bar{s}) \mid \iota.\mathtt{m}(\bar{\iota}, \mathcal{E}, \bar{s}) \mid \mathtt{ret}(\iota, \mathtt{m}, \mathcal{E}) \\
r & ::= & \iota.\mathtt{f} \mid \iota;s \mid \mathtt{new}\ \mathtt{C}(\bar{\iota}) \mid \iota\mathtt{!!S}(\bar{\iota}) \mid \mathtt{spawn}(\iota) \mid \iota.\mathtt{m}(\bar{\iota}) \mid \mathtt{ret}(\iota, \mathtt{m}, \iota) \mid \mathtt{unlock}(\iota.\mathtt{m}(\bar{\iota}))
\end{array}
$$

$$
\begin{array}{lcl}
\mathit{lockedBy}(e) & = & \{\iota \mid \mathtt{ret}(\iota, \cdots, \cdots) \text{ is a subexpression of } e \text{ and } \mathtt{unlock}(\iota.\cdots(\cdots)) \text{ is not a subexpression of } e\} \\
\mathit{lockedBy}(e_1 \cdots e_n) & = & \bigcup_{1 \le i \le n} \mathit{lockedBy}(e_i)
\end{array}
$$

**Reduction rules:**

$$\frac{\mathcal{H}(\iota) = \mathsf{o} \qquad \mathsf{o}(\mathtt{f}) = \iota'}{\bar{a}\,\mathcal{E}[\iota.\mathtt{f}]\,\bar{c},\ \mathcal{H} \longrightarrow \bar{a}\,\mathcal{E}[\iota']\,\bar{c},\ \mathcal{H}} \quad \text{(R-ATTR)}$$

$$\bar{a}\,\mathcal{E}[\iota;\ s]\,\bar{c},\ \mathcal{H} \longrightarrow \bar{a}\,\mathcal{E}[s]\,\bar{c},\ \mathcal{H} \quad \text{(R-SEQ)}$$

$$\frac{\mathtt{state\ class\ C} \cdots \{\mathtt{C}(\bar{\mathtt{C}}\ \bar{\mathtt{f}})\{\mathtt{this!!S}(\bar{\mathtt{f}})\} \cdots\} \qquad \mathsf{o} = [\mathsf{class} : \mathtt{C},\ \mathsf{state} : \mathtt{S},\ \bar{\mathtt{f}} : \bar{\iota}] \qquad \iota \notin \mathrm{Dom}(\mathcal{H})}{\bar{a}\,\mathcal{E}[\mathtt{new}\ \mathtt{C}(\bar{\iota})]\,\bar{c},\ \mathcal{H} \longrightarrow \bar{a}\,\mathcal{E}[\iota]\,\bar{c},\ \mathcal{H}[\iota : \mathsf{o}]} \quad \text{(R-NEW)}$$

$$\frac{\mathcal{H}(\iota)(\mathtt{class}) = \mathtt{C} \qquad \mathit{attributes}(\mathtt{C}, \mathtt{S}) = \bar{\mathtt{C}}\,\bar{\mathtt{f}} \qquad \mathsf{o} = [\mathsf{class} : \mathtt{C},\ \mathsf{state} : \mathtt{S},\ \bar{\mathtt{f}} : \bar{\iota}]}{\bar{a}\,\mathcal{E}[\iota\mathtt{!!S}(\bar{\iota})]\,\bar{c},\ \mathcal{H} \longrightarrow \bar{a}\,\mathcal{E}[\iota]\,\bar{c},\ \mathcal{H}[\iota : \mathsf{o}]} \quad \text{(R-TRANS)}$$

$$\bar{a}\,\mathcal{E}[\mathtt{spawn}(\iota)]\,\bar{c},\ \mathcal{H} \longrightarrow \bar{a}\,\mathcal{E}[\iota]\,\bar{c}\ \iota.\mathtt{run}(),\ \mathcal{H} \quad \text{(R-SPAWN)}$$

$$\frac{\iota \notin \mathit{lockedBy}(\bar{a}\,\bar{c}) \qquad \mathcal{H}(\iota) = [\mathsf{class} : \mathtt{D},\ \mathsf{state} : \mathtt{S},\ \cdots] \qquad \mathit{mDef}(\mathtt{m}, \mathtt{D}, \mathtt{S}) = \mathtt{C}\ \mathtt{m}\ (\bar{\mathtt{C}}\ \bar{\mathtt{x}})\ \{\mathtt{e}\}}{\bar{a}\,\mathcal{E}[\iota.\mathtt{m}(\bar{\iota})]\,\bar{c},\ \mathcal{H} \longrightarrow \bar{a}\,\mathcal{E}[\mathtt{ret}(\iota, \mathtt{m}, \mathtt{e}[\mathtt{this} := \iota, \bar{\mathtt{x}} := \bar{\iota}])]\,\bar{c},\ \mathcal{H}} \quad \text{(R-INVK-1)}$$

$$\frac{\iota \in \mathit{lockedBy}(\mathcal{E}[\iota.\mathtt{m}(\bar{\iota})]) \qquad \mathcal{H}(\iota) = [\mathsf{class} : \mathtt{D},\ \mathsf{state} : \mathtt{S},\ \cdots] \qquad \mathit{mDef}(\mathtt{m}, \mathtt{D}, \mathtt{S}) \text{ undefined}}{\bar{a}\,\mathcal{E}[\iota.\mathtt{m}(\bar{\iota})]\,\bar{c},\ \mathcal{H} \longrightarrow \bar{a}\,\mathcal{E}[\mathtt{unlock}(\iota.\mathtt{m}(\bar{\iota}))]\,\bar{c},\ \mathcal{H}} \quad \text{(R-INVK-2)}$$

$$\frac{\iota \notin \mathit{lockedBy}(\bar{a}\,\bar{c}) \qquad \mathcal{H}(\iota) = [\mathsf{class} : \mathtt{D},\ \mathsf{state} : \mathtt{S},\ \cdots] \qquad \mathit{mDef}(\mathtt{m}, \mathtt{D}, \mathtt{S}) = \mathtt{C}\ \mathtt{m}\ (\bar{\mathtt{C}}\ \bar{\mathtt{x}})\ \{\mathtt{e}\}}{\bar{a}\,\mathcal{E}[\mathtt{unlock}(\iota.\mathtt{m}(\bar{\iota}))]\,\bar{c},\ \mathcal{H} \longrightarrow \bar{a}\,\mathcal{E}[\mathtt{ret}(\iota, \mathtt{m}, \mathtt{e}[\mathtt{this} := \iota, \bar{\mathtt{x}} := \bar{\iota}])]\,\bar{c},\ \mathcal{H}} \quad \text{(R-UNLOCK)}$$

$$\bar{a}\,\mathcal{E}[\mathtt{ret}(\iota, \mathtt{m}, \iota_0)]\,\bar{c},\ \mathcal{H} \longrightarrow \bar{a}\,\mathcal{E}[\iota_0]\,\bar{c},\ \mathcal{H} \quad \text{(R-RET)}$$

**Fig. 4.** FSJ (simple) runtime expressions, evaluation contexts, redexes, auxiliary functions, and reduction rules.

specifies that a thread is currently holding the lock of the receiver $\iota$, in order to evaluate the expression $e$, which represents the body of the method m, and (2) $\mathtt{unlock}(\iota.\mathtt{m}(\bar{\iota}))$ specifies that the lock of $\iota$ has been released in order to give a chance to another thread to change the state of $\iota$ to a state in which m is defined. Note that, the definition of the syntax for runtime expressions implies that there can be nested `ret` expressions but only one `unlock`. The metavariables $a$, $b$, $c$, $d$, and $e$ range over runtime expressions. We write $\bar{a}$ as a shorthand for a possibly empty sequence $a_1 \cdots a_n$ and $\dot{a}$ as a shorthand for a possibly empty sequence of length almost one. The function $\mathit{lockedBy}(\bar{e})$, defined in Fig. 4, returns the set of addresses that are locked by the thread sequence $\bar{e}$.

The reduction relation has the form "$\bar{a}\,b_1\,\bar{c}, \mathcal{H}_1 \longrightarrow \bar{a}\,b_2\,\bar{c}\,\dot{d}, \mathcal{H}_2$", read "configuration $\bar{a}\,b_1\,\bar{c}, \mathcal{H}_1$ reduces to configuration $\bar{a}\,b_2\,\bar{c}\,\dot{d}, \mathcal{H}_2$ in one step". The (empty or singleton) sequence $\dot{d}$ indicates that a new thread might have been spawned because of the reduction of a `spawn` expression. We write $\longrightarrow^\star$ for the reflexive and transitive closure of $\longrightarrow$.

By using the definition of *evaluation context* and *redex* (see $\mathcal{E}$ and $r$ in Fig. 4), the reduction rules ensure that inside each thread the computation follows a call-by-value left-to-right reduction strategy. This implies that expressions such as `ret` and `unlock` can only be preceded by values and followed by simple runtime expressions, which do not contain `ret` and `unlock` (see the definition of $s$ and $e$ in Fig. 4).

The following property asserts that a context can be decomposed in a unique way in sub-contexts showing the activation stack of method calls.

*Property 1 (Unique Decomposition).* Every evaluation context $\mathcal{E}$ can be written as

$$\underbrace{\mathcal{E}_{1,1}[\texttt{ret}(\iota_1,\texttt{m}_{1,1},\cdots\mathcal{E}_{1,q_1}[\texttt{ret}(\iota_1,\texttt{m}_{1,q_1}}_{q_1}\cdots$$
$$\underbrace{\mathcal{E}_{p,1}[\texttt{ret}(\iota_p,\texttt{m}_{p,1},\cdots\mathcal{E}_{p,q_p}[\texttt{ret}(\iota_p,\texttt{m}_{p,q_p}}_{q_p},\mathcal{E}_0)]\cdots)]$$
$$\cdots)]\cdots)],$$

where $\mathcal{E}_{1,1}$, ..., $\mathcal{E}_{1,q_1}$, ..., $\mathcal{E}_{p,1}$, ..., $\mathcal{E}_{p,q_p}$ ($p \geq 0$, $q_1 \geq 1$, ..., $q_p \geq 1$) and $\mathcal{E}_0$ do not contain $\texttt{ret}(\cdots)$ subexpressions.

The reduction rules are given at the bottom of Fig. 4. Each reduction rule rewrites a configuration of the form "$\bar{a}\,\mathcal{E}[r]\,\bar{c}, \mathcal{H}_1$", where $\mathcal{E}$ is an evaluation context and $r$ is a redex, into a configuration of the form "$\bar{a}\,\mathcal{E}[e]\,\bar{c}\,\dot{d}, \mathcal{H}_2$". The metavariable $\mathsf{o}$ ranges over objects. We use $\mathcal{H}[\iota : \mathsf{o}]$ to denote the heap such that $\mathcal{H}[\iota : \mathsf{o}](\iota) = \mathsf{o}$ and $\mathcal{H}[\iota : \mathsf{o}](\iota') = \mathcal{H}(\iota')$, for $\iota' \neq \iota$.

The reduction rules for attribute selection, (R-ATTR), and sequential composition, (R-SEQ), are standard. The rule for object creation, (R-NEW), stores the newly created object at a fresh address of the heap and returns the address. The pseudo fields class and state, and the parameters of the initial state are initialized as specified by the class constructor. The rule for state transition, (R-TRANS), changes the current state of the object and returns its address. Rule (R-SPAWN) replaces the `spawn` expression with the address $\iota$ and adds a new thread evaluating the call of the method `run` on the object at $\iota$. Rule (R-INVK-1) is applied if the method `m` is defined in the current state of the receiver, $\iota$, and no other thread holds the lock of $\iota$. The expression produced replaces the call with $\texttt{ret}(\iota, \texttt{m}, e')$, indicating that the current thread holds the lock of $\iota$. The expression $e'$ is the body of the method `m` in which `this` and the formal parameters are replaced with the address $\iota$ and the actual parameters. Rule (R-INVK-2) is applied if the method `m` is not defined in the current state of the receiver and the current thread holds the lock of $\iota$. In this case, the lock of $\iota$ must be released and the thread must wait that some other thread changes the state of $\iota$ to a state in which `m` is defined. This is achieved by replacing the method call redex with the expression $\texttt{unlock}(\iota.\texttt{m}(\bar{\iota}))$. Note that, since the current thread had the lock of $\iota$, the newly introduced `unlock` expression is a subexpression of an expression $\texttt{ret}(\iota, \texttt{m}', e')$ for some $\texttt{m}'$ and $e'$. Rule (R-UNLOCK) replaces the expression $\texttt{unlock}(\iota.\texttt{m}(\bar{\iota}))$, if $\iota$ is not locked and the method `m` is defined in its state, with $\texttt{ret}(\iota, \texttt{m}, e')$, where $e'$ is the body of the method `m` in which `this` and the formal parameters are replaced with the address $\iota$ and the actual parameters. Rule(R-RET), that applies when the body of the method `m` on object $\iota$ has been evaluated completely, producing a value, releases the lock of $\iota$ by removing the $\texttt{ret}(\iota, \texttt{m}, \iota_0)$ subexpression.

*Example 1 (Application of the reduction rules).* First we define the following classes `CR` and `CW` representing the class of threads that have a shared access to a `ReaderWriter` object `rw` and the class of threads that have an exclusive access to it, respectively.

```
state class CR extends Object {
  CR(ReaderWriter rw) { this!!S(rw) }
```

```
  state S (ReaderWriter rw) {
    Object run () {
      rw.shared();
      ...
      rw.releaseShared();
      this.run() } }
}

state class CW extends Object {
  CW(ReaderWriter rw) { this!!S(rw) }
  state S (ReaderWriter rw) {
    Object run () {
      rw.exclusive();
      ...
      rw.releaseExclusive();
      this.run() } }
}
```

We consider as the *main expression* of the program, that is the expression to be evaluated,

$$\texttt{spawn}(\texttt{new}\,\texttt{CR}(\iota)); (\texttt{new}\,\texttt{CW}(\iota)).\texttt{run}(),$$

where $\iota$ is a `ReaderWriter` object, so the computation starts from the following *configuration*:

$$\texttt{spawn}(\texttt{new}\,\texttt{CR}(\iota)); (\texttt{new}\,\texttt{CW}(\iota)).\texttt{run}(),\ \mathcal{H}$$

where $\mathcal{H} = \iota : [\![\mathsf{class} : \texttt{ReaderWriterFair},\ \mathsf{state} : \texttt{FREE}]\!]$.

A possible computation is as in Fig. 5, where SH stands for SHARED, PW stands for PENDING_WRITER, EX stands for EXCLUSIVE, and PE stands for PRE_EXCLUSIVE. We adopt the following notations: (1) Threads $e_1, e_2$ being part of the *configuration* are written $\begin{pmatrix} e_1 \\ e_2 \end{pmatrix}$; (2) Redexes are underlined; (3) Redexes of suspended threads are underlined and written in grey; (4) the arrow $\Longrightarrow$ indicates one step of reduction for each thread of the sequence; (5) In `ret` expressions we omit method names. As we see in Fig. 5, in the example we assumed to have integers, decrement and if-statement. These are assumed, in line (#), to be reduced following the standard semantics.

## 4 From StateJ to Java

This section briefly illustrates a translation from StateJ to plain Java. The basic idea of the translation is to map a state class into a Java class using `synchronized` methods and the primitives `wait`() and `notify`(). A class contains a field indicating the current state of the object, and methods corresponding to the methods of the original StateJ class. The translation can be briefly described as follows.

**Method.** Methods defined in more than one state have more than one body. To be able to execute different bodies in different states our translation creates a unique `synchronized` method containing all the different bodies. At run-time, when the method is called,

$$\mathtt{spawn(new\ CR(\iota));\ (new\ CW(\iota)).run()},\ \mathcal{H} \longrightarrow \mathtt{spawn(\iota');\ (new\ CW(\iota)).run()},\ \mathcal{H}_1 \longrightarrow \quad \text{first by (R-New) and second by (R-Spawn)}$$

$$\begin{pmatrix}\mathtt{\iota';\ (new\ CW(\iota)).run()} \\ \mathtt{\iota'.run()}\end{pmatrix},\ \mathcal{H}_1 \Longrightarrow \quad \text{by (R-Seq) and (R-Invk-1)}$$

$$\begin{pmatrix}\mathtt{(new\ CW(\iota)).run()} \\ \mathtt{ret(\iota',\ \iota.shared();\ \ldots;\ \iota.releaseShared();\ \iota'.run())}\end{pmatrix},\ \mathcal{H}_1 \Longrightarrow \quad \text{by (R-New) and (R-Invk-1)}$$

$$\begin{pmatrix}\mathtt{\iota''.run()} \\ \mathtt{ret(\iota',\ ret(\iota,\ \iota!!SH(1));\ \ldots;\ \iota.releaseShared();\ \iota'.run())}\end{pmatrix},\ \mathcal{H}_2 \Longrightarrow \quad \text{by (R-Invk-1) and (R-Trans)}$$

$$\begin{pmatrix}\mathtt{ret(\iota'',\ \iota.exclusive();\ldots;\ \iota.releaseExclusive();\ \iota''.run())} \\ \mathtt{ret(\iota',\ ret(\iota,\ \iota);\ \ldots;\ \iota.releaseShared();\ \iota'.run())}\end{pmatrix},\ \mathcal{H}_3 \longrightarrow \quad \text{by (R-Ret)}$$

$$\begin{pmatrix}\mathtt{ret(\iota'',\ \iota.exclusive();\ \ldots;\ \iota.releaseExclusive();\ \iota''.run())} \\ \mathtt{ret(\iota',\ \iota;\ \ldots;\ \iota.releaseShared();\ \iota'.run())}\end{pmatrix},\ \mathcal{H}_3 \longrightarrow^{\star} \quad \text{by first applying (R-Invk-1) and (R-Seq)}$$

$$\begin{pmatrix}\mathtt{ret(\iota'',\ ret(\iota,\ \iota!!PW(1);\ \iota.pre\_exclusive();\ \iota!!EX);\ \ldots;\ \iota.releaseExclusive();\ \iota''.run())} \\ \mathtt{ret(\iota',\ \iota.releaseShared();\ \iota'.run())}\end{pmatrix},\ \mathcal{H}_3 \longrightarrow \quad \text{by (R-Trans)}$$

$$\begin{pmatrix}\mathtt{ret(\iota'',\ ret(\iota,\ \iota;\ \iota.pre\_exclusive();\ \iota!!EX);\ \ldots;\ \iota.releaseExclusive();\ \iota''.run())} \\ \mathtt{ret(\iota',\ \iota.releaseShared();\ \iota'.run())}\end{pmatrix},\ \mathcal{H}_4 \longrightarrow \quad \text{by (R-Seq)}$$

$$\begin{pmatrix}\mathtt{ret(\iota'',\ ret(\iota,\ \iota.pre\_exclusive();\ \iota!!EX);\ \ldots;\ \iota.releaseExclusive();\ \iota''.run())} \\ \mathtt{ret(\iota',\ \iota.releaseShared();\ \iota'.run())}\end{pmatrix},\ \mathcal{H}_4 \longrightarrow \quad \text{by (R-Invk-2)}$$

$$\begin{pmatrix}\mathtt{ret(\iota'',\ ret(\iota,\ unlock(\iota.pre\_exclusive());\ \iota!!EX);\ \ldots;\ \iota.releaseExclusive();\ \iota''.run())} \\ \mathtt{ret(\iota',\ \iota.releaseShared();\ \iota'.run())}\end{pmatrix},\ \mathcal{H}_4 \longrightarrow \quad \text{by (R-Invk-1)}$$

$$(\#) \quad \begin{pmatrix}\mathtt{ret(\iota'',\ ret(\iota,\ unlock(\iota.pre\_exclusive());\ \iota!!EX);\ \ldots;\ \iota.releaseExclusive();\ \iota''.run())} \\ \mathtt{ret(\iota',\ ret(\iota,\ n--;\ if\ (n=0)\ \iota!!PE);\ \iota'.run())}\end{pmatrix},\ \mathcal{H}_4 \longrightarrow^{\star}$$

$$\begin{pmatrix}\mathtt{ret(\iota'',\ ret(\iota,\ unlock(\iota.pre\_exclusive());\ \iota!!EX);\ \ldots;\ \iota.releaseExclusive();\ \iota''.run())} \\ \mathtt{ret(\iota',\ \iota;\ \iota'.run())}\end{pmatrix},\ \mathcal{H}_5 \Longrightarrow \quad \text{by (R-Unlock) and (R-Seq)}$$

$$\begin{pmatrix}\mathtt{ret(\iota'',\ ret(\iota,\ ret(\iota,\ \_);\ \iota!!EX);\ \ldots;\ \iota.releaseExclusive();\ \iota''.run())} \\ \mathtt{ret(\iota',\ \iota'.run())}\end{pmatrix},\ \mathcal{H}_5 \longrightarrow^{\star} \quad \text{by (R-Ret) and (R-Invk-1)}$$

$$\begin{pmatrix}\mathtt{ret(\iota'',\ ret(\iota,\ \iota.releaseExclusive();\ \iota''.run()))} \\ \mathtt{ret(\iota',\ ret(\iota',\ \ldots))}\end{pmatrix},\ \mathcal{H}_6 \longrightarrow^{\star} \cdots$$

where

$\mathcal{H} = \iota : [\text{class} : \mathtt{ReaderWriterFair},\ \mathtt{state} : \mathtt{FREE}]$

$\mathcal{H}_1 = \mathcal{H}[\iota' : [\text{class} : \mathtt{CR},\ \mathtt{state} : \mathtt{S},\ \mathtt{rw} : \iota]]$ $\quad \mathcal{H}_2 = \mathcal{H}_1[\iota'' : [\text{class} : \mathtt{CW},\ \mathtt{state} : \mathtt{S},\ \mathtt{rw} : \iota]]$

$\mathcal{H}_3 = \mathcal{H}_2[\iota : [\text{class} : \mathtt{ReaderWriterFair},\ \mathtt{state} : \mathtt{SH},\ \mathtt{n} : 1]]$ $\quad \mathcal{H}_4 = \mathcal{H}_3[\iota : [\text{class} : \mathtt{ReaderWriterFair},\ \mathtt{state} : \mathtt{PW},\ \mathtt{n} : 1]]$

$\mathcal{H}_5 = \mathcal{H}_4[\iota : [\text{class} : \mathtt{ReaderWriterFair},\ \mathtt{state} : \mathtt{PE}]]$ $\quad \mathcal{H}_6 = \mathcal{H}_5[\iota : [\text{class} : \mathtt{ReaderWriterFair},\ \mathtt{state} : \mathtt{EX}]]$

**Fig. 5.** An example of reduction

we have to check the current state of the object, and see whether the method was defined in this state or not. In case it is defined, then the corresponding body is executed, otherwise the thread calls a `wait()` putting it in hold. To keep the information on the methods defined in a certain state we use a hash table. Due to the limitation of the switch statement of JAVA, states are codified by the primitive type `int`. For example the following class

```
state class Ex extends Object {
  Ex() { this!!A(); }
  state A () {
    Object m() { /* body of m in A */ } }
  state B () {
    Object m() { /* body of m in B */ } } }
```

is translated into

```
class Ex extends Object {
  Ex() { ... }
  final static int A = 1;
  final static int B = 2;
  Hashtable stateMethods;
  int currentState;
```

```
  synchronized Object m() {
    while (!existsInCurrentState) wait();
    switch (currentState) {
      case A : /* body of m in A */ break;
      case B : /* body of m in B */ break;
} } }
```

where the existence of a method in a given state and its selection are done using the hash table of methods.

**State Transition.** The state transition expression `this!!A()` is translated into

```
   currentState = A;   notifyAll();
```

so in addition to change the state of the objects it notifies all the threads waiting for the lock of the current object. When the current thread will release the lock the notified threads will compete to get it to have a chance to see whether the method that caused the waiting is defined in the current state. If the method is defined, then the thread can proceed, otherwise it calls a `wait()`. Due to the non deterministic nature of JAVA scheduling we cannot insure the order in which notified threads will be waken up.

**Constructor.** The constructor of the translated class should initialize the hash table and then include the translation of the constructor of the original class.

**Inheritance.** A state class may extend another class (either state or not). In the subclass we inherit all the states and may add others. Therefore, we have to be careful to clashes of constants of state. Moreover, methods may be added/redefined. For instance method `exclusive()` of the example in Sect. 2, is defined in state `FREE` of `ReaderWriter`, and redefined in state `SHARED` of `ReaderWriterFair`. When a method is redefined in its translation we use the `default` clause as follows.

```
class ReaderWriterFairF
      extends ReaderWriter {
  ...
  synchronized void exclusive () {
    while (!existsInCurrentState) wait();
    switch (currentState) {
    case SHARED:
      currentState =PENDING_WRITER;
      notifyAll();
      pre_exclusive();
      currentState =EXCLUSIVE;
      notifyAll();
      break;
    default :
      super.exclusive;
      break; } } }
```

The current implementation of the translator (www.di.unito.it/~giannini/stateJimpl/) takes as input a program written in JAVA 1.4 extended with state classes *with attribute-free states* (attributes can be straightforwardly codified by class fields; however, their implementation would require to implement the type and effect analysis). The translation uses the tool for Language Recognition ANTLR, see [6], and the StringTemplate tecnology, see [7]. We first made a JAVA 1.4 to JAVA 1.4 translation taking advantage of the grammar defined by Parr and then modified the grammar to include our state related constructs. The use of ANTLR and StringTemplate makes the translator easily adaptable to different translation schemes and also to addition to the input language.

## 5 Related Work

According to [1] `states` provide a *boundary coordination* mechanism (we refer to Sect. 4.2 of [1] for a survey of several COOLs with boundary coordination). In particular, the state class construct is related to the *actor model* [8] and to the *behaviour abstraction* and *behaviour/enable sets* proposals [9,10].

At the best of our knowledge, the main novelties in our proposal are: the ability of states to carry values (thanks to the presence of attributes); the formalization of an abstract operational semantics of a notion of state for expressing coordination in JAVA-like languages; and the presence of a static type and effect system (presented in [2]) guaranteeing that during the execution there cannot be any access to undefined attributes of objects. Type systems for concurrent objects have been investigated in the literature, see, e.g., "regular object types" [11], the TYCO object calculus [12], and the $Fickle_{\mathsf{MT}}$ proposal [13].

Various improvements of the concurrency model of JAVA-like languages have been proposed. In JOIN JAVA [14] and POLYPHONIC C♯ [4] the synchronization mechanism relies on the *join pattern*, called *chord* in POLYPHONIC C♯, construct. Chords can be used to codify the state of an object through the pattern (illustrated, for instance, in [4]) of using private asynchronous method to carry object state. However, this pattern could be misused leading to deadlock or errors. In STATEJ the notion of object state is in the language definition, thus eliminating the possibility of many of such errors. In JEEG [15] the synchronization conditions on an object o are expressed with *linear temporal logic constraints* involving the value of fields and the method invocation history of o. These constraints could be used to codify the state of an object o. However, state attributes have to be mapped on object fields and there is no way to express the fact that some fields should be accessible only in some states.

STATEJ (as JOIN JAVA, POLYPHONIC C♯, and JEEG) focuses on a specific coordination mechanishm. The JR programming language [16] takes a different approach: it extends JAVA providing a rich concurrency model with a variety of mechanisms. None of this mechanisms directly models the notion of object state.

## 6 Future Work

The current prototypical implementation of STATEJ (www.di.unito.it/~giannini/stateJimpl/) is based on the translation scheme outlined in Sect. 4. It consists of a preprocessor that maps code written in JAVA 1.4 extended with state classes into plain JAVA. The current approach favors simplicity over efficiency. Its major drawback is that each state transition of an object o notifies *all* the threads waiting for *any* state of o. Note that, notifying just the threads waiting for the target state of the transition would not represent a significative improvement, since multiple state transitions may occur before the lock on `o` is released. A more significative improvement would be moving notification from state transition on `o` to lock release on `o`: this would allow notifying just the threads waiting for the current state of `o`. Note that, however, all but the first (according to the scheduling mechanism of JAVA) of such threads have to sleep again. We are currently investigating a quite different approach that support selective wakeups. It can be roughly described as follows:

- Each object `o` is equipped with a set of FIFO queues (one for each state).
- Whenever a thread invokes a method `m` on `o`, IF `o` is locked by some other thread OR `m` is not available in the current state of `o`
  - THEN the thread is suspended and enqueued in all the queues associated to the `states` of `o` where `m` is available, and the lock on `o` (if held by the suspended thread) is released
  - ELSE the method executed and the lock on `o` (if not already held by the invoking thread) is taken.
- Whenever the lock on `o` is released, IF the queue associated to the current state of `o` is not empty, THEN a thread `e` is extracted from the queue, removed from all the other queues, resumed, and it takes the lock on `o`.

Other future work includes: Refinement of the type and effect system given in [2]; Further investigations on the expressivity of the state class construct and on its integration in JAVA-like languages (by analyzing the interaction of state classes and their types with the advanced features of JAVA-like languages); Development of a new prototype (based on the translation scheme outlined above) including state attributes and the related type and effect analysis; and Development of benchmarks.

## References


1. Philippsen, M.: A Survey of Concurrent Object-Oriented Languages. Concurrency Computat.: Pract. Exper. 12, 917–980 (2000)
2. Damiani, F., Giachino, E., Giannini, P., Cameron, N., Drossopoulou, S.: A state abstraction for coordination in java-like languages. In: Electronic proceedings of FTfJP 2006 (2006), `www.cs.ru.nl/ftfjp/`
3. Birrel, A.D.: An introduction to programming with threads. Technical Report 35, DEC SRC (1989)
4. Benton, N., Cardelli, L., Fournet, C.: Modern Concurrency Abstractions for C♯. ACM TOPLAS 26, 769–804 (2004)
5. Igarashi, A., Pierce, B., Wadler, P.: Featherweight Java: A minimal core calculus for Java and GJ. ACM TOPLAS 23, 396–450 (2001)
6. Parr, T.: project group: ANTLR Reference Manual, Version 2.7.5 (2005), `http://www.antlr.org/doc/index.html`
7. Parr, T.: StringTemplate Documentation, (2003-2005), `http://www.stringtemplate.org/doc/doc.html`
8. Agha, G.A.: ACTORS: A Model of Concurrency Computation in Distribuited Systems. MIT Press, Cambridge (1986)

9. Kafura, D.G., Lavender, R.G.: Concurrent object-oriented languages and the inheritance anomaly. In: Casavant, T., Tvrdil, P., Plásil, F. (eds.) Parallel Computers: Theory and Practice, pp. 221–264. IEEE Press, Los Alamitos (1996)
10. Tomlinson, C., Singh, V.: Inheritance and synchronization with enabled-sets. In: OOPSLA 1989, pp. 103–112. ACM Press, New York (1989)
11. Nierstrasz, O.: Regular Types for Active Objects. In: OOPSLA 1993. ACM SIGPLAN Notices, vol. 28, pp. 1–15 (1993)
12. Ravara, A., Vasconcelos, V.T.: Typing Non-uniform Concurrent Objects. In: Palamidessi, C. (ed.) CONCUR 2000. LNCS, vol. 1877, pp. 474–488. Springer, Heidelberg (2000)
13. Damiani, F., Dezani-Ciancaglini, M., Giannini, P.: On re-classification and multithreading. JOT, 3, 5–30 (2004), `http://www.jot.fm`; Special issue: OOPS track at SAC 2004
14. Itzstein, G.S., Kearney, D.: Join Java: an alternative concurrency semantics for Java. Technical Report ACRC-01-001, Univ. of South Australia (2001)
15. Milicia, G., Sassone, V.: Jeeg: Temporal Constraints for the Synchronization of Concurrent Objects. Concurrency Computat.: Pract. Exper. 17, 539–572 (2005)
16. Keen, A.W., Ge, T., Maris, J.T., Olsson, R.A.: JR: Flexible distributed programming in an extended java. TOPLAS 26, 578–608 (2004)

# Software Implementation of the IEEE 754R Decimal Floating-Point Arithmetic

Marius Cornea, Cristina Anderson, and Charles Tsen

Intel Corporation

**Abstract.** The IEEE Standard 754-1985 for Binary Floating-Point Arithmetic [1] is being revised [2], and an important addition to the current text is the definition of decimal floating-point arithmetic [3]. This is aimed mainly to provide a robust, reliable framework for financial applications that are often subject to legal requirements concerning rounding and precision of the results in the areas of banking, telephone billing, tax calculation, currency conversion, insurance, or accounting in general. Using binary floating-point calculations to approximate decimal calculations has led in the past to the existence of numerous proprietary software packages, each with its own characteristics and capabilities. New algorithms are presented in this paper which were used for a generic implementation in software of the IEEE 754R decimal floating-point arithmetic, but may also be suitable for a hardware implementation. In the absence of hardware to perform IEEE 754R decimal floating-point operations, this new software package that will be fully compliant with the standard proposal should be an attractive option for various financial computations. The library presented in this paper uses the binary encoding method from [2] for decimal floating-point values. Preliminary performance results show one to two orders of magnitude improvement over a software package currently incorporated in GCC, which operates on values encoded using the decimal method from [2].

**Keywords:** IEEE 754R, IEEE 754, Floating-Point, Binary Floating-Point, Decimal Floating-Point, Basic Operations, Algorithms, Financial Computation, Financial Calculation.

## 1 Introduction

Binary floating-point arithmetic can be used in most cases to approximate decimal calculations. However errors may occur when converting numerical values between their binary and decimal representations, and errors can accumulate differently in the course of a computation depending on whether it is carried out using binary or decimal floating-point arithmetic.

For example, the following simple C program will not have in general the expected output b=7.0 for a=0.0007.

```
main () {
      float a, b;
      a = 7/10000.0;
      b = 10000.0 * a;
```

J. Filipe, B. Shishkov, and M. Helfert (Eds.): ICSOFT 2006, CCIS 10, pp. 97–109, 2008.

```
        printf ("a = %x = %10.10f\n",
        *(unsigned int *)&a, a);
        printf ("b = %x = %10.10f\n",
        *(unsigned int *)&b, b);
  }
```

(The value 7.0 has the binary encoding 0x40e00000.) The actual output on a system that complies with the IEEE Standard 754 will be:

```
a = 3a378034 = 0.0007000000
b = 40dfffff = 6.9999997504
```

Such errors are not acceptable in many cases of financial computations, mainly because legal requirements mandate how to determine the rounding errors - in general following rules that humans would use when performing the same computations on paper, and in decimal. Several software packages exist and have been used for this purpose so far, but each one has its own characteristics and capabilities such as precision, rounding modes, operations, or internal storage formats for numerical data. These software packages are not compatible with each other in general. The IEEE 754R standard proposal attempts to resolve these issues by defining all the rules for decimal floating-point arithmetic in a way that can be adopted and implemented on all computing systems in software, in hardware, or in a combination of the two. Using IEEE 754R decimal floating-point arithmetic, the previous example could then become:

```
main () {
        decimal32 a, b;
        a = 7/10000.0;
        b = 10000.0 * a;
        printf ("a = %x = %10.10fd\n",
        *(unsigned int *)&a, a);
        printf ("b = %x = %10.10fd\n",
        *(unsigned int *)&b, b);
}
```

(The hypothetical format descriptor %fd is used for printing decimal floating-point values.) The output on a system complying with the IEEE Standard 754R proposal would then represent the result without any error:

```
a = 30800007 = 0.0007000000
b = 32800007 = 7.0000000000
```

(The IEEE 754R binary encoding for decimal floating-point values was used in this example.) The following section summarizes the most important aspects of the IEEE 754R decimal floating-point arithmetic definition.

## 2 IEEE 754R Decimal Floating-Point

The IEEE 754R standard proposal defines three decimal floating-point formats with sizes of 32, 64, and 128 bits. Two encodings for each of these formats are specified: a decimal-based encoding which is best suited for certain possible hardware implementations of the decimal arithmetic [4], and a binary-based encoding better suited for software implementations on systems that support the IEEE 754 binary floating-point arithmetic in hardware [5]. The two encoding methods are otherwise equivalent, and a simple conversion operation is necessary to switch between the two.

As defined in the IEEE 754R proposal, a decimal floating-point number n is represented as

$$n = \pm C \cdot 10^e$$

where C is a positive integer coefficient with at most p decimal digits, and e is an integer exponent. A precision of p decimal digits will be assumed further for the operands and results of decimal floating-point operations.

Compared to the binary single, double, and quad precision floating-point formats, the decimal floating-point formats denoted here by decimal32, decimal64, and decimal128 cover different ranges and have different precisions, although they have similar storage sizes. For decimal, only the wider formats are used in actual computations, while decimal32 is defined as a storage format only. For numerical values that can be represented in these binary and decimal formats, the main parameters that determine their range and precision are shown in Table 1.

**Table 1.** IEEE 754 binary and IEEE 754R decimal floating-point format parameters

| | Binary Formats | | |
|---|---|---|---|
| | single | double | quad |
| Prec. | n=24 | n=53 | n=113 |
| $E_{min}$ | –126 | –1022 | –16382 |
| $E_{max}$ | +127 | +1023 | +16383 |
| | Decimal Formats | | |
| | decimal32 | decimal64 | decimal128 |
| Prec. | p=7 | p=16 | p=34 |
| $E_{min}$ | –101 | –398 | –6178 |
| $E_{max}$ | +90 | +369 | +6111 |

The following sections will present new algorithms that can be used for an efficient implementation in software of the decimal floating-point arithmetic as defined by the IEEE 754R proposal. Mathematical proofs of correctness have been developed, but will not be included here for brevity. Compiler and run-time support libraries could use the implementation described here, which addresses the need to have a good software solution for the decimal floating-point arithmetic.

## 3 Conversions between Decimal and Binary Formats

In implementing the decimal floating-point arithmetic defined in IEEE 754R, conversions between decimal and binary formats are necessary in many situations.

For example, if decimal floating-point values are encoded in a decimal-based format (string, BCD, IEEE 754R decimal encoding, or other) they need to be converted to binary before a software implementation of the decimal floating-point operation can take full advantage of the existing hardware for binary operations. This conversion is relatively easy to implement, and should exploit any available instruction-level parallelism.

The opposite conversion, from binary to decimal format may have to be performed on results before writing them to memory, or for printing in string format decimal numbers encoded in binary.

Another reason for binary-to-decimal conversion could be for rounding a decimal floating-point result to a pre-determined number of decimal digits, if the exact result was calculated first in binary format. The straightforward method for this is to convert the exact result to decimal, round to the destination precision and then, if necessary, convert the coefficient of the final result back to binary. This step can be avoided completely if the coefficients are stored in binary.

The mathematical property presented next was used for this purpose. It gives a precise way to 'cut off' x decimal digits from the lower part of an integer C when its binary representation is available, thus avoiding the need to convert C to decimal, remove the lower x decimal digits, and then convert the result back to binary. This property was applied to conversions from binary to decimal format as well as in the implementation of the most common decimal floating-point operations: addition, subtraction, multiplication, fused multiply-add, and in part, division.

For example if the decimal number C = 123456789 is available in binary and its six most significant decimal digits are required, Property 1 specifies precisely how to calculate the constant $k_3 \approx 10^{-3}$ so that $\lfloor C \cdot k_3 \rfloor = 123456$, with certainty, while using only the binary representation of C. The values $k_x$ are pre-calculated. (Note: the floor(x), ceiling(x), and fraction(x) functions are denoted here by $\lfloor x \rfloor$, $\lceil x \rceil$, and $\{x\}$ respectively.)

**Property 1.** Let $C \in \mathbf{N}$ be a number in base b = 2 and

$C = d_0 \cdot 10^{q-1} + d_1 \cdot 10^{q-2} + \ldots + d_{q-2} \cdot 10^1 + d_{q-1}$ its representation in base B=10, where $d_0, d_1, \ldots d_{q-1} \in \{0, 1, \ldots, 9\}$ and $d_0 \neq 0$.

Let $x \in \{1, 2, 3, \ldots, q-1\}$ and $\rho = \log_2 10$.

If $y \in N$, $y \geq \lceil \{\rho \cdot x\} + \rho \cdot q \rceil$ and $k_x$ is the value of $10^{-x}$ rounded up to y bits (the subscript RP,y indicates rounding up y bits in the significand), i.e.:

$$k_x = (10^{-x})_{RP,y} = 10^{-x} \cdot (1 + \varepsilon) \qquad 0 < \varepsilon < 2^{-y+1}$$

then $\lfloor C \cdot k_x \rfloor = d_0 \cdot 10^{q-x-1} + d_1 \cdot 10^{q-x-2} + d_2 \cdot 10^{q-x-3} + \ldots + d_{q-x-2} \cdot 10^1 + d_{q-x-1}$

Given an integer C represented in binary, this property specifies a method to remove exactly x digits from the lower part of the decimal representation of C, without actually converting the number to a decimal representation. The property specifies the minimum number of bits y that are necessary in an approximation of $10^{-x}$, so that the integer part (or 'floor') of $C \cdot k_x$ will be precisely the desired result. The property states that $y \geq \lceil \{\rho \cdot x\} + \rho \cdot q \rceil$. However, in practice it is sufficient to take $y = \lceil 1 + \rho \cdot q \rceil = 1 + \lceil \rho \cdot q \rceil$ where $\lceil \rho \cdot q \rceil$ is the 'ceiling' of $\rho \cdot q$ (e.g. $\lceil 33.3 \rceil = 34$). Note that $\rho = \log_2 10 \approx 3.3219\ldots$ and $2^\rho = 10$. For example if we want to remove the x lower decimal digits of a 16-digit decimal number, we can multiply the number with an approximation of $10^{-x}$ rounded up to $y = 1 + \lceil \rho \cdot 16 \rceil = 55$ bits, followed by removal of the fractional part in the product.

The relative error ε associated with the approximation of $10^{-x}$ which was rounded up to y bits satisfies $0 < \varepsilon < 2^{-y+1} = 2^{-\lceil \rho \cdot q \rceil}$.

The values $k_x$ for all x of interest are pre-calculated and are stored as pairs $(K_x, e_x)$, with $K_x$ and $e_x$ positive integers:

$$k_x = K_x \cdot 2^{-ex}$$

This allows for implementations exclusively in the integer domain of some decimal floating-point operations, in particular addition, subtraction, multiplication, fused multiply-add, and certain conversions.

## 4 Decimal Floating-Point Addition

It will be assumed that

$$n1 = C1 \cdot 10^{e1} \qquad C1 \in \mathbf{Z}, 0 < C1 < 10^p$$
$$n2 = C2 \cdot 10^{e2} \qquad C2 \in \mathbf{Z}, 0 < C2 < 10^p$$

are two non-zero decimal floating-point numbers with coefficients having at most p decimal digits stored as binary integers and that their sum has to be calculated, rounded to p decimal digits using the current IEEE rounding mode (this is indicated by the subscript rnd,p).

$$n = (n1 + n2)_{rnd,p} = C \cdot 10^e$$

The coefficient C needs to be correctly rounded, and is stored as a binary integer as well. For simplicity, it will be assumed that $n1 \geq 0$ and $e1 \geq e2$. (The rules for other combinations of signs or exponent ordering can be derived from here.)

If the exponent e1 of n1 and the exponent e2 of n2 differ by a large quantity, the operation is simplified and rounding is trivial because n2 represents just a rounding error compared to n1. Otherwise if e1 and e2 are relatively close the coefficients C1 and C2 will 'overlap', the coefficient of the exact sum may have more than p decimal digits, and so rounding may be necessary. All the possible cases will be quantified next.

If the exact sum is n', let C' be the exact (not yet rounded) sum of the coefficients:

$$n' = n1 + n2 = C1 \cdot 10^{e1} + C2 \cdot 10^{e2} =$$
$$(C1 \cdot 10^{e1-e2} + C2) \cdot 10^{e2}$$
$$C' = C1 \cdot 10^{e1-e2} + C2$$

Let q1, q2, and q be the numbers of decimal digits needed to represent C1, C2, and C'. If not zero, the rounded coefficient C will require between 1 and p decimal digits. Rounding is not necessary if C' represented in decimal requires at most p digits, but it is necessary otherwise.

If $q \leq p$, then the result is exact:

$$n = (n')_{rnd,p} = (C' \cdot 10^{e2})_{rnd,p} =$$
$$(C')_{rnd,p} \cdot 10^{e2} = C' \cdot 10^{e2}$$

Otherwise, if $q > p$ let $x = q - p \geq 1$. Then:

$$n = (n')_{rnd,p} = (C' \cdot 10^{e2})_{rnd,p} =$$
$$(C')_{rnd,p} \cdot 10^{e2} = C \cdot 10^{e2+x}$$

If after rounding $C = 10^p$ (rounding overflow), then $n = 10^{p-1} \cdot 10^{e2+x+1}$.

A simple analysis shows that rounding is trivial if $q1 + e1 - q2 - e2 \geq p$. If this is not the case, i.e. if

$$|q1 + e1 - q2 - e2| \leq p - 1$$

then the sum C' has to be calculated and it has to be rounded to p decimal digits. This case can be optimized by separating it in sub-cases as shall be seen further.

The algorithm presented next uses Property 1 in order to round correctly (to the destination precision) the result of a decimal floating-point addition in rounding to nearest mode, and also determines correctly the exactness of the result by using a simple comparison operation. First, an approximation of the result's coefficient is calculated using

Property 1. This will be either the correctly rounded coefficient, or it will be off by one ulp (unit-in-the-last-place). The correct result as well as its exactness can be determined directly from the calculation, without having to compute a remainder through a binary multiplication followed by a subtraction for this purpose. This makes the rounding operation for decimal floating-point addition particularly efficient.

**Decimal Floating-Point Addition with Rounding to Nearest.** The straightforward method to calculate the result is to convert both coefficients to a decimal encoding, perform a decimal addition, round the exact decimal result to nearest to the destination precision, and then convert the coefficient of the final result back to binary. It would also be possible to store the coefficients in decimal all the time, but then neither software nor hardware implementations could take advantage easily of existing instructions or circuitry that operate on binary numbers. The algorithm used for decimal floating-point addition in rounding to nearest mode is Algorithm 1, shown further.

If the smaller operand represents more than a rounding error in the larger operand, the sum $C' = C1 \cdot 10^{e1-e2} + C2$ is calculated. If the number of decimal digits q needed to represent this number does not exceed the precision p of the destination format, then no rounding is necessary and the result is exact. If $q > p$, then $x = q - p$ decimal digits have to be removed from the lower part of C', and C' has to rounded correctly to p decimal digits. For correct rounding to nearest, 0.5 ulp is added to C': $C'' = C' + 1/2 \cdot 10^x$. The result is multiplied by $k_x \approx 10^{-x}$ ($C^* = C'' \cdot k_x$), where the pre-calculated values $k_x$ are stored for all $x \in \{1, 2, \ldots, p\}$. A test for midpoints follows ($0 < f^* < 10^{-p}$, where f* is the fractional part of C*) and if affirmative, the result is rounded to the nearest even integer. (For example if the exact result 4567.5 has to be rounded to nearest to four decimal places, the rounded result will be 4568.) Next the algorithm checks for rounding overflow (p+1 decimal digits are obtained instead of p) and finally it checks for exactness.

Note that the straightforward method for the determination of midpoints and exactness is to calculate a remainder $r = C' - C \cdot 10^x \in [0, 10^x)$. Midpoint results could be identified by comparing the remainder with $1/2 \cdot 10^x$, and exact results by comparing the remainder with 0. However, the calculation of a remainder – a relatively costly operation – was avoided in Algorithm 1 and instead a single comparison to a pre-calculated constant was used. This simplified method to determine midpoints and exactness along with the ability to use Property 1 make Algorithm 1 more efficient for decimal floating-point addition than previously known methods.

**Algorithm 1. Calculate the Sum of Two Decimal Floating-point Numbers Rounded to Nearest to p Decimal Digits, and Determine its Exactness.**

q1, q2 = number of decimal digits needed to
represent C1, C2 // from table lookup
if $|q1 + e1 - q2 - e2| \geq p$ then
// assuming that $e1 \geq e2$ round the result
// directly as $0 < C2 < 1$ ulp $(C1 \cdot 10^{e1-e2})$;
the result $n = C1 \cdot 10^{e1}$ or
$n = C1 \cdot 10^{e1} \pm 10^{e1+q1-p}$ is inexact
else // if $|q1 + e1 - q2 - e2| \leq p - 1$
$C' = C1 \cdot 10^{e1-e2} + C2$ // binary integer

```
    // multiplication and addition;
    // 10^(e1–e2) from table lookup
q = number of decimal digits needed to
    represent C' // from table lookup
if q ≤ p the result n = C' · 10^e2 is exact
else if q ∈ [p+1, 2·p] continue
x = q – p, number of decimal digits to be
    removed from lower part of C', x ∈ [1, p]
C'' = C' + 1/2 · 10^x // 1/2 · 10^x
    // pre-calculated, from table lookup
k_x = 10^(–x) · (1 + ε), 0 < ε < 2^(–⌈2·ρ·p⌉)
    // / pre-calculated as specified in Property 1
C* = C'' · k_x = C'' · K_x · 2^(–Ex)
    // binary integer multiplication with
    // implied binary point
f* = the fractional part of C*
    // consists of the lower Ex bits of the
    // product C'' · K_x
if 0 < f* < 10^(–p) then
    if ⌊C*⌋ is even then C = ⌊C*⌋
        // logical shift right;
        // C has p decimal digits,
        // correct by Property 1
    else if ⌊C*⌋ is odd then C = ⌊C*⌋ – 1
        // logical shift right; C has p dec.
        // digits, correct by Property 1
else C = ⌊C*⌋ // logical shift right; C has p
    // decimal digits, correct by Property 1
n = C · 10^(e2+x)
if C = 10^p then n = 10^(p–1) · 10^(e2+x+1)
    // rounding overflow
if 0 < f* – 1/2 < 10^(–p) then the result is exact
else it is inexact
```

Note that conditions $0 < f^* < 10^{-p}$ and $0 < f^* - 1/2 < 10^{-p}$ from Algorithm 1 for midpoint detection and exactness determination hold also if $10^{-p}$ is replaced by $10^{-x}$ or even by $k_x = 10^{-x} \cdot (1 + \varepsilon)$. These comparisons are fairly easy in practice. For example, since $C''{\cdot}k_x = C''{\cdot}K_x{\cdot}2^{-ex}$, in $f^*<10^{-p}$ the bits shifted to the right out of $C''{\cdot}k_x$, representing $f^*$ can be compared with a pre-calculated constant that approximates $10^{-p}$ (or $10^{-x}$).

**Decimal Floating-Point Addition when Rounding to Zero, Down, or Up.** The method to calculate the result when rounding to zero or down is similar to that for rounding to nearest. The main difference is that the step for calculating $C'' = C' + 1/2 \cdot 10^x$ is not necessary anymore, because midpoints between consecutive floating-point numbers do not have a special role here. For rounding up, the calculation of the result and the determination of its exactness are identical to those for rounding down. However, when the result is inexact then one ulp has to be added to it.

## 5 Decimal Floating-Point Multiplication

It will be assumed that the product

$$n = (n1 \cdot n2)_{rnd,p} = C \cdot 10^{e}$$

has to be calculated, where the coefficient C of n is correctly rounded to p decimal digits using the current IEEE rounding mode, and is stored as a binary integer. The operands $n1 = C1 \cdot 10^{e1}$ and $n2 = C2 \cdot 10^{e2}$ are assumed to be strictly positive (for negative numbers the rules can be derived directly from here). Their coefficients require at most p decimal digits to represent and are stored as binary integers, possibly converted from a different format/encoding.

Let q be the number of decimal digits required to represent the full integer product $C' = C1 \cdot C2$ of the coefficients of n1 and n2. Actual rounding to p decimal digits will be necessary only if $q \in [p+1, 2\cdot p]$, and will be carried out using Property 1. In all rounding modes the constants $k_x \approx 10^{-x}$ used for this purpose, where $x = q - p$, are pre-calculated to y bits as specified in Property 1. Since $q \in [p+1, 2\cdot p]$ for situations where rounding is necessary, all cases are covered correctly by choosing $y = 1+ \lceil 2\cdot\rho\cdot p \rceil$. Similar to the case of the addition operation, the pre-calculated values $k_x$ are stored for all $x \in \{1, 2, \ldots, p\}$.

**Decimal Floating-Point Multiplication with Rounding to Nearest.** The straightforward method to calculate the result is similar to that for addition. A new and better method for decimal floating-point multiplication with rounding to nearest that uses existing hardware for binary computations is presented in Algorithm 2. It uses Property 1 to avoid the need to calculate a remainder for the determination of midpoints or exact floating-point results, as shall be seen further. The multiplication algorithm has many similarities with the algorithm for addition.

**Algorithm 2. Calculate the Product of Two Decimal Floating-point Numbers Rounded to Nearest to p Decimal Digits, and Determine its Exactness.**

$C' = C1 \cdot C2$ // binary integer multiplication
q = the number of decimal digits required to
represent C' // from table lookup
if $q \le p$ then the result $n = C' \cdot 10^{e1+e2}$ is exact else if $q \in [p+1, 2\cdot p]$ continue
$x = q - p$, the number of decimal digits to be removed from the lower part of C', $x \in [1, p]$
$C'' = C' + 1/2 \cdot 10^{x}$ // $1/2 \cdot 10^{x}$ pre-calculated
$k_x = 10^{-x} \cdot (1 + \varepsilon)$, $0 < \varepsilon < 2^{-\lceil 2\cdot\rho\cdot p \rceil}$ // pre-calculated
// as specified in Property 1
$C^* = C'' \cdot k_x = C'' \cdot K_x \cdot 2^{-Ex}$ // binary integer
// multiplication with implied binary point
$f^*$ = the fractional part of $C^*$ // consists of the
// lower $E_x$ bits of the product $C'' \cdot K_x$
if $0 < f^* < 10^{-p}$ then // since $C^* = C'' \cdot K_x \cdot 2^{-Ex}$,
// compare $E_x$ bits shifted out of $C^*$ with 0
// and with $10^{-p}$
if $\lfloor C^* \rfloor$ is even then $C = \lfloor C^* \rfloor$ // logical right
  // shift; C has p decimal digits, correct by

```
      // Property 1
   else C = ⌊C*⌋ – 1 // if ⌊C*⌋ is odd // logical
      // right shift; C has p decimal digits, correct
      // by Property 1
   else
   C = ⌊C*⌋ // logical shift right; C has p
      // decimal digits, correct by Property 1
   n = C · 10^(e1+e2+x) // rounding overflow
   if 0 < f* – 1/2 < 10^-p then the result is exact
   else the result is inexact
   // C* = C'' · K_x · 2^-Ex ⇒ compare E_x bits
   // shifted out of C* with 1/2 and 1/2+10^-p
```

If $q \geq p + 1$ the result is inexact unless the x decimal digits removed from the lower part of $C'' \cdot k_x$ were all zeros. To determine whether this was the case, just as for addition, the straightforward method is to calculate a remainder $r = C' - C \cdot 10^x \in [0, 10^x)$. Midpoint results could be identified by comparing the remainder with $1/2 \cdot 10^x$, and exact results by comparing the remainder with 0. However, the calculation of a remainder – a relatively costly operation – was avoided in Algorithm 2 and instead a single comparison to a pre-calculated constant was used.

The simplified method to determine midpoints and exactness along with the ability to use Property 1 make Algorithm 2 better for decimal floating-point multiplication than previously known methods.

**Decimal Floating-Point Multiplication when Rounding to Zero, Down, or Up.** The method to calculate the result when rounding to zero or down is similar to that for rounding to nearest. Just as for addition, the step for calculating $C'' = C' + 1/2 \cdot 10^x$ is not necessary anymore. Exactness is determined using the same method as in Algorithm 2. For rounding up, the calculation of the result and the determination of its exactness are identical to those for rounding down. However, when the result is inexact then one ulp has to be added to it.

## 6 Decimal Floating-Point Division

It will be assumed that the quotient

$$n = (n1 / n2)_{rnd,p} = C \cdot 10^e$$

has to be calculated where $n1 > 0$, $n2 > 0$, and q1, q2, and q are the numbers of decimal digits needed to represent C1, C2, and C (the subscript rnd,p indicates rounding to p decimal digits, using the current rounding mode). Property 1 cannot be applied efficiently for the calculation of the result in this case because a very accurate approximation of the exact quotient is expensive to calculate. Instead, a combination of integer operations and floating-point division allow for the determination of the correctly rounded result. Property 1 is used only when an underflow is detected and the calculated quotient has to be shifted right a given number of decimal positions. The decimal floating-point division algorithm is based on Property 2 presented next.

**Property 2.** If a, b are two positive integers and $m \in \mathbf{N}$, $m \geq 1$ such that $b < 10^m$, $a/b < 10^m$ and $n \geq \lfloor m \cdot \log_2 10 \rfloor$, then $| a/b - \lfloor ((a)_{rnd,n}/(b)_{rnd,n})_{rnd,n} \rfloor | < 8$.

The decimal floating-point division algorithm for operands $n1 = C1 \cdot 10^{e1}$ and $n2 = C2 \cdot 10^{e2}$ follows. While this algorithm may be rather difficult to follow without working out an example in parallel, it is included here for completeness. Its correctness, as well as that of all the other algorithms presented here has been verified.

**Algorithm 3. Calculate the Quotient of Two Decimal Floating-point Numbers, Rounded to p Decimal Digits in any Rounding Mode, and Determine its Exactness.**

```
if C1 < C2
  find the integer d > 0 such that (C1/C2)·10^d ∈
[1, 10).
// compute d based on the number
// of decimal digits q1, q2 in C1, C2
   C1' = C1·10^(d+15), Q = 0
   e = e1 – e2 – d – 15  // expected res. expon.
else
    a = (C1 OR 1)_rnd,n, b = (C2)_rnd,n // logical OR
    Q = ⌊((a/b)_rnd,n)⌋
    R = C1 – Q · C2
    if R < 0
      Q = Q – 1
      R = R + C2
    if R = 0 the result n = Q · 10^(e1–e2) is exact
    else continue
    find the number of decimal digits for Q:  d >
     0 such that Q ∈ [10^(d–1), 10^d)
    C1' = R · 10^(16–d)
    Q = Q · 10^(16–d)
    e = e1 – e2 – 16 + d
Q2 = ⌊((C1')_rnd,n/(C2)_rnd,n)_rnd,n⌋
R = C1' – Q2 · C2
Q = Q + Q2
if R ≥ 4 · C2
    Q = Q + 4
    R = R – 4 · C2
if R ≥ 2 · C2
    Q = Q + 2
    R = R – 2 · C2
if R ≥ C2
    Q = Q + 1
    R = R – C2
if e ≥ minimum_decimal_exponent
    apply rounding in desired mode by
```

```
    comparing R and C2
    // e.g. for rounding to nearest add 1 to Q
    // if 5 · C2 < 10 · R + (Q AND 1)
    the result n = Q ·10^e is inexact
else
    result underflows
    compute the correct result based on Prop. 1
```

## 7 Decimal Floating-Point Square Root

Assume that the square root

$$n = (\sqrt{n1})_{rnd,p} = C \cdot 10^e$$

has to be calculated (where the subscript rnd,p indicates rounding to p decimal digits using the current rounding mode). The method used for this computation is based on Property 3 and Property 4, shown next. A combination of integer and floating-point operations are used. It will be shown next that the minimum precision n of the binary floating-point numbers that have to be used in the computation of the decimal square root for decimal64 arguments (with p = 16) is n = 53, so the double precision floating-point format can be used. The minimum precision n of the binary floating-point numbers that have to be used in the computation of the square root for decimal128 arguments (with p = 34) is n = 113, so the quad precision floating-point format can be used safely.

Properties 3 and 4 as well as the algorithm for square root calculation are included here for completeness.

**Property 3.** If $x \in (1, 4)$ is a binary floating-point number with precision n and $s = (\sqrt{x})_{RN,n}$ is its square root rounded to nearest to n bits, then $s + 2^{-n} < x$.

**Property 4.** Let m be a positive integer and $n = \lfloor m \cdot \log_2 10 + 0.5 \rfloor$. For any integer $C \in [10^{2 \cdot m-2}, 10^{2 \cdot m})$, the inequality $| \sqrt{C} - \lfloor \sqrt{((C)_{RN,n})} \rfloor | < 3/2$ is true.

The round-to-nearest decimal square root algorithm can now be summarized as follows:

**Algorithm 4. Calculate the Square Root of a Decimal Floating-point Number n1 = C·10^e, Rounded to Nearest to p Decimal Digits, and Determine its Exactness.**

```
if e is odd then
    e' = e – 1
    C' = C · 10
else
    e' = e
    C' = C
let S = ⌊√(( C')RN,n)⌋
if S * S = C'
    the result n = S · 10^(e'/2) is exact
else
    q = number of decimal digits in C
```

$C'' = C' \cdot 10^{2 \cdot p-1-q}$ and $Q = \lfloor \sqrt{((C'')_{RN,n})} \rfloor$
if $(C'' - Q \cdot Q < 0)$ sign = –1 else sign = 1
M = 2·Q + sign // will check against this
// midpoint for rounding to nearest
if $(M \cdot M - 4 \cdot C'' < 0)$ sign_m = –1
else sign_m = 1
if sign ≠ sign_m Q' = Q + sign else Q' = Q
the result $n = Q' \cdot 10^{e'/2}$ is inexact

**Decimal Floating-Point Square Root when Rounding to Zero, Down, or Up.** The algorithm shown above can be easily adapted for other rounding modes. Once Q is computed such that $|\sqrt{C''} - Q| < 1.5$, one needs to consider rounding the result coefficient to one of the following values: Q–2, Q–1, Q, Q+1, Q+2, and only two of these values need to be considered after the sign of $(\sqrt{C''} - Q)$ has been computed.

## 8 Conclusions

A new generic implementation in C of the basic operations for decimal floating-point arithmetic specified in the IEEE 754R standard proposal was completed, based on new algorithms presented in this paper. Several other operations were implemented that were not discussed here for example remainder, fused multiply-add, comparison, and various conversion operations. Performance results for all basic operations were in the expected range, for example the latency of decimal128 operations is comparable to that of binary quad precision operations implemented in software.

It was also possible to compare the performance of the new software package for basic operations with that of the decNumber package contributed to GCC [6]. The decNumber package represents the only other implementation of the IEEE 754R decimal floating-point arithmetic in existence at the present time. It should be noted that decNumber is a more general decimal arithmetic library in ANSI C, suitable for commercial and human-oriented applications [7]. It allows for integer, fixed-point, and decimal floating-point computations, and supports arbitrary precision values (up to a billion digits).

Tests comparing the new decimal floating-point library using the algorithms described in this paper versus decNumber showed that the new generic C implementations for addition, multiplication, division, square root, and other operations were faster than the decNumber implementations, in most cases by one to two orders of magnitude.

Table 2 shows the results of this comparison for basic 64-bit and 128-bit decimal floating-point operations measured on a 3.4 GHz Intel® EM64t system with 4 GB of RAM, running Microsoft Windows Server 2003 Enterprise x64 Edition SP1. The code was compiled with the Intel(R) C++ Compiler for Intel(R) EM64T-based applications, Version 9.0. The three values presented in each case represent minimum, median, and maximum values for a small data set covering operations from very simple (e.g. with operands equal to 0 or 1) to more complicated, e.g. on operands with 34 decimal digits in the 128-bit cases. For the new library, further performance improvements can be attained by fine-tuning critical code sequences or by optimizing simple, common cases.

**Table 2.** New Decimal Floating-Point Library Performance vs. decNumber on EM64t (3.4 GHz Xeon). Minimum-median-maximum values are listed in sequence, after subtracting the call overhead.

| Operation | New Library [clock cycles] | decNumber Library [clock cycles] | dec Number /New Library |
|---|---|---|---|
| 64-bit ADD | 14-140-241 | 99-1400-1741 | 4-10-14 |
| 64-bit MUL | 21-120-215 | 190-930-1824 | 6-8-9 |
| 64-bit DIV | 172-330-491 | 673-2100-3590 | 4-6-11 |
| 64-bit SQRT | 15-288-289 | 82-16700-18730 | 7-58-107 |
| 128-bit ADD | 16-170-379 | 97-2300-3333 | 4-13-14 |
| 128-bit MUL | 19-300-758 | 95-3000-4206 | 5-10-18 |
| 128-bit DIV | 153-250-1049 | 1056-2000-7340 | 4-8-9 |
| 128-bit SQRT | 16-700-753 | 61-42000-51855 | 4-60-152 |

For example for the 64-bit addition operation the new implementation, using the 754R binary encoding for decimal floating-point, took between 14 and 241 clock cycles per operation, with a median value around 140 clock cycles. For the same operand values decNumber, using the 754R decimal encoding, took between 99 and 1741 clock cycles, with a median around 1400 clock cycles. The ratio shown in the last column was between 4 and 14, with a median of around 10 (probably the most important of the three values).

It is also likely that properties and algorithms presented here for decimal floating-point arithmetic can be applied as well for a hardware implementation, with re-use of existing circuitry for binary operations. It is the authors' hope that the work described here will represent a step forward toward reliable and efficient implementations of the IEEE 754R decimal floating-point arithmetic.

## References


1. IEEE Std. 754, IEEE Standard 754-1985 for Binary Floating-Point Arithmetic. IEEE (1985)
2. IEEE Std. 754R Draft. Draft of the Revised IEEE Standard 754-1985 (2006), `http://754r.ucbtest.org/drafts/754r.pdf`
3. Cowlishaw, M.: Decimal Floating-Point: Algorism for Computers. In: 16th IEEE Symposium on Computer Arithmetic (2003)
4. Erle, M., Schwarz, E., Schulte, M.: Decimal Multiplication with Efficient Partial Product Generation. In: 17th Symposium on Computer Arithmetic (2005)
5. Peter Tang, BID Format, IEEE 754R Draft (2005), `http://754r.ucbtest.org/subcommittee/bid.pdf`
6. Grimm, J.: Decimal Floating-Point Extension for C via decNumber, IBM, GCC Summit. decNumber (2005), `http://www.alphaworks.ibm.com/`
7. `/tech/decnumber`

# PART II

# Software Engineering

# Bridging between Middleware Systems: Optimisations Using Downloadable Code

Jan Newmarch

Faculty of Information Technology, Monash University, Melbourne, Australia
jan.newmarch@infotech.monash.edu.au

**Abstract.** There are multiple middleware systems and no single system is likely to become predominant. There is therefore an interoperability requirement between clients and services belonging to different middleware systems. Typically this is done by a bridge between invocation and discovery protocols. In this paper we introduce three design patterns based on a bridging service cache manager and dynamic proxies. This is illustrated by examples including a new custom lookup service which allows Jini clients to discover and invoke UPnP services. There is a detailed discussion of the pros and cons of each pattern.

**Keywords:** Middleware, UPnP, Jini, Service oriented architecture, downloadable code, proxies.

## 1 Introduction

There are many middleware systems which often overlap in application domains. For example, UPnP is designed for devices in zero-configuration environments such as homes [16], Jini is designed for adhoc environments with the capability of handling short as well as long-lived services [19] while Web Services are designed for long running services across the Web [20]. There are many other middleware systems such as CORBA, Salutation, HAVi etc each with their own preferred application space, and these different application spaces will generally overlap to some extent[1].

It is unlikely that any single middleware will become predominant, so that the situation will arise where multiple services and clients exist but belonging to different middleware systems. To avoid middleware "silos", it is important to examine ways in which clients using one middleware framework can communicate with services using another.

This issue is not new: the standard approach is to build a "bridge" which is a two-sided component that uses one middleware on one side and another middleware on the other. Examples include Jini to CORBA [13], Jini to UPnP [1], SLP to UPnP, etc. These essentially replace an end-to-end communication between client and service by an end-to-middle-to-end communication, where the middle (the bridge) performs translation from one protocol to the other.

[1] The middleware systems we are interested in involve discovery of services, rather than just transport-level middleware such as HTTP connecting web browsers and HTTP servers.

J. Filipe, B. Shishkov, and M. Helfert (Eds.): ICSOFT 2006, CCIS 10, pp. 113–126, 2008.

Newmarch [14] has investigated how a Jini lookup service can be embedded into a UPnP device to provide an alternative to the bridging architecture. However, in practical terms this is an invasive mechanism which requires changes to the UPnP device and cannot be easily retro-fitted into devices.

Jini [2] is apparently unique in production-quality middleware sytems with service discovery in that rather than giving some sort of remote reference to clients it downloads a proxy object into the client (the proxy is a Java object). Many of the obvious security issues in this have already been addressed by Jini. It has also been claimed that this will lead "to the end of protocols" [18]. In this paper we investigate the implications of downloadable code for bridging systems, and show that it can lead to many optimisations.

Some of our work can be applied to middleware systems which support downloadable code but not discovery, such as JavaScript in HTML pages.

We illustrate some of these optimisations with a Jini-to-Web Services bridge and others with Jini- to-UPnP bridge.

The principal contribution of this paper is that it proposes and demonstrates a number of optimisations that could be considered to be additional architectural patterns that can sometimes be applied to bridge between different middleware systems. The validity of these patterns are demonstrated by discussion of several example systems and through an implementation for bridging between UPnP services and Jini clients. However, the patterns do have strong requirements on the client-side middleware: it must be possible to dynamically download code to clients and to dynamically determine the content of this downloaded code.

The structure of this paper is as follows: the next section discusses some general properties of bridging systems and the following section discusses downloadable code in this context. Section 4 introduces the first of three optimisations, one for transport-level bridging. Section 5 considers service cache management and the following section applies this to the second optimisation, for service-level transport. This is followed by a section on device-level optimisation. Successive sections deal with event handling and the implementation of a Jini-UPnP bridge based on these principles. We then assess the proposals and consider the value and generality of our work, before a concluding section.

Background knowledge of Jini may be found in Newmarch [13] and on the UPnP home site [16].

## 2 Bridging

Nakazawa et al [12] discuss general properties of middleware bridges. They distinguish three features

- Transport-level bridging concerns translation between two invocation protocols where a client makes a request of a service. Examples of invocation protocols include SOAP and CORBA's IIOP. Transport-level bridging is concerned with translating from the invocation of a request to its delivery, and also between any replies.

- Service-level bridging involves the advertisement and discovery of services. Examples of discovery protocol include CORBA's use of a Naming service and UPnP's Simple Service Discovery Protocol.
- Device-level bridging concerns the semantics of services.

Transport level bridging includes translating between the data-types carried by each protocol. For example for Web Services using SOAP, these are XML data-types while for Java RMI using JRMP these are serialisable Java objects. There are usually problems involved in such conversions. Vinoski [17] points to the mismatch between Java data-types and XML data-types. While he goes on to examine the consequences for JAX-RPC, the same issues cause problems converting from SOAP data-types to Java objects on JRMP. Newmarch [14] discusses the mismatch between UPnP data-types and Java objects and concludes that the UPnP to Java mapping is generally okay but the opposite direction is not. There is no general solution to the data-mapping problem, and indeed the use of the so-called "language independent" XML in some middleware systems appears to have exacerbated this. Services where the data-types are not convertable cannot be bridged. This paper does not address this issue.

While the transport protocol is usually end-to-end, the discovery protocol may be either end-to-end as in UPnP or involve a third party. Dabrowski and Mills [6] term this third-party a service cache manager (SCM). Examples of such a manager are the Jini lookup service, the CORBA and RMI Naming service and UDDI (although this does not seem to be heavily used). The implications for service-level bridging involve the discovery protocol: in an end-to-end discovery system the service-level bridge will need to understand how to talk directly to services and/or clients, while with a service cache manager the bridge will need to understand how to talk to the service cache manager.

Device-level bridging concerns the meaning of "service" in different middleware systems, and how services (and devices) are represented.

In general a bridge system will look like Figure 1.

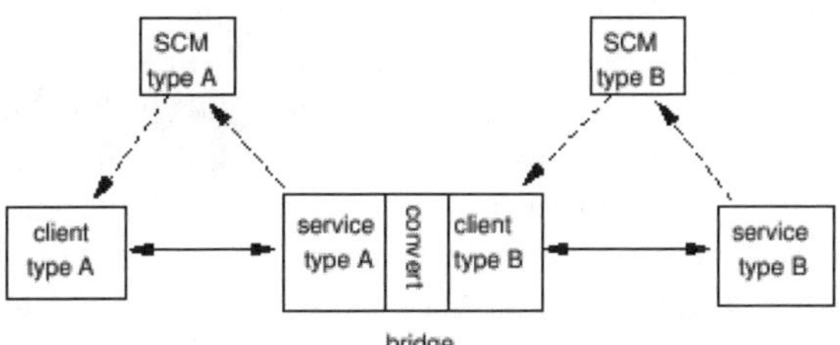


**Fig. 1.** Typical bridge system

## 3 Downloadable Code

There are many examples where code is downloaded from one computer to execute in another. These include JavaScript in HTML pages, Safe-Tcl [11] and Erlang [4]. Jini as a service-oriented architecture makes use of RMI to download a proxy object representing a service into a client. This changes the nature of the client/service transport protocol since that is now managed by the proxy object, not by the client-the client just makes local calls on the proxy. The Java Extensible Remote Invocation framework (Jeri) in Jini 2.0 allows the proxy and service to use any protocol that they choose.

Proxy/service communication in Jini can be represented in Figure 2.

**Fig. 2.** Proxy communication

The pattern of communication of Figure 2 can also be employed by JavaScript using the Ajax extensions [8], and is used by Google Maps and Google Mail for example, although the communication is restricted to HTTP calls.

## 4 Optimising Transport-Level Bridging

Transport-level bridging involves the bridge receiving messages from a client using the client's transport protocol, translating them into messages for the service and sending them using the service's transport protocol. Responses are handled in a similar way.

Many internet protocols specify all components of the interaction between clients, services and service cache managers. For example, UPnP specifies the search and discovery protocols and also the protocol for procedure call interaction between client and service as SOAP. However, as was shown by Java RMI over CORBA's IIOP instead of JRMP, and also by CORBA's use of Naming and Trader services, there is no necessary link between discovery and invocation. As long as a client and service are using the same invocation protocol they can interact directly.

For UPnP and many systems there is little choice since the invocation protocol is fixed by the middleware specification. However, Jini 2.0 allows a "pluggable" communications protocol. While most systems would require the client to have the communications protocol "hard coded" (or loadable from local files), Jini allows a service proxy to be downloaded from a lookup service (service cache manager) to a client, and this can carry code to implement any desired communicaration protocol.

In a similar but less flexible way, the Ajax `XMLHttpRequest` object can exchange any type of data with its originating service. Usually this is XML data, but could be other types such as JSON [9]

In the most common situations, the service proxy communicates with its bridge service. However, a transport-level bridge is just there to translate and communicate between the client and the service. If the code to do this translation is moved into the proxy, then the transport-level component of the bridge service becomes redundant. That is, the client makes local calls on the proxy, which makes calls directly to the service using the service's transport protocol. One leg of the middleware has been removed. This is illustrated in Figure 3.

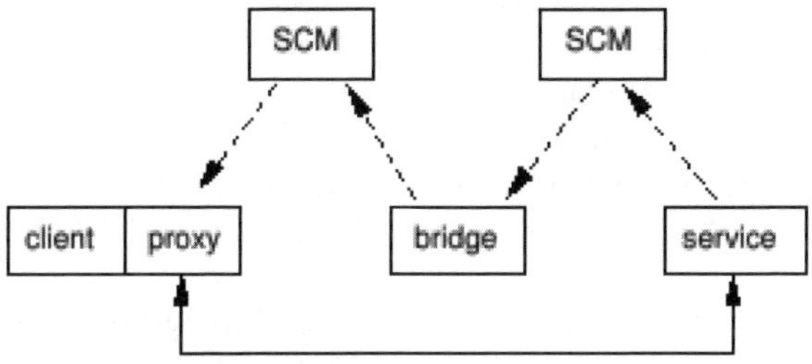


**Fig. 3.** Removing one transport step

This optimisation improves performance by

- removing one serialisation step
- removing one deserialisation step
- removing one network transport leg

In addition, the conversion to the destination protocol is performed once at the client-side. There are some systems such as that of Nakazawa et al [12] in which the bridge performs conversion from source data-types to an intermediate "standard" type and from there to the destination type. This (or even just conversion from source transport data-types to destination transport data-types) introduce possibilities for semantic problems which are mitigated by a single conversion step at the client-side.

This pattern has been used by Newmarch [15] to show how a Jini client can communicate with a Web Service. The proxy uses SOAP, the transport protocol for the Web Service. The conversion from Java data-types to XML data-types is performed by the JAX-RPC package (which cannot do a perfect conversion job, as mentioned earlier). The role of the bridge is just there to advertise the Web Service to the Jini federation and to upload a proxy to the Jini lookup service.

This pattern can also be used by Jini clients to talk to CORBA services, since Jini can directly generate proxies that use IIOP.

Casati [5] shows how JavaScript downloaded into a browser can talk directly to Web Services instead of the more usual HTML-Servlet-Web Service (or similar) bridge (as typified by the web site `www.xmethods.com`). Casati employs the `XMLHttpRequest` object which allows a browser to communicate with an HTTP server asynchronously. This is usually used to exchange data between the browser and original page server. But as Web Services typically use SOAP over HTTP, Casati gives JavaScript for the object to be used as a proxy to talk directly to the Web Service.

In a later section we discuss how we use this pattern for a Jini client to talk to a UPnP service.

## 5 Service Cache Manager

Service cache managers are expected to store "services" in some format and deliver them to clients. The stored service can be a simple name/address pair as in naming systems such as Java RMI or CORBA, complex XML structures linked to WSDL URLs for Web Services in UDDI directories, or other possibilities. The Jini lookup service stores service proxy objects, along with type information to locate them.

When clients and services are trying to locate a service cache manager, there is often an assumed symmetry, that the client and service are searching for the same thing. In our examples above, this occurs in all of naming services, UDDI registries and Jini lookup services.

Once found though, clients and service do different things: services register whereas clients look for services. The Jini `ServiceRegistrar` for example contains two sets of methods, one for services (`register()`) and one for clients (`lookup()`). UDDI similarly has two sets of messages, but there are more of them since UDDI has a more complex structure [3]. Conceptually, there should be one protocol for services discovering caches and another for clients discovering them, with different interfaces exposed to each.

## 6 Optimising Service-Level Bridging

The standard bridge acts as a client to one discovery protocol and as service to the other. For example, in a Jini/UPnP bridge [1] UPnP device advertisements are heard by a bridge acting as a UPnP control point, which re-advertises the service as a Jini service. In addition, it also acts as a transport-level bridge.

As a second optimisation we propose folding the service cache managers into the bridge, to just leave service-level bridging as in Figure 4.

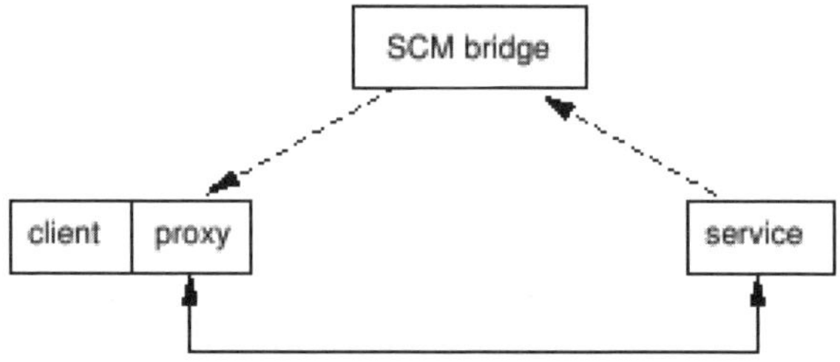


**Fig. 4.** Optimised service-level bridging

As an illustration of this, we have built a lookup service as a service-level bridge which listens for UPnP device advertisements on one side. It can handle device registration and device farewells and will deal with device renewals, timing out if they are not received. In this respect it acts like a UPnP control point, but unlike a control point it does not send any action calls to the UPnP device or register itself for events. The other side of the service-level bridge handles requests from Jini clients, primarily a discovery request for the lookup service.

The lookup service will act like a normal Jini lookup service as far as the Jini client is concerned and return a lookup service proxy. The Jini client will be a normal Jini client and uses the lookup service to search for a service using the standard Jini API.

If the lookup service knows of UPnP devices that deliver the service, it will prepare a proxy for the UPnP device and send it back to the Jini client.

This optimisation is only useful in conjunction with the first one. Transport-level bridging or its replacement will still need to be in place. If there is no replacement then little is gained by separating transport-level and service-level bridging. However, when the transport-level bridge is replaced by a smart proxy then it is possible to just keep the service-level bridge.

This is at present a practical restriction on the applicability of this pattern, since there do not appear to be many middleware systems in practical use apart from Jini that support both downloadable proxies and discovery services. However, this could be expected to change with future development of more advanced service oriented frameworks (for example, see Edwards [7]).

## 7 Device-Level Bridging

Different middleware systems have different basic ideas of services. Many systems such as CORBA, Jini and WebServices only have the notion of services. Others like UPnP and Bluetooth have devices. UPnP devices contain a number of services (and possibly other devices, recursively).

The different systems give different meanings to discovery. For example, the UPnP on/off light is a `BinaryLight` device containing a `SwitchPower` service. Jini has no concept of `BinaryLight`'s and can only look for a `SwitchPower` service. So a Jini client cannot search for a binary light device but only some subset (as a collection) of the service interfaces offered. On the other hand, UPnP advertises the binary light device and the services, but with separate messages for each service, rather than as a group. UPnP devices usually only have one service although some may have more. For example, an internet gateway device may have several services and embedded devices. This device has a total service list of `Layer3Forwarding`, `WANCommonInterfaceConfig`, `WANDSLLinkConfig` and `WANPPPConnection`. In general, a Jini service may implement a number of service interfaces, and a Jini client may request a service that simultaneously implements a number of interfaces.

In the case of UPnP, services are described by XML documents, while Jini services are described by Java interfaces. We have defined a standard mapping from UPnP services to Jini services. For example, the UPnP service description for a `SwitchPower` service is

```
<?xml version="1.0"?>
<scpd xmlns="urn:schemas-upnp-org:...">
   ...
   <actionList>
    <action>
    <name>SetTarget</name>
      <argumentList>
        <argument>
          <name>newTargetValue</name>
          <relatedStateVariable>Target
```

```
          </relatedStateVariable>
          <direction>in</direction>
        </argument>
      </argumentList>
    </action>
    ...
  </actionList>
  ...
</scpd>
```

Our mapping translates this into the Java interface (along with a suitable definition of Target)

```
public interface SwitchPower
       extends Remote {
   void SetTarget(Target newTargetValue)
           throws RemoteException;
   ...
}
```

The service-level bridge will need to be able to translate from one representation to the other. A direct approach is to store a table mapping each service. In the case of UPnP and Jini, the table would just hold UPnP service names matched to Java class files. At this stage, the service-level bridge will be responsible for creating the proxy, and to do this it needs the class files for the service interfaces. While it would be fine for the bridge to have class files for the set of "standard" devices and services maintained by the UPnP Consortium, it would not allow for new, unknown services to be managed.

New UPnP services would require the service-transport bridge to examine in detail the UPnP service description and generate source code for the Java interface. Then compile this on the fly using a local Java compiler (such as `javac` or Kirby's dynamic compiler [10]). This is similar to dynamic compilation of JSP and servlets by servlet engines such as Tomcat. The resultant class files can be cached against repeated use.

Similar mechanisms may be needed for bridging between any middleware systems where new and unknown services may be presented to the service-level bridge. This will depend on what information is required by the bridge in order to create a proxy.

## 8 Optimising Device-Level Bridging

In the architecture proposed so far, the service-level bridge needs to be able to generate a proxy to represent the original service. For a Jini client, this requires class files on the lookup service for the Java interfaces, and for unknown service types these will need to be generated by the bridge. This will involve detailed introspection of the service descriptions and use of a Java compiler. While dynamic compilation of JSP pages demonstrates that this is feasible, it nevertheless has overheads.

The Jini client on the other hand has to know the service interface, otherwise it cannot ask for a service proxy. So if knowledge of the Java interfaces can be deferred to the client side, then it just becomes a lookup of already instantiated classes. The name of the interface is all that is required for the client to find the interface class[2] .

The Jini lookup service already downloads a proxy to the client to represent it. This has not been shown in the figures so far as it is a Jini-specific (but standard) detail. Usually this proxy just makes remote calls back to the lookup service. However, just like any downloaded code, the proxy can be designed to perform any functions on the client side (subject to security constraints). In particular, on a lookup operation the proxy could just pass back to the lookup service enough to allow a match to be made, and on success the lookup service could pass back just enough for the lookup service's proxy to create a proxy for the original service. In the case of a UPnP/Jini bridge, the minimal information is the names of the interfaces required, and the returned information just needs to be the URL of the UPnP device description. These are enough for a proxy to be created on the client-side that can talk to the UPnP service. See Figure 5 for the final system.

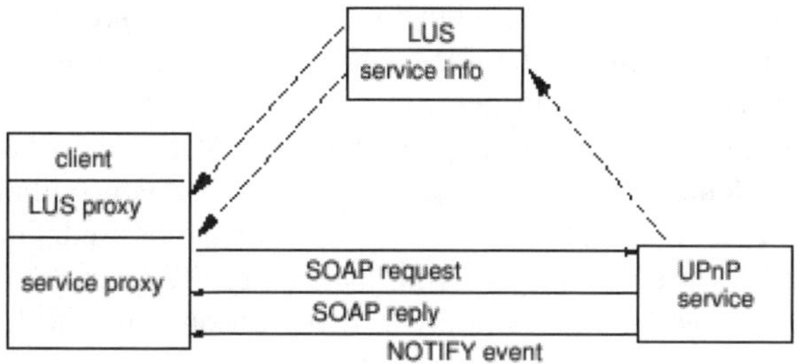


**Fig. 5.** Optimised service-level bridging

## 9 Event Handling

The discussion so far has used the remote procedure call paradigm. However, there are other possibilities such as an asynchronous callback mechanism where the service makes calls back to the client. This is easily handled by the proposed systems, as the proxy just registers itself as the callback address.

## 10 Implementation of Optimised Jini-UPnP Bridging

There is an open source implementation of UPnP devices and control points by CyberGarage (Konno, 2006). This is very closely modelled on the UPnP Device Architecture specification (UPnP Consortium, 2006a). It exposes an API to allow a client to create a `ControlPoint` which can listen for device announcements, to determine the services within the device and it has methods to prepare parameters and make action calls on UPnP services. It also supports getting device information such as friendly name and registering as listener for state variable change events.

[2] The client has to know the interfaces it is interested in. It should not know the implementation classes. This is addressed in the implementation section.

We use this in our lookup service to monitor UPnP devices and keep track of the services that are available, as well as device information.

The CyberGarage API treats UPnP devices and services using a DOM-oriented model, unlike the SOA-oriented manner of Jini. We use the UPnP to Java mapping discussed earlier to translate between the two representations.

In our implemention, we use the Java `Proxy class` to give a dynamic proxy. This proxy implements all of the services on a UPnP device that are requested by the client. The proxy is supplied with the device URL so that it can access the device description. This description contains the URLs for action calls, for registering listeners and for the presentation. The Jini proxy requires an invocation handler. We use the CyberGarage classes to build a generic handler to deal with SOAP calls to the device. The CyberGarage classes and this handler are downloaded from the bridge to the client. This avoids the need for the bridge to know the service interfaces at all and allows the client to only know the service interfaces.

The proxy implementation uses the CyberGarage library, but only for the control components of the CyberGarage ControlPoint. That is, it is used to prepare and make SOAP action calls and to register and listen for UPnP events. However, it does not listen for devices, since that is done by the bridging lookup service. When a method call is made on the service proxy it uses the control point to make a SOAP remote procedure call.

Our current implementation relies heavily on the CyberGarage library, but only on the control point code. The device advertisement code is not used. Only a part of the control point code is used by the bridging lookup service to monitor devices while another part is used by the service proxy to make action calls and listen for events. However, the CyberGarage code is tightly interwoven, and it was not possible to use only the relevant parts. The lookup service has to import almost all of the library, as does the service proxy. It should be possible to produce a lighter-weight version for each with only the required partial functionality.

## 11 Assessment

Any "optimisation" often has both positive and negative sides. We try to offer a balanced viewpoint on the advantages and disadvantages of our pattern.

### 11.1 Transport-Level Optimisation

In transport-level optimisation, we place the code to perform service invocation directly in the proxy downloaded to the client. The principal advantages of this are

- perfomance improvement by removing one serialisation step
- perfomance improvement by removing one deserialisation step
- perfomance improvement by removing one network transport leg
- reducing the risk of semantic mismatches between client and service data-types by reducing the number of data conversion steps.

The ma jor disadvantage is that code has to be downloaded to the client that is capable of talking directly to the service. This is generally downloaded from an HTTP server. Some examples follow

- Casati[5] gives JavaScript that can be downloaded to a web browser such as Firefox or IE that can make function calls on Web Services. This requires just 10kbytes of JavaScript source code. This relies on the extensive libraries and support within the browsers for many of the library calls made.
- Newmarch [15] discusses a Jini proxy that can make function calls on Web Services. The particular implementation used there makes use of the Apache Axis objects `Call`, `QName` and `Service`. These classes and all the classes they depend on are substantial in size–over 900kbytes. There are clear redundancies in this: for example, there are many classes which deal with WSDL document processing, and this is not needed by the proxy.
- For the Jini UPnP proxy discussed here, the CyberGarage classes are used. These classes are 270kbytes in size. However, the jar file also contains the source code for the package. Removing these reduces the size to 160kbytes and a specialised version could be even smaller. CyberGarage also requires an XML parser to interpret SOAP responses. The default parser (Xerces) and associated XML API package are over 1Mbyte in size which is substantial for an HTTP download. The kXML package can be used instead, and this is a much more reasonable 20kbytes and there is even a light version of this. This gives a total of 180kbytes which is acceptable for any Jini client–the reference implementation of Sun's lookup service takes 50kbytes just by itself.

The actual amount of code downloaded depends on the complexity of the proxy and the degree of support that already exists in the client. These three examples show variations from 10kbytes to nearly 1Mbyte.

## 11.2 Service-Level Optimisation

The standard bridge requires upto two service cache managers, one for each discovery protocol. In addition, the bridge has to act as a client to discover the original service and as a service to advertise to the original client. Service-level optimisation reduces this to two halves of two SCMs: one half to listen to service adverts, the other half for the original client to discover the service. UPnP does not have an SCM and control points listen directly to service adverts, which reduces the savings somewhat.

On the downside, it is necessary to write parts of service cache managers. Although this is not inherently difficult, knowledge of how to do this and API support by middleware systems is not so widespread as for writing simple clients and services. Jini has the necessary classes, but there are no tutorials on how to write a lookup service. CyberGarage has support for control points, but this is tightly woven with the device code and so contains redundant code.

In addition, the need to possibly perform introspection on service descriptions, to generate appropriate client-side definitions and to compile them are disadvantages.

## 11.3 Device-Level Optimisation

This optimisation gains by removal of some code (introspection, generation of interfaces and compilation) completely. On the other hand, code to generate the proxy is just moved into the client. In the case of Jini, most of this code is already present in the client from the Jini libraries and does not represent much of an overhead. For other systems it may be more costly.

### 11.4 Generality

The design patterns discussed in this paper rely on a number of properties of the two middleware systems in order to be applicable

- it must be possible for a service cache manager to be used in each middleware system. In practise this is not an onerous provision and it can be applied even to systems such as UPnP which do not require an SCM.
- There must be a (sufficiently good) mapping of the datatypes from service system to client system. This allows UPnP services to be called from Jini clients, but would limit the scope of Jini services that could be invoked by UPnP clients. As another example, the flexibility of XML data-types means that it should be possible to mix Jini clients with Web Services, and Jini services with Web Service clients.
- It must be possible to download code from the SCM to run in either the client or service. In our case study, we have downloaded code to the client that understands the service invocation protocol, but it would work equally well if code could be downloaded to the service that understands the client invocation protocol. Without this, the recipient would already need to know how to deal with a foreign invocation protocol, which would largely defeat the value of the pattern.

The third point is the most difficult to realise in practise. Many languages support dynamic code execution: most interpreted languages have an equivalent of the `eval()` mechanism, through to dynamic linking mechanisms such as dynamic link libraries of compiled, relocatable code.However, the only major language supporting dynamic downloads of code across a network appears to be Java, and the principal middleware system using this is Jini. Given some level of dynamic support, adding network capabilities to this is not hard: the author wrote a few pages of code as proof of concept to wrap around the Unix C `dlopen()` call to download compiled code across the network into a C program.

## 12 Value of Work

The value of mixing different middleware systems can be seen by a simple example. Through UPnP, various devices such as hardware-based clocks and alarms can be managed. A stock exchange service may be available as a Web Service. A calendar and diary service may be implemented purely in software as a Jini service. Using the techniques described in this paper, a Jini client could access all of these. Acting on events from UPnP clocks to trigger actions from the Jini diary the client could query the Web Service stock exchange service and ring UPnP alarms if the value of the owner's shares has collapsed.

In addition to extending the use of clients and services, there are also some side benefits:

- Jini has suffered by a lack of standards work for Jini devices and device services, with a corresponding lack of actual devices. This work allows Jini to "piggyback" on the work done now and in the future by the UPnP Consortium and to bring a range of standardised devices into the Jini environment. Jini clients will be able to invoke UPnP services in addition to services specifically designed for Jini.
- UPnP is a device-centric service architecture. It allows clients to use services on devices, but has no mechanism for UPnP clients to deal with software-only services since they cannot be readily expressed in UPnP. Work is ongoing within the UPnP Consortium to bring WSDL descriptions into the UPnP world. Jini clients on the other hand are agnostic to any hardware or software base, and can mix services of any type.

Both middleware systems have limitations–in the case of Jini, in the types of services that can be accessed, and in the case of UPnP, in the range of services that can be offered. Other middleware systems will have similar limitations. For example, Web Services tend to deal with long-lived services at well-known addresses whereas Jini can handle transient services

## 13 Conclusions

We have proposed a set of alternative architectures to bridge between different middleware systems which uses a service cache bridge and a downloadable proxy understanding the service or client invocation protocol. In addition, we have used this between Jini and UPnP and we have automated the generation and runtime behaviour of this proxy from a UPnP specification. This has been demonstrated to give a simple solution for UPnP services and Jini clients. The techniques are applicable to any client protocol which supports downloadable code and any service protocol.

## References


1. Allard, J., Chinta, V., Gundala, S., Richard, III.G.G.: Jini meets upnp. In: Proceedings of the Applications and the Internet (SAINT) (2003)
2. Arnold, K.: The Jini Specification. Addison-Wesley, Reading (2001)
3. Bellwood, T.: Uddi version 2.04 api specification. (2002) (Retrieved, July 7, 2006), `http://uddi.org/pubs/ProgrammersAPI-V2.04-Published-20020719.htm`
4. Brown, L., Sablin, D.: Extending erlang for safe mobile code execution. In: Varadharajan, V., Mu, Y. (eds.) ICICS 1999. LNCS, vol. 1726, pp. 39–53. Springer, Heidelberg (1999)
5. Casati, M.: Javascript soap client. (2006) (Retrieved, July 7, 2006), `http://www.codeproject.com/Ajax/JavaScriptSOAPClient.asp`
6. Dabrowski, C., Mills, K.: Analyzing properties and behavior of service discovery protocols using an In: Proc. Working Conference on Complex and Dynamic Systems Architecture (2001)
7. Edwards, W.K., Newman, M.W., Smith, T.F., Sedivy, J., Izadi, S.: An extensible set-top box platform for home media applications. IEEE Transactions on Consumer Electronics 4(51) (2005)

8. Garrett, J.J.: Ajax: a new approach to web applications. (2005) (Retrieved, July 7, 2006), http://www.adaptivepath.com/publications/essays/archives/000385.php
9. JSON, Json in javascript. (2006) (Retrieved, July 7, 2006), http://www.json.org/js.html
10. Kirby, G.: Dynamic compilation in java. (2005) (Retrieved, July 7, 2006), http://www-ppg.dcs.st-and.ac.uk/Java/DynamicCompilation
11. Levy, J.Y., Ousterhout, J.K., Welch, B.B.: The safe-tcl security model. Technical report, Sun Microsystems. (1997) (Retrieved, July 7, 2006), http://research.sun.com/technical-reports/1997/abstract-60.html
12. Nakazawa, J., Edwards, W., Tokuda, H., and Ramachandran, U.: A bridging framework for universal interoperability in pervasive systems. In: ICDCS (2006) (Retrieved, July 7, 2006), www-static.cc.gatech.edu/ keith/pubs/ icdcs06-bridging.pdf
13. Newmarch, J.: A Programmers Guide to Jini. APress (2001)
14. Newmarch, J.: Upnp services and jini clients. In: ISNG, Las Vegas (2005)
15. Newmarch, J.: Foundations of Jini 2 Programming. APress (2006)
16. UPnP Consortium, Upnp home page. (2006) (Retrieved, July 7, 2006), http://www.upnp.org
17. Vinoski, S.: Rpc under fire. IEEE Internet Computing (2005)
18. Waldo, J.: The end of protocols. (2000) (Retrieved July 7, 2006), http://java.sun.com/developer/technicalArticles/jini/protocols.html
19. Waldo, J.: An architecture for service oriented architectures. (2005) (Retrieved July 7, 2006), http://www.jini.org/events/0505NYSIG/WaldoNYCJUG.pdf
20. Consortium, W.W.W.: Web services home page. (2002) (Retrieved July 7, 2006), http://www.w3.org/2002/ws

# MDE for BPM: A Systematic Review

Jose Manuel Perez, Francisco Ruiz, and Mario Piattini

Alarcos Research Group, University of Castilla-La Mancha
Paseo de la Universidad, 4, 13071, Ciudad Real, Spain
JoseM.Perez2@alu.uclm.es, Francisco.RuizG@uclm.es
Mario.Piattini@uclm.es

**Abstract.** Due to the rapid change in the business processes of organizations, Business Process Management (BPM) has come into being. BPM helps business analysts to manage all concerns related to business processes, but the gap between these analysts and people who build the applications is still large. The organization's value chain changes very rapidly; to modify simultaneously the systems that support the business management process is impossible. MDE (Model Driven Engineering) is a good support for transferring these business process changes to the systems that implement these processes. Thus, by using any MDE approach, such as MDA, the alignment between business people and software engineering should be improved. To discover the different proposals that exist in this area, a systematic review was performed. As a result, the OMG's Business Process Definition Metamodel (BPDM) has been identified as the standard that will be the key for the application of MDA for BPM.

**Keywords:** Business process management, Model driven engineering, Model driven architecture, Systematic review.

## 1 Introduction

There is a need for today's business to create and modify value chains rapidly. This brings about continuous growth and change in business processes. The goal of Business Process Management (BPM) is to help business people to manage these changes.

Business process management is defined as the capability to discover, design, deploy, execute, interact, operate, optimize and analyze process in a way that is complete, doing it at the business design level and not at the technical implementation level [1].

BPM offers numerous benefits to organizations such as improving the speed of business, giving increased customer satisfaction, process integrity and accountability. It promotes process optimization, at the same time eliminating unnecessary tasks. It also includes customers and partners alike in the business processes and provides organizational agility.

BPM represents a "third wave" in business process engineering. The first wave was guided by process papers that reorganized human activity. The second wave focused on reengineering of business processes and the use of Enterprise Resource Planning (ERP). The third wave centers on formal business process models and the

J. Filipe, B. Shishkov, and M. Helfert (Eds.): ICSOFT 2006, CCIS 10, pp. 127–135, 2008.

ability to modify and combine those models so as to align business process with organizational needs [2].

BPM starts with process modeling. Process modeling is a business-driven exercise in which current and proposed process flows are documented in detail, linked to quantifiable performance metrics, and optimized through simulation analysis. Standards for process modeling languages are the key to the attaining of BPM's goal as well as in achieving the platform independence of the process models. Platform independence is one of the principles on which Model Driven Engineering (MDE) is based. The combination of both concepts, MDE and BPM, is the target of this systematic review.

MDE was conceived in an effort to solve several problems that have arisen in the last decade. On one hand, the growth of platform complexity, there being thousands of classes and methods with very complicated dependencies. On the other hand, we can observe the continuous technological evolution of the systems, forcing programmers to modify the system code every time a new requirement is given.

In the MDE paradigm, every concept must be modeled. Thus, any change in the system must be shown in the model that represents that system. To model the systems, MDE proposes using Domain-Specific Modeling Languages (DSML). By means of these languages, different modeling notations for each kind of system are achieved. Thus, the software engineer has specific tools for modeling all kind of systems.

Another important concept in MDE is model transformation. By transforming models, the evolution of the systems is facilitated. A model could be transformed to another model or to a XML specification as well as to the source code that implements the model functionality.

The OMG group has developed Model Driven Architecture (MDA) as an example of MDE. MDA emerged with the established idea of separating the business logic specification of a system from the platform specific details in which the system is implemented [3].

MDA adds some concepts to the MDE philosophy. MDA defines three level of abstraction. The Computational Independent Model (CIM), the Platform Independent Model (PIM) and the Platform Specific Model (PSM).

The key technology in MDA is MOF, as it is as in the definition of metamodels, which are MOF instances (figure 1) [4]. The transformations among these models are the basis of MDA philosophy.

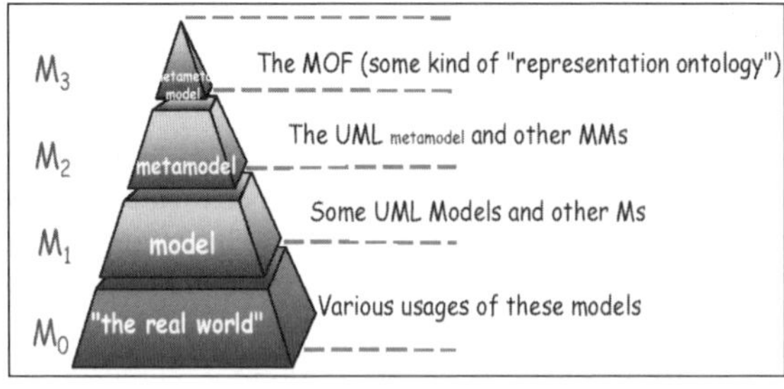


**Fig. 1.** MOF metamodels structure [4]

The structure of this paper is as follows. In section 2, systematic reviews are introduced. In section 3, the carrying out of the review is shown in part, presenting the selection of studies and the classification of these. The information analysis is described in section 4 by summarizing the different authors' proposals about the MDE for BPM application. Section 5 presents the conclusions extracted from the systematic review along with future work, taking into account the different views found.

## 2 Systematic Reviews

A systematic review of the literature is a means of identifying, evaluating and interpreting all available research relevant to a particular research question, or topic area, or phenomenon of interest [5].

Systematic review is a scientific methodology that can be used to integrate empirical research on software engineering [6].

Some of the characteristics that make the above methodology different from a conventional review are that a systematic review starts by defining a review protocol that specifies the research question, along with the methods and the criteria to drive the review. Added to all this, a systematic review is based on a search strategy that aims to detect as much relevant literature as possible. Moreover, performing a systematic review is needed in order to document the whole search strategy so that another researcher can replicate the same review with identical results.

There are three main phases that organize the different stages of the review process.

The phase called "planning the review" has as its purpose to identify the need for this study and to see through the development of a review protocol. A researcher may need a systematic review to be able to draw more general conclusions about a phenomenon or as a prelude to further research activities.

The protocol specifies the methods that will be used to undertake a specific systematic review. A pre-defined protocol is needed to avoid the possibility of researcher bias. Without a protocol, the selection of individual studies might possibly be driven by the expectations of the researcher.

When the whole planning is done, the review can start. This is the second phase, called "conducting the review". This phase lies in the identification of research, the selection of primary studies, the quality assessment study, data extraction and monitoring, together with data synthesis.

Firstly, the researcher must search the documents by using the strings specified in the protocol. When a first potential set of primary studies is obtained, the researcher must perform a selection by assessing the studies' actual relevance. Quality assessment must be done over the selected studies. As the result of assessing the information quality, according to the criteria defined in the protocol, a new set of studies is generated.

Finally, the data synthesis provides researchers with the results of the systematic review. The synthesis may be either quantitative or descriptive.

The last phase lies in the communication of the results. Usually the systematic review is reported in at least two formats: In a technical report or in a section of a PhD thesis as well as in a journal or conference paper.

**Table 1.** Studies Selection

| Author, date | Study name | Source |
|---|---|---|
| Roser and Bauer (2005) | A Categorization of Collaborative Business Process Modeling Techniques | IEEE Digital Library |
| Zeng, et al. (2005) | Model-Driven Business Performance Management | IEEE Digital Library |
| Pfadenhauer, et al (2005) | Comparison of Two Distinctive Model Driven Web Service Orchestration Proposals | IEEE Digital Library |
| Rosen (2004) | SOA, BPM and MDA | ACM Digital Library |
| Frankel (2005) | BPMI and OMG: The BPM Merger | Business Process Trends |
| Harmon (2004) | The OMG's Model Driven Architecture and BPM | Business Process Trends |
| Frankel (2003) | BPM and MDA: The Rise of Model-Driven Enterprise Systems | Business Process Trends |
| Smith (2003) | BPM and MDA: Competitor, Alternatives or Complementary | Business Process Trends |
| Kano, et al. (2005) | Analysis and simulation of business solutions in a service-oriented architecture | Wiley Digital Library |
| MEGA & Standard Bodies (2004) | Business Process Modeling and Standardization | bpmg.org |

# 3 Review Results

This section presents the selected works in the searches performed in the digital libraries, journals and internet sites related to the issue in hand. Moreover, a classification of studies is given. This has used aspects which are of relevance to the goal of the review as a basis for this classification

## 3.1 Studies Selection

The first step was to search in the predefined information sources. Those sources are: ACM digital library, IEEE digital library, Science Direct Digital Library, Business Source Premier, Wiley InterScience, www.BPTrends.com, www.bpmg.org.

The result of this search was a first set, composed of 22 studies. With the aim of tuning the set of studies, the selection criteria were applied. The studies had to contain information about the application of model driven engineering or model driven architecture in business process management. The issue of the systematic review is MDE for BPM, but because MDA is currently so widespread in the model engineering world, MDA was included in the selection criteria.

As the result of the application of selection criteria, the new set of studies was composed of 10 works (Table 1).

### 3.2 Classification of Studies

The selected studies have been classified according to several aspects that have been chosen to satisfy the goal of the systematic review (Table 2).

First of all, the author's opinion about the issue of systematic review is the most important aspect to take into account in classifying the studies. Another important aspect is whether the study offers a proposal about the use of CIM, PIM and PSM (MDA models) within the business process context. This means that the author suggests a specific utilization of MDA models, pointing out the possible modeling standards used in each model. Finally, the different standards proposed by authors for modeling business process are also aspects that are taken into account.

**Table 2.** Classification of the selected studies

| Author, date | MDE for BPM | Propose CIM, PIM, & PSM | UML | BPML | BPMN | BPDM | BPEL | J2EE | Others |
|---|---|---|---|---|---|---|---|---|---|
| Roser and Bauer (2005) | Yes | Yes | | X | X | X | X | X | ebXML, AIRIS, WS-CDL |
| Zeng, et al. 2005 | Yes | No | X | | | | | X | |
| Pfadenhauer, et al. (2005) | Yes | Partially | X | | | X | X | | |
| Rosen (2004) | Yes | Yes | | X | X | X | X | | |
| Frankel (2005) | Yes | No | X | | X | X | X | | SBVR |
| Harmon (2004) | Yes | Yes | X | X | X | X | X | | SBVR |
| Frankel (2003) | Yes | No | X | X | X | | X | | |
| Smith (2003) | No | No | | X | X | | | | |
| Kano, et al. (2005) | Yes | Yes | X | | | | X | X | |
| MEGA & Standard Bodies (2004) | Yes | No | | X | X | X | X | | XPDL |

## 4 Findings and Analysis

This systematic review goal is to identify studies that can provide an approach for the application of the MDE paradigm to business process management. Note that from here on in the text, MDA will be the modeling approach that will always be mentioned, whereas MDE will not. This is because MDA is the most widely-seen example of MDE application, and because all the papers deal specifically with MDA, and not with MDE in general.

The article “BPM and MDA: Competitors, Alternatives or Complementary” [7], does not share the optimism of the rest of the authors. In Smith’s opinion, BPM and MDA are very different. He declares that MDA must be used by software engineers and that BPM must be used by business people. He also affirms that the latter are not interested in a new approach for developing more software, but rather in a design-driven architecture based on processes and on a business process management system (BPMS) that interprets such designs, in the same way that RDBMS interprets a relational model. Although he does not deny the possibility that in the future the two philosophies may work together, at the moment he advocates the separation of both approaches.

The work “Model-Driven Business Performance Management” [8] proposes a technical approach for developing a complete application related to the BPM context. This study presents a relation between the two important concepts of this systematic review, using a model-driven approach to build the solution. The technical approach is based on the observation metamodel and its transformations. When the models are transformed, the approach suggests compiling the operational aspects of the model and finally developing a runtime engine that interprets the model and executes the generated code.

The study ”Comparison of Two Distinctive Model Driven Web Service Orchestration Proposals” [9] focuses on the way to generate a set of web services that implement the organization business processes. By applying the MDA approach, and using some of the business process standards, the final solution is generated. This document mentions the BPDM standard as the MDA BPM connection.

The article “Analysis and simulation of business solutions in a service-oriented architecture” [10], offers a four-layer model architecture, in which the first two layers, when viewed together, are similar to the CIM layer in MDA from the business point of view rather than from the software system point of view. The last two layers correspond directly to the MDA PIM and PSM layers. By separating the independent platform concerns of a solution from the specific platform concerns and their associated code by means of MDA, the reuse of solution components is supported. Furthermore, the system is more flexible and adaptable to the changes in business requirements.

The work “A Categorization of Collaborative Business Process Modeling Technique” [11], provides a proposal for applying MDA within the collaborative business process framework. Collaborative business processes are performed among different enterprises, which could have different business process development methodologies. Therefore, the creation of a common framework in which the organizations could communicate to each other in terms of business process would be ideal. The authors have spoken about MDA as the common framework for integrating business process from different organizations. They propose to create the business process CIMs, PIMs and PSMs in every organization, by using their own model language for each kind of model. These model languages must be MOF metamodels. Thus, transformations among metamodels can be done. The communication among the enterprises in terms of business process will be done by means of the common CIMs, PIMs and PSMs. These common models are written by using a common metamodel (one for each kind of model) and contain a view for the models of each organization from their CIMs, PIMs and PSMs. Thus, the common framework is well-known for all the organizations.

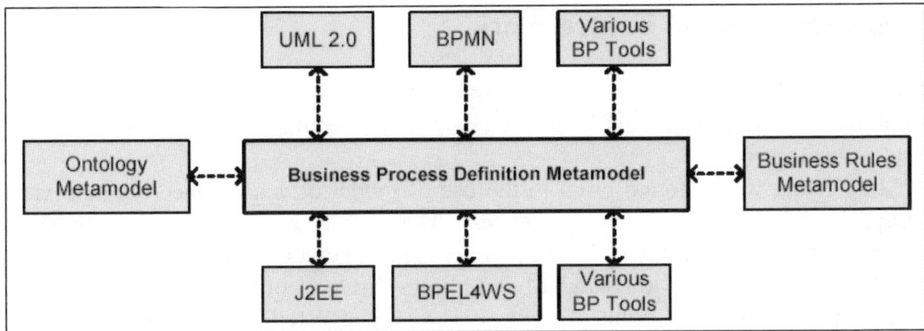


**Fig. 2.** Use of OMG BPDM [15]

The study "Business Process Modeling and Standardization" [12], is a review concerning all of the standards existing around business process, from languages to modeling notations. It provides a whole view of the state of standards (as it stood on September 2004), as well as their coverage within the BPM context. Moreover, it reports on the capacities of versions of new standards that are about to come out.

The study "SOA, BPM and MDA" [13] does not offer a specific proposal for using MDE within some business process management areas, but provides an abstract vision about the role that both MDE and BPM play. The article points out how MDA can help business process automation, reuse and maintenance.

The two works by Frankel selected in the systematic review, concerning MDA and BPM, [2], [14], point to the use of MDA as the methodology that guides business process design, implementation, maintenance and management. Frankel's theory is that BPM joined to MDA is stronger than BPM alone, and MDA together with BPM is stronger than MDA alone. Moreover, he gives a wide classification of the different business process standards that currently exist. He aims at the aligning of the business process modeling notation (BPMN) with the OMG metamodel BPDM. This would provide portability utility by means of the XMI format and the power of the MDA transformations, in line with the well-known BPMN standard. Although Frankel is optimistic about the application of MDA in BPM, he also warns us about the wide gap that exists between the abstraction represented by a business process model and the specific models that represent the implementation of the business process.

The study "The OMG's Model Driven Architecture and BPM" [15], has as its goal the use of MDA within the BPM. Harmon puts BPDM at the centre of business process modeling (Figure 2). The rest of business process modeling standards should be transformed directly to BPDM, even BPMN. He proposes a way to use the different kinds of MDA models (CIM, PIM and PSM) for business process design and implementation. Thus, CIM will be specified in terms of business process by using BPDM; the business rules by means of business rules metamodel (BRM). These models are used by business analysts. PIM are a transformation from previous CIM, specified in a software system metamodel, for example UML. These models are used by software architects. Finally, PSM are built by transforming PIM to the platform specific language in which the business process will be implemented, for example the J2EE UML Profile.

## 5 Conclusions and Future Work

The systematic review performed provides a complete view of the proposals and opinions existing in the recent literature about MDE paradigm application in business process management.

Most of the works found point to the use of model driven engineering as a valid approach for business process management. There are proposals for the use of MDA in the context of collaborative business process management, where the model driven plays the role of integration standard and allows different organizations to cooperate from a business process point of view. It is also suggested, on the other hand, that MDA is the methodology that drives the organization business process design, implementation, maintenance and management.

Although most authors are in favor of the use of MDE in business process management, there is some rejection of this idea, throwing into relief how far apart both concepts are, and how difficult it is to obtain cooperation to achieve better results.

Business process modeling standards become the key issue for the MDA application in the context of BPM. These standards must be metamodels, which are instances of meta-metamodel MOF. OMG propose the business process definition metamodel (BPDM) as the standard for business process modeling, which has no final version yet [16]. BPDM is a semantic description of the logical relations among several elements of any business process description. It is not a notation. Its advantage is that it is a MOF metamodel. Thus, any other notation language, such as BPMN, can be transformed to BPDM. As BPDM is a MOF metamodel, this can be transported via XMI to any business process tool that knows such a metamodel. The companies only have to define MDA transformations from the BPDM metamodel to executable languages like J2EE or BPEL.

BPMN is the notation standard most frequently used to define business process at a high level. So some authors are quite adamant in their assertions that the next version of BPDM will take on the BPMN standard. Thus, any high level BPMN model will be able to be shared via XMI and transformed to follow the MDA methodology.

In future research, we will monitor the evolution of BPDM and its convergence with the BPMN standard. We will propose a QVT transformation from BPMN to BPDM, as well as from BPDM to a web services metamodel. To do this, the model management framework MOMENT will be used [17].

## Acknowledgements

This work has been partially financed by the FAMOSO project (Ministerio de Industria, Turismo y Comercio, FIT-340000-2005-161), ENIGMAS project (Junta de Comunidades de Castilla-La Mancha, PBI-05-058), ESFINGE project (Ministerio de Educación y Ciencia, TIN2006-15175-C05-05), COMPETISOFT project (Programa Iberoamericano de Ciencia y Tecnología para el Desarrollo, 506PI0287), including the support of the "Fondo Europeo de Desarrollo Regional (FEDER)", European Union.

## References

1. Smith, H.: The emergence of Business Process Management. CSC's Research Services (2002)
2. Frankel, D.S.: BPM and MDA. The rise of model-driven enterprise systems. Business Process Trends (2003)
3. Miller, J., Mukerji, J.: MDA Guide Version 1.0.1, OMG (2003)
4. Bézivin, J.: MDA: From hype to hope, and reality. In: UML 2003. 6th International Conference. ATLAS Group (2003)
5. Kitchenham, B.: Procedures for performing systematic reviews. NICTA Joint Technical Report (2004)
6. Travassos, G.H., Biolchini, J., Gomes, P., Cruz, A.C.: Systematic review in software engineering. Technical Report (2005)
7. Smith, H.: BPM and MDA: Competitors, alternatives or complementary. Business Process Trends (2003)
8. Zeng, L., Lei, H., Dikun, M., Chang, H., Bhaskaran, K.: Mode-driven business performance management. In: ICEBE 2005. Proceedings of the 2005 IEEE International Conference on e-Business Engineering. IEEE, Los Alamitos (2005)
9. Pfadenhauer, K., Dustdar, S., Kittl, B.: Comparision of two distinctive model driven web service orchestration proposals. In: CECW 2005. Proceedings of the 2005 Seventh IEEE International Conference on E-Commerce Technology Workshops. IEEE, Los Alamitos (2005)
10. Kano, M., Koide, A., Liu, T., Ramachandran, B.: Analysis and simulation of business solutions in a service oriented architecture. IBM Systems Journals 44(4) (2005)
11. Roser, S., Bauer, B.: A categorization of collaborative business process modeling techniques. In: CECW 2005. Proceedings of the 2005 Seventh IEEE International Conference on E-Commerce Technology Workshops. IEEE, Los Alamitos (2005)
12. MEGA & Standard Bodies, Business process modeling and standarization. Bpmg.org (2004)
13. Rosen, M.: SOA, BPM and MDA. CTO Azora Technologies (2004)
14. Frankel, D.S.: BPMI and OMG: The BPM Merger. Business Process Trends (2005)
15. Harmon, P.: The OMG's model driven architecture and BPM. Business Process Trends (2004)
16. O.M.G.: Business Process Definition Metamodel, Request for Proposal, OMG document (June 1, 2003), http://www.omg.org/docs/bei/03-01-06.pdf
17. Boronat, A., Carsí, J.A., Gómez, A., Ramos, I.: Utilización de Maude desde Eclipse Modeling Framework para la gestión de modelos. Departament de Sistemes Informatics i Computacio. Universitat Politécnica de Valencia (2005)

# Exploring Feasibility of Software Defects Orthogonal Classification

Davide Falessi and Giovanni Cantone

Univ. of Roma "Tor Vergata", DISP, Via del Politecnico 1, Rome, Italy
falessi@ing.uniroma2.it, cantone@uniroma2.it

**Abstract.** Defect categorization is the basis of many works that relate to software defect detection. The assumption is that different subjects assign the same category to the same defect. Because this assumption was questioned, our following decision was to study the phenomenon, in the aim of providing empirical evidence. Because defects can be categorized by using different criteria, and the experience of the involved professionals in using such a criterion could affect the results, our further decisions were: (i) to focus on the IBM Orthogonal Defect Classification (ODC); (ii) to involve professionals after having stabilized process and materials with students. This paper is concerned with our basic experiment. We analyze a benchmark including two thousand and more data that we achieved through twenty-four segments of code, each segment seeded with one defect, and by one hundred twelve sophomores, trained for six hours, and then assigned to classify those defects in a controlled environment for three continual hours. The focus is on: Discrepancy among categorizers, and orthogonality, affinity, effectiveness, and efficiency of categorizations. Results show: (i) training is necessary to achieve orthogonal and effective classifications, and obtain agreement between subjects, (ii) efficiency is five minutes per defect classification in the average, (iii) there is affinity between some categories.

**Keywords:** Software engineering, Experimental software engineering, Orthogonal Defect Classification, Defect class affinity, Fault detection, Effectiveness, Efficiency.

## 1 Introduction

Defect classification plays an important role in software quality. In fact, software quality is strictly related to the number and types of defects present in software artifacts and eventually in software code.

The analysis of defect data can help to better understand the quality of software products and the related processes, and how they evolve.

An invalid defect categorization would obviously imply wrong data, which could lead analysts to wrong conclusions, concerning the product, development process or phase, methods, and/or tools.

For instance, in order to define the best mix of code testing and inspection techniques for given application domain and development environment, it is crucial to collect valid defect-category data [1], [3], [4].

J. Filipe, B. Shishkov, and M. Helfert (Eds.): ICSOFT 2006, CCIS 10, pp. 136–152, 2008.

### 1.1 Related Works

The Orthogonal Defect Classification (ODC) is a schema [12] that IBM proposed in the aim of capturing semantics of software defects (see Section 1.3 for further details concerning ODC). ODC was originally published on 1992; because in the mean time the software world changed, the IBM provided to update ODC regularly. The classification adopted in this work is ODC v5.11, i.e. the last version of ODC, to the best of our knowledge. ODC is defined as a technology-independent (software process, programming language, operative system, etc.) classification schema. This is based on eight different kinds of attributes, each of them having its own categories.

Khaled El Emam and Isabella Wieczorek (1960), and Kennet Henningsson and Claes Wohlin (2004) investigated ODC empirically by focusing on subjectivity of defect classification. In order to evaluate the level of cohesion among classifications that different subjects enacted, both studies used "Kappa statistics" [5], and worked on their own variations of ODC. In particular, El Emam and Wieczorek involved various combinations of three subjects who performed in the role of defect categorizers on an actual software artifact, during the development process; they hence collected and eventually analyzed "real inspection data" [7]. Eight subjects, each having at least a Master's degree but with limited experience in defect classification, participated to the experiments conducted by Henningsson and Wohlin, where objects were utilized that included thirty defects selected from a repository. Concerning results from those studies, the former presents high level of cohesion with respect to standards utilized by medical studies, the latter shows that there might be subjectivity in classification. [6] used the ODC as initial defect categorization framework and afterwards faults were classified in a detailed manner according to the high-level constructs where the faults reside and their effects in the program. The analysis of field data on more than five hundred real software faults shows a clear trend in fault distribution across ODC classes. Moreover, results show that a smaller subset of specific fault types is clearly dominant regarding fault occurrence.

### 1.2 Study Motivations and View

We can count a significant number of empirical works from many authors worldwide, whose conclusions are based on categorization of software defects. A common assumption of all those works (see Section 8 for few samples of them: [3], [4], [9], [10]) is that in large extent defects can be classified objectively, whatever the classification model might be. In the absence of enough evidence for such an assumption, all those empirical results could be questioned. Consequently, the basic question of this study is whether software practitioners can uniformly categorize defects.

In this paper we focus on the ODC attribute "Defect Type" (DT), which role is to catch the semantics of defects, that is the nature of the actual correction that was made to remove a defect from a software code. DT categorization hence follows defect detection, identification and fixing: in fact, the real nature of a defect can be understood (and than suitably categorized) only after the code is fixed, in the ODC approach.

DT includes seven defect categories [12], [13]:

1. Assignment/Initialization: value(s) assigned incorrectly or not assigned at all.
2. Checking: errors caused by missing or incorrect validation of parameters or data in conditional statements. It might be expected that a consequence of checking for a value would require additional code such as a do while loop or branch.

3. Algorithm/Method: efficiency or correctness problems that affect the task and can be fixed by re-implementing an algorithm or local data structure without the need for requesting a design change; problems in the procedure, template, or overloaded function that describes a service offered by an object.
4. Function/Class/Object: the defect should require a formal design change, as it affects significantly capability, end-user interfaces, product interfaces, interface with hardware architecture, or global data structure(s); defect occurred when implementing the state and capabilities of a real or an abstract entity.
5. Interface/O-O Messages: communication problems between modules, components, device drivers, objects or functions.
6. Relationship: problems related to associations among procedures, data structures and objects.
7. Timing/Serialization: necessary serialization of shared resource was missing, the wrong resource was serialized, or the wrong serialization technique was employed.

In the remaining, we present, analyze, and discuss a benchmark including two-thousand and more data that we achieved through an experiment based on twenty-four segments of code, each segment seeded with one defect, and one hundred twelve sophomores, trained for six hours and then assigned to classify those defects in a controlled environment for three continual hours. In particular, Section 2 presents the experiment problem and goal definition. Section 3 shows the experiment planning and operation. Section 4 and 5 present and discuss results. Some final remarks and further intended works conclude the paper.

## 2 Goal and Experiment Hypotheses

The goal [3] of this paper is to analyze the (ODC)'s DT attribute from the point of view of the researcher, in the context of an academic course on "OO thinking and programming with Java" for sophomores, for the purpose of evaluating dependences of software defect categorizations on: i) defect ($d \in DD$): ii) subjectivity of practitioners ($s \in S$); iii) expertise in defect detection (X), and (iv) Programming language (PL) utilized to code artifacts, by focusing on: a) Effectiveness (E), i.e. in what extent a defect is associated to its most frequent categorization (MFC); b) Efficiency (Ec), i.e. the number of (MFC)s per time unit; c) Orthogonality (O), i.e. in what extent a defect is assigned to just one category; d) Affinity (A), i.e. in what extent a defect category looks like other categories, and e) Discrepancy (D), i.e. in what extent subjects assign a defect different categories (see Sections 3 for quantitative definitions of all those variables).

Based on that goal, the hypotheses of our work concern the impact of expertise ($h_X$), defect category ($h_C$), and programming language ($h_L$) on orthogonality ($h_O$), effectiveness ($h_E$), and discrepancy ($h_D$).

The null ($h_0$) and alternative ($h_1$) hypotheses for expertise versus orthogonality (resp. effectiveness, and discrepancy) are:

– $h_{XO0}$: Expertise does not significantly impact on orthogonality (resp. $h_{XE0}$, and $h_{XD0}$).
– $h_{XO1}$: Expertise impacts significantly on orthogonality (resp. $h_{XE1}$, and $h_{XD1}$).

Hypotheses concerning programming language ($h_{LO0}$, $h_{LO1}$, $h_{LE0}$, $h_{LE1}$, $h_{LD0}$, $h_{LD1}$), and defect category ($h_{CO0}$, $h_{CO1}$, $h_{CE0}$, $h_{CE1}$, $h_{CD0}$, $h_{CD1}$) have similar formulations. In the remaining, while we evaluate the impact of defect category, expertise, and programming language on outcomes, our reasoning mainly focuses on expertise. In fact, in our expectation, in case of significant dependence of defect categorizations from the categorizers' subjectivity, expertise should play the most important role and behave as the main discriminating factor; consequently, our planning and training emphasis was in providing variable expertise.

## 3 Experiment Planning and Operation

Whoever the participant subject, three items characterize our elementary experiment: a defect, as seeded and fixed in a program segment, the programming language of that segment, and dissimilarity of that defect.

In order to average on differences among participant subjects, our planning decisions was to utilize subjects with the same level of experience; in particular: i) one hundred or more subjects from the same academic class, ii) subjects showing the same OOP class frequency record, iii) subjects who would be attending all the training sessions. Moreover, in order to manage the impact of learning effect on results, we kept further planning decisions, which also helped to prevent exchange of information among participant subjects: iv) to arrange four master files, where experiment artifacts are located in different order, v) to assign subjects seats randomly, and give neighbors copies of different master files, and vii) to ask subjects to handle artifacts in sequence, staring from the first artifact their assigned.

We hence developed and saved into repository defected artifacts. An artifact consists in a less than twenty ELOC segment of code, plus comments to ensure easy and valid understanding; one defect is seeded per code segment, and fixed through specific comments. Let us note that while we used our understanding of DT ODC to generate defected artifacts, we no further utilize such understanding in the remaining of this study, where categorizations are utilized as enacted by subjects.

In parallel with repository construction, we called for participation, and trained subjects through three two-hour lectures, which presented the role and importance of defect categorization, defined categories of the ODC DT attribute, and explained extensively two or more exemplar cases for each defect category. Subsequently, we evaluated in Low (L), Average (A), and High (H) the dissimilarity between defects in the experiment artifacts and defects in the examples given for training (see Expertise in Section 3.1.4 for further details). Finally, we ruled the random selection of experiment artifacts from the repository, as in the following: (i) Get as many C++ as Java coded artifacts; (ii) Get two or more artifacts for each defect category; (iii) Get 20% of artifacts for each value of Dissimilarity, and remaining (40%) at random.

### 3.1 Independent Variables: Parameters, Blocking Variables and Factors

**Subjects.** As already mentioned, one hundred twelve sophomores participated to the experiment, who were attending the course of Object-Oriented Programming, their fourth CS course at least. All of them had attended all the training lectures and, in term of experience, they can be considered as novice programmers. Subjects' participation was part of a course test; they worked individually in the same 250 seats room, in the continual presence of two or more observers; communication among subjects was not allowed. Other one hundred subjects, who had not fully attended the training or the OOP course, were located in an adjacent room: their data will be no further considered in the present paper.

**Objects.** Experiment artifacts, twelve C++ coded and twelve Java coded, were assigned to all subjects, each artifact seeded by one defect. All quadruples of neighbor subjects handled the same artifacts but in different order.

**Experiment Duration.** Subjects had up to three hours assigned to enact their task. They were allowed to quit the experiment any time, after the start and before the formal end.

**Factors and Treatments.** Factors of the basic experiment and their levels are:
- **Programming Language** (PL), levelled at C++ and Java, respectively.
- **Defect Category** (Ctg). Six defect categories are utilized, i.e. all the DT ODC less Timing/Serialization: in fact, subjects had not yet been exposed to concurrent programming concepts, constructs, and mechanisms, when they participated to the experiment.
- **Expertise** (X). It is analogous to Dissimilarity but scale is reversed; it hence relates to quantity of examples given per defect during training. In fact, for each defect type, we set artificially the subjects expertise by dosing the explanation time, and the numbers of examples given per defect. (0, 1, 2) are the values of the ordinal scale we use to measure the subjects expertise, where: 0 means that training did not include examples showing that specific instance of the defect category (hence, the defect shows low level of similarity with the explained defects, and its Dissimilarity measure is H); 1 means that training exposed subjects just one time to that specific instance of the defect category (Dissimilarity measures A); 2, means that subjects trained with two or more instances concerning that specific defect category (the defect shows high similarity with the explained ones, and its Dissimilarity measure is L). Concerning this point, let us finally note that, because subjects had already attended two CS courses in C++ and were attending a Java course, trainers gave more emphasis to defected artifacts coded in the latter.

### 3.2 Dependent Variables

We directly measured:
- Completion Time: Actual task duration per subject (duration of all the elementary experiments assigned to the same subject).
- Categorization: ODC per elementary experiment and subject. A subject, whether sure about his understanding, assigns a defect just one category, else zero or two categories.

Based on such direct measures, we derive the variables described in the followings, which characterize the DT attributes of the OD Classification. Let us note that measures in the following are given to each specific defect, and then applied in the same way in each defect category, each programming language, and so on..

- **Effectiveness** (E): percentage of the most frequent categorization with respect to the universe of categorizations given by subjects for this defect.
- **Efficiency** (Ec): how many (MFC)s occur per time unit, in the average, for this defect. Because of the experiment infrastructure that we choose (paper supports for data collection; data registration enacted by subjects), our decision was to collect the task Completion time only, rather than the time duration of each elementary experiment. Consequently, data from the basic experiment are not enough to investigate efficiency in deep.
- **Orthogonality** (O): what percentage of subjects assigned this defect just one category (rather than zero or two).
- **Discrepancy** (D): this does measure the average distance in percentage related to the entire population for the same categorization, and is a variant of the Agreement's [8], [7] one complement. In other word, discrepancy is the average probability that a given categorization is different from those given by other subjects for the same defect.
- **Affinity** (A): this expresses a relationship of a category with respect to one more category, and is a variant of the Confusion's [8] one complement. Given two categories, the source category CS and the destination category CD, let us take in consideration defects, which MFC is CS. The affinity of CS *with respect to* CD, A_WRT(CS,CD), measures the percentage of CS or CD categorizations given for those defects. Formally:
- $\forall d \in DD(Exp)$: $MFC(d) = CS$, $\exists A\_WRT \in [0..100]$: $(100 * p(d) \in \{CS, CD\} = A\_WRT)$; (1)

where: d is any of the defect set DD in the experiment Exp, and p is the probability function averaged on all instances of the argument defect. A_WRT is not commutative (sometimes A_WRT(C1,C2) ≠ A_WRT(C1, C2)), and its reflexive closure, A_WRT(C,C), is the Effectiveness with respect to category C.

The affinity *between* CS and CD, A_Btw(CS, CD), is then defined as:

$$\forall d \in DD(Exp): MFC(d) \in \{CS, CD\}, \exists\ A\_Btw \in [0..100]: (100 * p(d) \in \{CS, CD\} = A\_Btw); \quad (2)$$

Note that ∀(CS, CD), A_Btw(CS,CD) = A_Btw(CD, CS), i.e. A_Btw is commutative:.

Definitions above can be extended to three or more categories.

## 4 Results and Data Analysis

At experiment conduction time, subjects registered more than two thousand six hundred data fields, which we eventually deposited in a database. Two subjects provided exorbitantly distant data from the most frequent ones; data analysis identified those data as outliers, and consequently we excluded them form further analysis.

In this study all categorizations given by subjects, are evaluated, null ones included: in our evaluation, null categorizations candidate IBM DT definitions for further clarification, or our training for improvement.

### 4.1 Descriptive Statistics

Let us consider now orderly relationships between each response variable and factors.

**Effectiveness.** We want to describe the evolution of the most frequent categorizations as a whole and versus expertise, programming languages, and defect categories involved, and eventually with respect to the task completion time.

Fig. 1 shows subjects given categorizations, as averaged on the whole available data. Concerning the abscissa, "0" stands for not categorized defects (null); "1_MFC" (resp. "1_NMFC") denotes that the subject assigned this defect just the most frequent categorization (resp. one category, but different from the MFC); "Others" stands for assignment of two categories to this defect. Effectiveness (see MFC in Fig. 1) is 0.69, and variance is 8.

Fig. 2 and Fig. 3 relate effectiveness with expertise and specific defects, respectively.

Table 1 shows effectiveness versus ODC categories, and related variances. Table 2 relates effectiveness to the programming language of the defected segments.

Fig. 4 shows the evolution of effectiveness in time.

**Efficiency.** Fig. 5 presents efficiency with respect to completion time. Table 3 shows statistical summary for efficiency.

**Orthogonality.** Fig. 6 and Fig. 7 relate orthogonality with expertise and specific defects, respectively. Table 4 and Table 5 present orthogonality versus ODC categories, and programming language, respectively. Table 6 shows statistical summary for Orthogonality, and Fig. 8 presents the evolution of orthogonality in time.

**Discrepancy.** Table 7 shows statistical summary for discrepancy. In the remaining, this Section presents discrepancy with respect to ODC categories (Table 8), programming languages (Table 9), expertise (Fig. 9), and seeded defects ( Fig. 10), respectively.

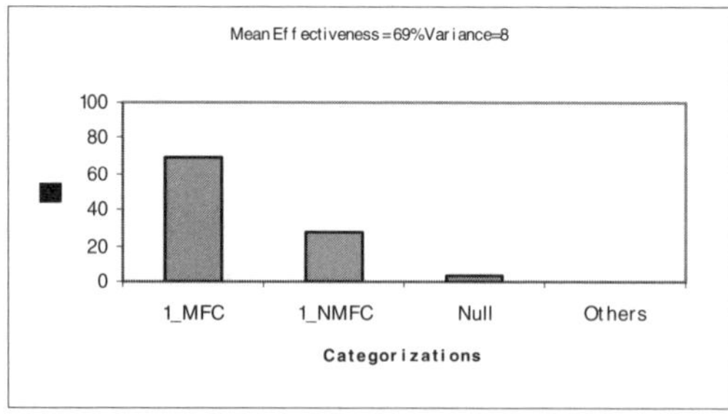


**Fig. 1.** Categorizations and Effectiveness (with respect to the whole data collected)

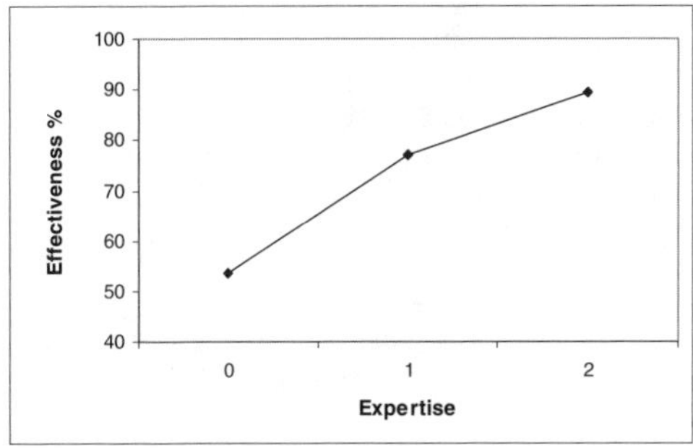


**Fig. 2.** Effectiveness versus Expertise

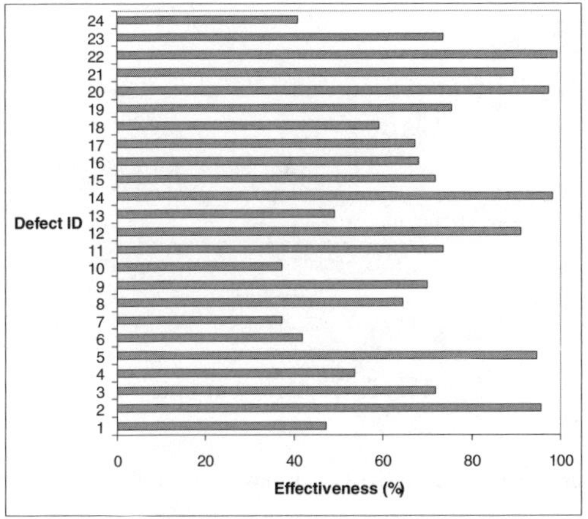


**Fig. 3.** Effectiveness per Defect

**Table 1.** Effectiveness versus ODC Categories

| Effectiveness \ Category | 1 | 2 | 3 | 4 | 5 | 6 |
|---|---|---|---|---|---|---|
| Average (%) | 77 | 83 | 48 | 75 | 54 | 82 |
| Variance | 204 | 156 | 18 | 470 | 237 | 151 |

**Table 2.** Effectiveness versus Programming Language

| Effectiveness \ Language | Java | C++ |
|---|---|---|
| Average (%) | 61 | 78 |
| Variance | 402 | 248 |

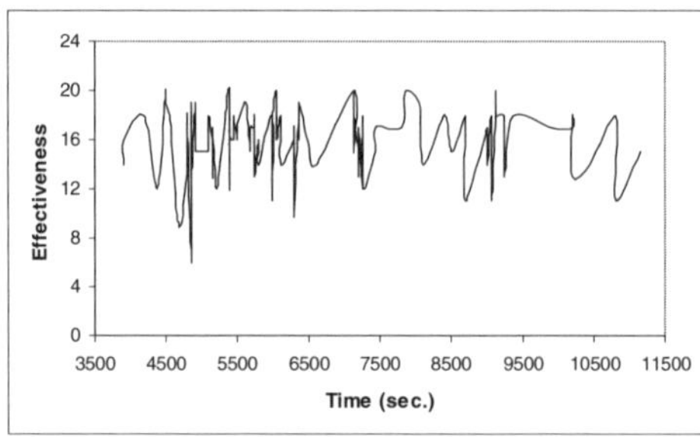


**Fig. 4.** Effectiveness in time

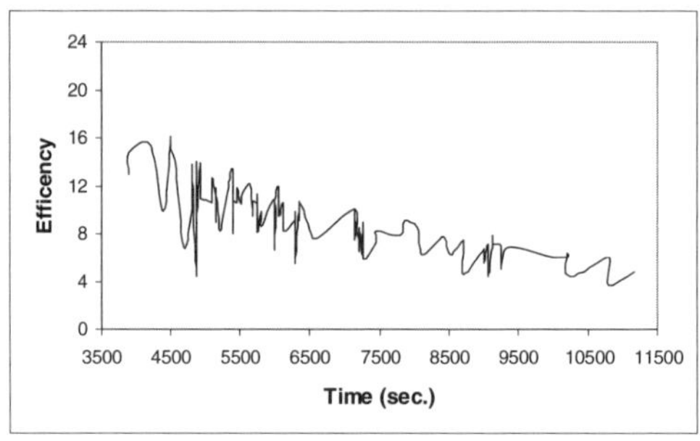


**Fig. 5.** Efficiency in time

**Table 3.** Statistical summary for efficiency

| Efficiency | |
|---|---|
| Average (MFC/h) | 9 |
| Variance | 7,31 |

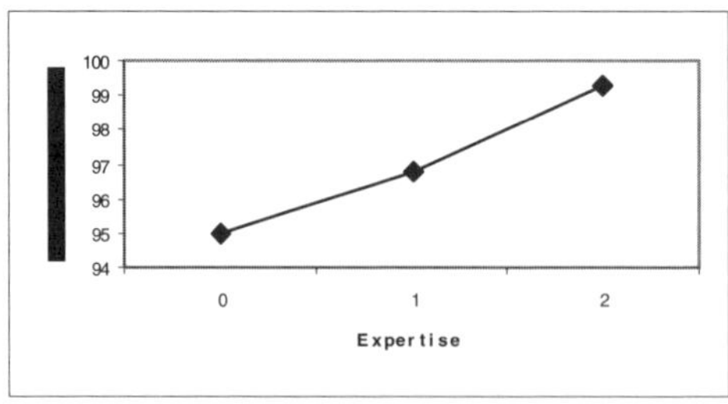


**Fig. 6.** Orthogonality versus Expertise

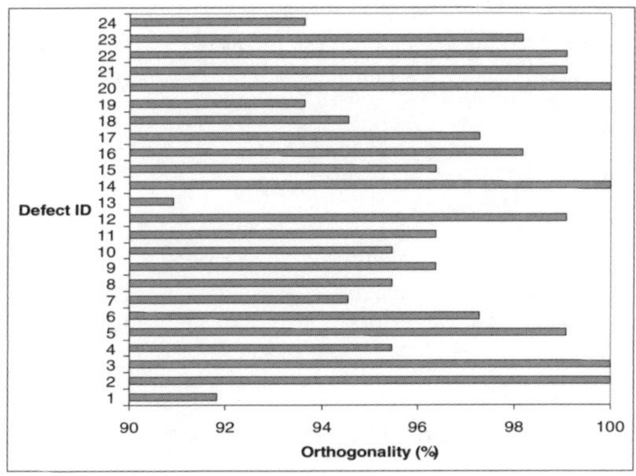


**Fig. 7.** Orthogonality per defect

**Table 4.** Orthogonality versus ODC categories

| Category / Orthogonality | 1 | 2 | 3 | 4 | 5 | 6 |
|---|---|---|---|---|---|---|
| Average (%) | 98 | 97 | 94 | 98 | 95 | 98 |
| Variance | 4 | 5 | 7 | 4 | 3 | 2 |

**Table 5.** Orthogonality versus programming language

| Language / Orthogonality | Java | C++ |
|---|---|---|
| Average (%) | 96 | 97 |
| Variance | 6 | 7 |

**Table 6.** Statistical summary for orthogonality

| Orthogonality | |
|---|---|
| Average (%) | 97 |
| Variance | 2 |

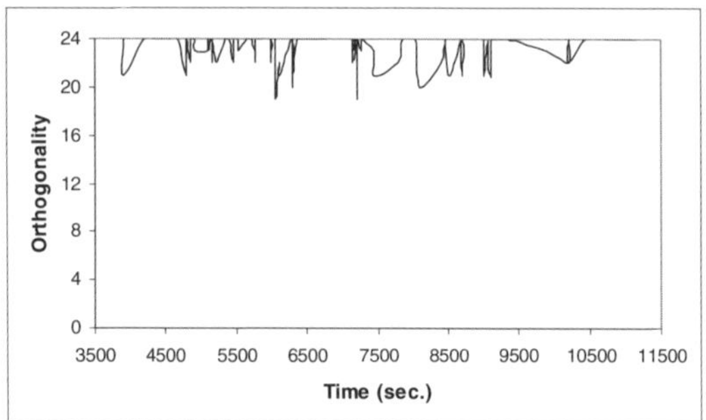


**Fig. 8.** Orthogonality in time

**Table 7.** Statistical summary for Discrepancy

| Discrepancy | |
|---|---|
| Average (%) | 43 |
| Variance | 1010 |

**Table 8.** Discrepancy versus ODC Categories

| Category / Discrepancy | 1 | 2 | 3 | 4 | 5 | 6 |
|---|---|---|---|---|---|---|
| Average (%) | 39 | 28 | 65 | 34 | 60 | 29 |
| Variance | 44 | 379 | 13 | 591 | 171 | 326 |

**Table 9.** Discrepancy versus Programming Language

| Language / Discrepancy | Java | C++ |
|---|---|---|
| Average (%) | 50 | 50 |
| Variance | 453 | 478 |

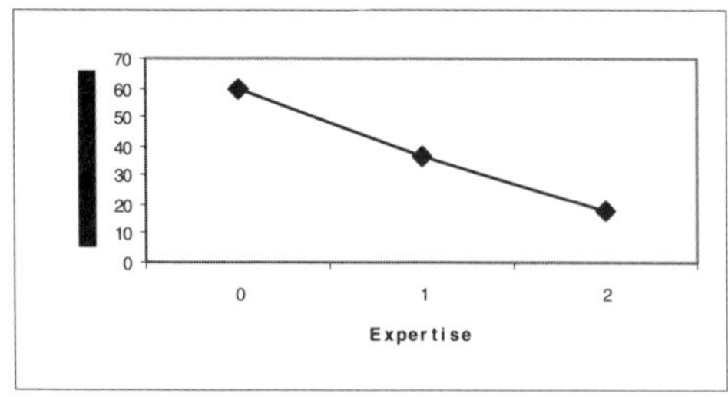


**Fig. 9.** Discrepancy versus Expertise

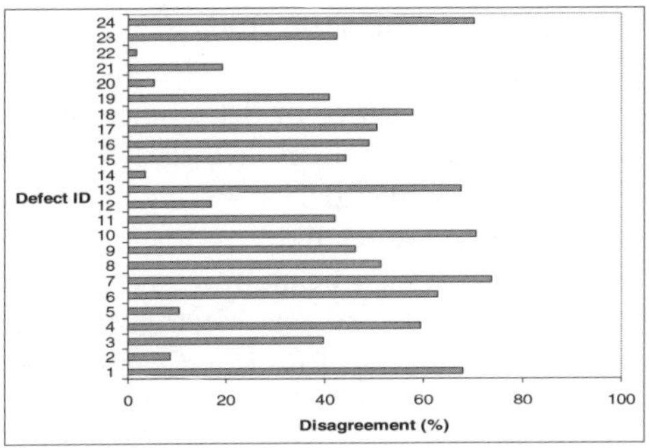


**Fig. 10.** Discrepancy per defect

**Affinity.** Based on the average effectiveness shown above (E=0.69), the number of categorizations that differ from their (MFC)s is around 818.

While it is not possible to include all those categorizations in this paper, we can describe their tendencies, based on definitions given for Affinity in Section 3.2 above: according to expressions (1), A_WRT(6, 5)= 90; A_WRT(2, 3)= 95; according to expression (2), A_Btw(1, 3, 5)= 90.

In words, when the MFC is 6 (Relationship) then 90% of categorizations provided by subjects are 6 (Relationship) or 5 (Interface/ OO Messages).

Moreover, when the MFC is 2 (Checking) then 95% of categorizations provided by subjects are 2 (Checking) or 3 (Algorithm/Method).

Furthermore, when MFC is 1 (Assignment/ Initialization), 3, or 5 then 90% of categorizations provided by subjects are 1, 3, or 5.

Finally, let us spread on data "Others" in Fig. 1, which concern affinity. Columns Ctg1 and Ctg2 in Table 10 present the alternative categorizations that doubtful subjects assigned to defects; the Ocs columns show the occurrences of those double categorizations.

**Table 10.** Defect's double categorizations (as provided by doubtful subjects)

| Ctg1 | Ctg2 | Ocs | Ctg1 | Ctg2 | Ocs |
|---|---|---|---|---|---|
| 1 | 3 | 3 | 3 | 6 | 1 |
| 1 | 5 | 2 | 4 | 5 | 1 |
| 2 | 3 | 3 | 4 | 6 | 2 |
| 3 | 4 | 2 | 5 | 6 | 1 |
| 3 | 5 | 2 | Others | | 0 |

## 4.2 Hypothesis Testing

In order to test hypotheses concerning expertise, we separate cases where the involved expertise is null (0) from remaining ones (expertise measures 1 or 2), so having the seeded defects partitioned in two groups, $G_{X=0}$, and $G_{X\neq0}$, respectively.

**Testing $h_{XO0}$. Expertise does Insignificantly Impact on Orthogonality: $O(G_{X=0}) \cong O(G_{X\neq0})$.** The number of subjects, who assigned one category to $G_{X=0}$ defects, are:

(100, 101, 103, 104, 104, 105, 105, 105, 106, 106, 107, 108), respectively; those for $G_{X\neq0}$ are: (103, 106, 107, 108, 109, 109, 109, 109, 110, 110, 110, 110). Fig. 11 shows the Box-and-Whisker plots for such series of data. Since the latter cannot fit under normal curve at 99% of confidence level (in fact, its lowest P-value from Shapiro-Wilks test is 0.0051, which is less than 0.01), we applied the Mann-Whitney (Wilcoxon) W test to compare medians. Since the W test's P-value is 0.000919, which is less than 0.05, there is a statistically significant difference between the medians at the 95.0% confidence level. Consequently, we can reject the null hypothesis $h_{XO0}$ at 95% of significance level. In other words, expertise significantly impacts on orthogonality of defect categorizations.

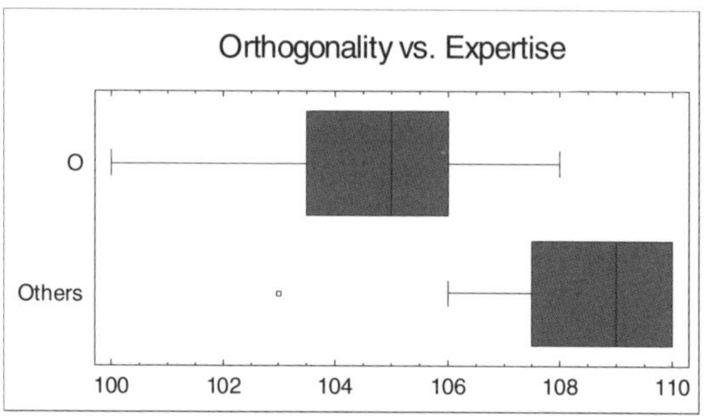


**Fig. 11.** Orthogonal classifications versus expertise

**Testing $h_{XE0}$. Expertise does Insignificantly Impact on Effectiveness: $E(G_{X=0}) \cong E(G_{X\neq0})$.** The effectiveness values for categorizing $G_{X=0}$ defects are (41, 41, 45, 46, 52, 54, 59, 65, 71, 75, 79, 81), respectively; those for $G_{X\neq0}$ are (74, 77, 79, 81, 83, 98, 100, 104, 105, 107, 108, 109). Fig. 12 shows the Box-and-Whisker plots for such series of data. Since the latter cannot fit under normal curve at 95% of confidence level (in fact, its lowest P-value from Shapiro-Wilks test is 0.037, which is less than 0.05), we applied the W test. Since the W test's P-value is 0.000194, which is less than 0.05, there is a statistically significant difference between the medians at the 95.0% confidence level. Consequently, we can reject the null hypothesis $h_{XE0}$ at 95% of significance level. In other words, expertise significantly impacts on effectiveness of defect categorizations.

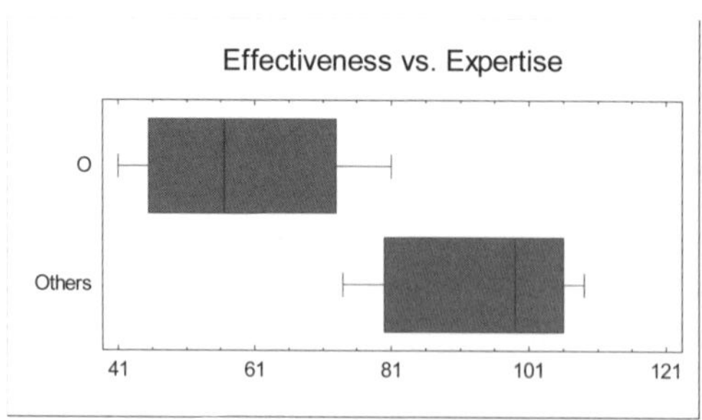


**Fig. 12.** Effectiveness versus expertise

**Testing $h_{XD0}$. Expertise does Insignificantly Impact on Discrepancy: $D(G_{X=0}) \cong D(G_{X\neq0})$.** The discrepancy values related to categorizations of $G_{X=0}$ defects are (42, 44, 49, 52, 58, 59, 63, 68, 68, 70, 71, 74), respectively; those for $G_{X\neq0}$ are (2, 4, 5, 9, 10, 17, 19, 40, 41, 42, 46, 51). Fig. 13 shows the Box-and-Whisker plots for such series of data. Since the latter cannot fit under normal curve (in fact, its lowest P-value from Chi-Square test is 0.022, which is less than 0.05), we applied the W test. Since the W test's P-value is 0.000137, which is less than 0.05, there is a statistically significant difference between the medians at the 95.0% confidence level. Consequently, we can reject the null hypothesis $h_{XD0}$ at 95% of significance level. In other words, expertise significantly impacts on discrepancy between defect categorizations.

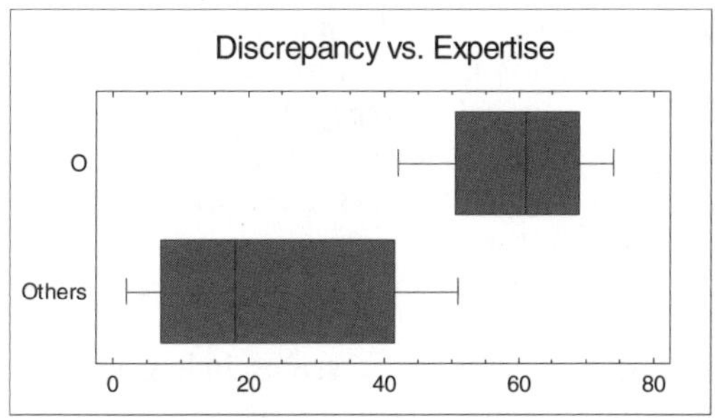


**Fig. 13.** Level of Discrepancy, treated by experience

# 5 Discussion

## 5.1 Experiment Results

**Effectiveness.** Based on Fig. 1, the percentage of most frequent categorizations is in average 69%. This seems quite a small value for effectiveness, which also means that there seems to be high subjectivity in defect categorization when trained/untrained novice programmers are involved. Again Fig. 1 shows that those programmers perform quite dissimilarly, since variance (8) is very high - one third of the seeded defects (24) - as also shown by Fig. 3 and Table 1.

Fig. 1 also shows that single non-MFC classifications (1_NMFC) are in number ten times greater than the doubtful ones (Null + Others). In our understanding, this means that novices seem unconscious of consequences that their limited knowledge of ODC DT could have. Another view is that IBM should improve the presentation of ODC DT, in order to help practitioners to distinguish among categories more easily.

Based on Fig. 2, it seems that effectiveness is strongly related to expertise. In fact, effectiveness grows from 54% up to 89% as the given training grows. Based on that slope, the trend for effectiveness is 100%, which expert professionals should be able to approach. The impact of expertise on results explains, in our understanding, the variance previously observed with aggregated data. This also asserts that data in Table 1 should not be utilized to evaluate the impact of defect category on effectiveness, and, similarly, data in Table 2 should not be used to evaluate the impact of programming languages on effectiveness.

Finally, based on Figure 4, effectiveness seems independent from the completion time, when this is limited to 3 hours.

**Efficiency.** The amount of time a subject employed to enact a categorization is around 5 minutes in average.

Based on date in Table 3, the mean time for an MFC categorization is 6.66 minutes (9 MFC/hour), and variance is 7.3 MFC/hour.

Since variance is similar to the average, it seems that efficiency is highly subjective with novice programmers. Let us recall that it was not possible to collect the duration time during the basic experiment; consequently, we cannot investigate efficiency more deeply.

**Orthogonality.** Based on Table 6, which data are again not yet disaggregated with respect to expertise, orthogonality is 97%, while variance is 2. This expresses that, in the average, programmers commonly percept ODC with respect, and tend to provide just one classification per defect, whatever is their expertise. However, taking in consideration data disaggregated by expertise (Fig. 6), with novices, orthogonality grows from 95% up to 99.3% as expertise grows.

Based on Table 5, aggregated data show no difference of C++ and Java versus orthogonality.

Finally, based on Fig. 8, orthogonality seems independent from the completion time, when this is limited to 3 hours.

**Discrepancy.** Subjects had to select a category out of seven (including null). In theory, the maximum value for discrepancy is 86%, which occurs when all selections are equally probable; it is the probability that six categories are selected out seven (less scale factor 100). The minimum of discrepancy is 0%, which occurs in case of complete agreement between subjects for each categorization. Table 7 shows 43% discrepancy (and 1010 variance!), as registered in average for our basic experiment, again with respect to data not yet disaggregated by expertise. That value is exactly the mean between the discrepancy's minimum and maximum theoretic values; as a result, ODC seems to be quite dependent from the categorizers' subjectivity, when trained/untrained novices are involved.

Based on Fig. 9, wich relates to data disaggregated by expertise, it seems that discrepancy is strongly related to expertise. In fact, discrepancy decreases from 60% up to 17% as the given training grows. Based on that slope, the trend for discrepancy is the theoretic minimum (0%), which expert professionals should be able to approach. The impact of expertise on results explains, in our understanding, the very large value previously observed for variance, when aggregated data were considered.

Based on Table 9, aggregated data show no difference of C++ and Java versus discrepancy. Again, discrepancy seems independent from the completion time, when this is limited to 3 hours.

**Affinity.** Based on data elaboration that we presented above (see Section 0), it seems that categories "Assignment/ Initialization", "Algorithm/Method", and "Interface/OO Message" are one each other strongly affine. Moreover, category "Interface/OO Message" is frequently provided in place of "Relationship", and the same for "Algorithm/Method" with respect to "Checking". This, in our understanding, calls for training improvement by emphasizing on dissimilarities among those categories.

### 5.2 Threats to Validity

This empirical study has a number of limitations that should be taken into account when interpreting its results.

Concerning the internal validity [11] (i.e. the degree to which conclusions can be drawn about the causal relationship between independent variables and dependent variables), it should be noted that we utilized a very limited number of defect samples: 12 per language, hence two defects per category. Moreover, while the task completion time assigned was quite small, and subjects were continually in control of observers during the conduction of the experiment, we cannot guaranty absence of interactions between participants; in fact, these were student, who we partially graded for their performance; in the experiment cultural context, a student is appreciated, who passes his solutions to colleagues. Furthermore, our training emphasized on Java language, and the real experience and expertise of subjects with C++ was not in control.

Another limitation of this study is related to the external validity [11], i.e. the degree to which the results from this study can be generalized. It cannot be assumed a priori that the results of a study generalize beyond the specific environment and context in which it was conducted. In fact, subjects involved with the basic experiment are sophomores in OO Programming, who should not be considered as novice professional programmers. Moreover, the experiment software artifacts that we utilized in the basic experiment are small segments of code, which should not be taken to represent real software. Finally, we utilized paper supports both for experiment artifacts and forms, while realism asked for electronic-supported code, and electronic-network-supported form distribution, and data collection.

## 6 Conclusions and Future Works

This paper has presented an empirical investigation on the (IBM)'s ODC-DT attribute for software defect categorization. Foci of the investigation have been the classification effectiveness, efficiency, orthogonality, discrepancy, and affinity with respect to practitioners' subjectivity (110 students performing in the role of experiment subjects), defects individuality (6 DT categories of seeded defects), and software artifacts' coding language (Java and C++). Results shown include averages for time for defect categorization (≅5 minutes), effectiveness (69%), and orthogonality (97%). Results also show that subject's expertise seems to impact very significantly on all the results, and subjects with enough expertise should be able to easily approach the theoretic best value for effectiveness, as for orthogonality and discrepancy. Our consequent expectation is that there should be objectivity in defect categorization, whether enacted by software practitioners. However, such an expectation still needs empirical evidence. Further results show that, when time spent in categorizing defects lasts between 1 and 3 hours, the effectiveness, orthogonality, and discrepancy are not affected by the time duration of the classification section. Moreover, results show that the programming language of coded artifacts, and the defect nature seem to impact insignificantly on effectiveness, orthogonality, and discrepancy. Finally, our results show that there are some categories that tend to confuse subjects; this, in our understanding, calls for improving definitions of those ODC DT categories, as actually given by IBM. Namely, those categories are "Interface/OO Message" and "Relationships".

Further confusing categories are "Assignment/Initialization" and "Algorithm/Method" on one side, and "Algorithm/Method" and "Checking" on the other side, which confirm previous results [8].

Our plan for the future is first to extend the size of our defect repository, place the material in electronic format, and contact IBM experts in the aim of receiving their categorizations of our defect samples (to use as the reference "correct" categorizations), and then to proceed with replicating the experiment with professionals both in a controlled environment, and through the Web. This should also provide the precise timing of each categorization, and help to investigate efficiency in deep.

## References

1. Abdelnabi, Z., Cantone, G., Ciolkowski, M., Rombach, D.: Comparing Code Reading Techniques Applied to Object-oriented Software Frameworks with regard to Effectiveness and Defect Detection Rate. In: Proceedings of the 2004 International Symposium on Empirical Software Engineering, Redondo Beach (CA) (2004)
2. Basili, V.R., Caldiera, G., Rombach, H.D.: Goal Question Metric Paradigm. In: Marciniak, J.J. (ed.) Encyclopaedia of Software Engineering, vol. 1, pp. 528–532. John Wiley & Sons, Chichester (1994)
3. Basili, V.R., Selby, R.: Comparing the Effectiveness of Software Testing Strategies. In: IEEE Transactions on Software Engineering, December 1987, pp. 1278–1296. CS Press (1987)
4. Cantone, G., Abdulnabi, Z.A., Lomartire, A., Calavaro, G.: Effectiveness of Code Reading and Functional Testing with Event-Driven Object-Oriented Software. In: Conradi, R., Wang, A.I. (eds.) ESERNET 2001. LNCS, vol. 2765, pp. 166–193. Springer, Heidelberg (2003)
5. Cohen, J.: A Coefficient of Agreement for Nominal Scales. Educational and Psychological Measurement 20, 37–46 (1960)
6. Durães, J., Madeira, H.: Definition of Software Fault Emulation Operators: a Field Data Study. In: Proc. of 2003 International Conference on Dependable Systems and Networks (2003)
7. El Emam, K., Wieczorek, I.: The Repeatability of Code Defect Classifications. In: Proceedings of International Symposium on Software Reliability Engineering, pp. 322–333 (1998)
8. Henningsson, K., Wohlin, C.: Assuring Fault Classification Agreement – An Empirical Evaluation. In: Proceedings of the 2004 International Symposium on Empirical Software Engineering (2004)
9. Juristo, N., Vegas, S.: Functional Testing, Structural Testing, and Code Reading: What Fault Type Do They Each Detect? In: Conradi, R., Wang, A.I. (eds.) ESERNET 2001. LNCS, vol. 2765, pp. 208–232. Springer, Heidelberg (2003)
10. Myers, G.J.: A Controlled Experiment in Program Testing and Code Walkthroughs/Reviews. Communications of ACM 21(9), 760–768 (1978)
11. Wohlin, C., Runeson, P., Höst, M., Ohlsson, M.C., Regnell, B., Wesslén, A.: Experimentation in Software Engineering: An Introduction. The International Series in Software Engineering (2000)
12. IBM a, Details of ODC v 5.11, (last access May 02, 2006), `http://www.research.ibm.com/softeng/ODC/DETODC.HTM`
13. IBM b, ODC Frequently Asked Questions, (last access May 02, 2006), `http://www.research.ibm.com/softeng/ODC/FAQ.HTM`

# Mapping Medical Device Standards Against the CMMI for Configuration Management

Fergal McCaffery[1], Rory V O'Connor[2], and Gerry Coleman[3]

[1] Lero – The Irish Software Engineering Research Centre, University of Limerick, Ireland
Fergal.McCaffery@dkit.ie
[2] School of Computing, Dublin City University, Dublin, Ireland
roconnor@computing.dcu.ie
[3] Department of Computing, Dundalk Institute of Technology, Dundalk, Ireland
gerry.coleman@dkit.ie

**Abstract.** This paper outlines the development of a Configuration Management model for the MEDical device software industry (CMMED). The paper details how medical device regulations associated with Configuration Management (CM) may be satisfied by adopting less than half of the practices from the CM process area of the Capability Maturity Model Integration (CMMI). It also investigates how the CMMI CM process area may be extended with additional practices that are outside the remit of the CMMI, but are required in order to satisfy medical device regulatory guidelines.

**Keywords:** Configuration Management, Medical device, Software Process Improvement, CMMI.

## 1 Introduction

Software is becoming an increasingly important aspect of medical devices and medical device regulation. Medical devices can only be marketed if compliance and approval from the appropriate regulatory bodies of the Food and Drug Administration [8], and the European Commission under its Medical Device Directives [6] is achieved. Medical device companies must produce a design history file detailing the software components and processes undertaken in the development of their medical devices. Due to the safety-critical nature of medical device software it is important that a highly efficient CM process is in place within medical device companies.

CM is the discipline of *coordinating software development and controlling the change and evolution of software products and components* [13]. It involves the 'unique identification, controlled storage, change control, and status reporting of selected intermediate work products, product components and products during the life of a system' [17]. Such CM procedures are needed to manage the vast number of elements (source code, documentation, change requests, etc) that are created and updated over the lifetime of a software system.

For many software companies, who report CM problems, CM is the first major process weakness that they are required to address. For example, as the company expands, it must fulfil the task of acquiring new customers whilst satisfying the demands of existing

J. Filipe, B. Shishkov, and M. Helfert (Eds.): ICSOFT 2006, CCIS 10, pp. 153–164, 2008.

customers. Often these demands include product customisations which many young companies, lacking reliable revenue streams, do not feel they can ignore. In many situations this results in companies having to support multiple code bases and product versions with very limited resources. Ultimately, a detailed CM process is the only way this problem can be solved.

A study of a small Danish software firm shows how it was forced to review the number of products it developed, and the amount of work it accepted, because of CM difficulties [2]. But CM is equally important in large software companies as a case study of Netscape and Microsoft's development practices shows [5]. Therefore, in a software company or department without CM to control product development, there is no process to assess and no basis for measurement [7]. To succeed in this area Humphrey [14] proposes that a CM plan be developed in conjunction with the establishment of a configuration control board to manage changes to all of the baseline configuration items and to ensure that configuration control procedures are followed.

A number of 'best practice' software process improvement (SPI) models such as ISO/IEC 15504 (also known as 'SPICE') and Capability Maturity Model Integration (CMMI) have been designed to help companies manage their software development activity. For example, CMMI is an SPI improvement model which specifies recommended practices in specific process areas – including CM - that have been shown to enhance software development and maintenance capability [4].

This paper will investigate how thorough current medical device regulations are in relation to specifying what CM practices medical device companies should adopt when developing software. This will be achieved through comparing current medical device regulations and guidelines for CM against the formally documented software engineering 'best practices' of the CMMI for the CM process area.

## 2 Medical Device Industry

Medical device companies have to adhere to medical device regulations in relation to CM. Therefore the main area of concern for medical device companies in relation to CM is to ensure that the checklist of CM elements required by Food and Drug Administration (FDA) are in place rather than trying to improve their overall CM practices. GAMP [12] details CM practices that medical device companies may adopt in order to comply with medical device regulations, however no documentation exists within the medical device domain in relation to how such practices could be improved by incorporating practices from formal software engineering SPI models for CM.

However, if we investigate other regulated industries such as the automotive and space industries we realise that these domains are not content with satisfying regulatory standards, but have proactively developed SPI models specifically for their domain so that they may continuously improve the development of their information systems to achieve higher levels of safety, greater efficiency, and a faster time to market, whilst seamlessly satisfying regulatory quality requirements.

The major process improvement frameworks that currently exist, namely ISO/IEC 15504 and CMMI, do not address the regulatory requirements of either the medical device, automotive or space industries. Therefore, a new SPI model [1] was developed specifically for the automotive industry, this model was based upon ISO/IEC15504 [15] and is referred to as 'Automotive SPICE'. Likewise, a new ISO/IEC15504 based SPI model was developed specifically for the space industry, this model is known as SPiCE for SPACE [3]. Both of these models contain reference and assessment information in relation to how companies may improve their configuration management practices within their domain.

This paper will not address the issue of developing an entire SPI model for the medical device industry ([19] for full discussion), but shall instead focus upon the individual process area of CM. This work addresses an opportunity to integrate the regulatory issues and SPI mechanisms to achieve improvements that are critical to the CM of software for medical devices.

## 3 CMMED Development

The CMMED (Configuration Management model for the MEDical device software industry) was initiated by work that one of the authors performed whilst performing research for the Centre for Software Process Technologies at the University of Ulster, Northern Ireland. This work is now progressing with Lero – the Irish Software Engineering Research Centre. The initial research work was assisted by the involvement of a steering group with a pilot of 5 medical device companies and a notified standards body (all based in Northern Ireland). Each of the five companies expressed a desire to have access to a CM model that would incorporate software process improvement practices and could fulfil the relevant medical device regulatory requirements. However, this work is now being extended to include medical device companies in the Republic of Ireland.

The CMMED may be defined as a set of activities that if performed at a base level will satisfy the CM guidelines specified in the medical device standards. However, CMMED also enables medical device companies to follow a SPI path to achieving CMMI certification. The CMMED will be flexible in that relevant elements of the model may be adopted as required to provide the most significant benefit to the business. The model is based on the CMMI, however another model is also being developed that is based upon ISO/IEC15504. The regulations used to extend the CMMI framework will be those of the FDA and the ANSI/AAMI SW68:2001 (SW68) standard (Medical device software – Software life cycle processes).

The CMMED will provide a means of assessing the software engineering capability for the configuration management process area in relation to software embedded in medical devices [9], [10], [11]. The CMMED is being developed to promote SPI practices into the CM process adopted by medical device companies. This is an attempt to improve the effectiveness and efficiency of CM within medical device companies through investigating the mapping of medical device regulatory guidelines against the CMMI CM process area.

The mappings between the medical device standards and the CMMI specific practices for the CM process result in the CMMED being composed of a number of goals, practices and activities. The CMMED determines what parts of the CMMI CM process area are required to satisfy medical device regulations. It also investigates the possibility of extending the CMMI process areas with additional practices that are outside the remit of CMMI, but are required in order to satisfy medical device regulatory guidelines.

The following section will detail a mapping of existing software development and regulatory guidelines for the medical device industry against the CMMI for the CM process area.

# 4 Guideline Mapping

The FDA provides little insight into how CM should be performed other than to state that a CM plan should exist and that this should be adopted to manage configuration items for medical device software. Therefore in order to gain a greater understanding of the CM guidelines that medical device companies follow in order to achieve regulatory compliance we referred to the medical device software life cycle processes (SW68) standard. This standard was drafted for use in the medical device sector based on the lifecycle requirements of ISO/IEC 12207 [16]. This section illustrates the CMMED structure for the CM process area. In order to achieve this, FDA regulations & SW68 guidelines (for the rest of the paper we refer to these together as medical device standards) were mapped against the goals and practices of the CMMI CM process area.

This mapping is presented as follows: Firstly, we identify the goals that exist within the CMMI CM process area. Next the CMMI CM practices are identified within each CM goal. Then the CM activities (associated with the current practice) that have to be performed in order to comply with medical device regulations are listed. We then identify the activities that have to performed in order in to adhere to the CMMI in relation to the current practice. Finally we lists the CMMI CM activities that are required in order to meet the medical device regulatory requirements associated with the current practice. The composition of the resulting CMMED is illustrated in figure 1.

It should be noted however, in some instances the CMMI CM activities associated with the current practice may not provide full coverage of the medical device standards and therefore these additional activities have to be added in order to achieve the full list of activities required to fulfil the objectives of CMMED.

The CMMED has three goals: Goal 1: *Establish Baselines*, Goal 2: *Track and Control Changes* and Goal 3: *Establish Integrity*. To meet each of these goals it is necessary for a number of practices and activities to be performed. Each of the following subsections will present the CM activities required for each of the 3 goals.

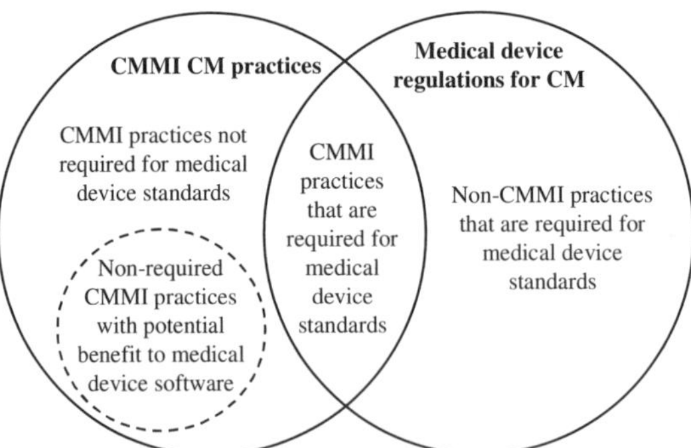


**Fig. 1.** Composition of the CMMED

## 4.1 Goal 1: Establish Baselines

In order to fulfil Goal 1 *Establish Baselines* the following practices have to be performed: *Identify Configuration Items*, *Establish a CM System* and *Create or Release Baselines*.

**Identify Configuration Items.** The 4 activities that have to be performed in order to achieve regulatory compliance in relation to identifying configuration items are:

1. Select the configuration items and the work products that compose them based on documented criteria
2. Assign unique identifiers to configuration items
3. Specify when each configuration item is placed under CM
4. Identify Off the Shelf Components

The 5 activities that have to be performed in order to satisfy the CMMI practice for identifying configuration items are:

1. Select the configuration items and the work products that compose them based on documented criteria
2. Assign unique identifiers to configuration items
3. Specify the important characteristics of each configuration
4. Specify when each configuration item is placed under CM
5. Identify the owner responsible for each configuration item

The 3 activities that are common to both the CMMI and the medical device standards for identifying configuration items are:

1. Select the configuration items and the work products that compose them based on documented criteria
2. Assign unique identifiers to configuration items
3. Specify when each item is placed under CM

Therefore, in order to adhere to the medical device standards only 3 out of the 5 activities required for the CMMI in relation to identifying configuration items are necessary. However an additional activity is required in order to identify Off-the-Shelf (OTS) components as this is not included in the CMMI. Therefore 4 CMMED activities are required for identifying configuration items are:

1. Select the configuration items and the work products that compose them based on documented criteria
2. Assign unique identifiers to configuration items
3. Specify when each configuration item is placed under CM
4. Identify Off the Shelf Components. *Note: this activity is not present in the CMMI but is required in order to fulfil the requirements specified in the medical device standards.*

**Establish a CM System.** The 2 activities that have to be performed in order to achieve regulatory compliance in relation to establishing a configuration management system (CMS) are:

1. Store and retrieve configuration items in the CM system
2. Store, update, and retrieve CM records

The 8 sub-practices that have to be performed in order to satisfy the CMMI practice for establishing a CMS are:

1. Establish a mechanism to manage multiple control levels of CM
2. Store / retrieve configuration items in the CMS
3. Share and transfer configuration items between control levels within the CMS
4. Store and recover archived versions of configuration items

5. Store, update, and retrieve CM records
6. Create CM reports from the CMS
7. Preserve the contents of the CMS
8. Revise the CM structure as necessary

There are 2 activities that are common to both the CMMI and the medical device standards for establishing a CMS. Therefore, in order to adhere to the medical device standards, only 2 of the 8 activities required by the CMMI for establishing a CMS are necessary. The main differences are that CMMI requests the usage of multiple control levels of CM, as well as archiving and restoration procedures to be in place. The 2 CMMED activities for establishing a CMS are:

1. Store and retrieve configuration items in the CM system
2. Store, update, and retrieve CM records

**Create or Release Baseline.** There is only a single activity that has to be performed in order to adhere to the medical device standards in relation to creating or releasing baselines - *Document the set of configuration items that are contained in a baseline*. Whereas there are 4 activities that have to be performed in order to satisfy the CMMI practice for creating or releasing baselines:

1. Obtain authorisation from the CCB before creating or releasing baselines of configuration items
2. Create or release baselines only from configuration items in the CM system
3. Document the set of configuration items that are contained in a baseline
4. Make the current set of baselines readily available

There is only single CMMED activity that is common to both the CMMI and medical device standards for creating or releasing baselines. Therefore, in order to adhere to the medical device standards only one of the 4 activities - *Document the set of configuration items that are contained in a baseline* – is required for the associated CMMI practice is necessary.

**Summary of CMMED Goal 1.** Table 1 summarises goal 1 of CMMED (*Establish Baselines*). It may be observed from table 1 that not all of activities of the CMMI have to be performed in order to satisfy the medical device regulations (in fact only 6 of the 17 CMMI activities have to be performed). However, in order to satisfy the objectives of the CMMED 1 additional (medical device specific) activity had to be added (i.e. to satisfy goal 1 of the CMMED).

**Table 1.** Summary of CMMED Goal 1

| Practice | CMMI activities | CMMI activities to meet medical device standards | Additional activities to meet medical device standards |
|---|---|---|---|
| Identify CM items | 5 | 3 | 1 |
| Establish a CMS | 8 | 2 | 0 |
| Create or delete Baselines | 4 | 1 | 0 |
| **Total** | **17** | **6** | **1** |

### 4.2 Goal 2: Track and Control Changes

In order to adhere to the CMMED goal 2 of tracking and controlling changes, the following specific practices have to be performed: *Track Change Requests* and *Control Configuration Items*.

**Track Change Requests.** The 5 activities that have to be performed in order to achieve regulatory compliance in relation to tracking change requests:

1. Initiate and record change requests in the change request database
2. Analyse the impact of changes and fixes proposed in the change requests.
3. Review change requests that will be addressed in the next baseline with those who will be affected by the changes and get their agreement.
4. Track the status of change requests to closure.
5. Each upgrade, bug fix, or patch for OTS software shall be evaluated, and the evaluation shall be documented

There are 4 activities that have to be performed in order to satisfy the CMMI practice for tracking change requests:

1. Initiate and record change requests in the change request database
2. Analyse the impact of changes and fixes proposed in the change requests.
3. Review change requests that will be addressed in the next baseline with those who will be affected by the changes and get their agreement.
4. Track the status of change requests to closure.

There are 4 activities that are common to both the CMMI and the medical device standards for tracking change requests:

1. Initiate and record change requests in the change request database
2. Analyse the impact of changes and fixes proposed in the change requests.
3. Review change requests that will be addressed in the next baseline with those who will be affected by the changes and get their agreement.
4. Track the status of change requests to closure.

Therefore, in order to adhere to the medical device standards all of the activities required for this CMMI practice are necessary, but not always to the same level of detail. However an additional practice is required in order to ensure that each upgrade, bug fix, or patch for OTS software is identified and evaluated, and that the evaluation is documented, as this is not included in the associated CMMI practice.

The CMMED activities for tracking change requests are:

1. Initiate and record change requests in the change request database
2. Analyse the impact of changes and fixes proposed in the change requests.
3. Review change requests that will be addressed in the next baseline with those who will be affected by the changes and get their agreement.
4. Track the status of change requests to closure.
5. Each upgrade, bug fix, or patch for OTS software shall be evaluated, and the evaluation shall be documented. *Note: this activity is not present in the CMMI but is required in order to fulfil the requirements specified in the medical device standards.*

**Control Configuration Items.** The 4 activities that have to be performed in order to achieve regulatory compliance in relation to controlling configuration items are:

1. Control changes to configuration items throughout the life of the product
2. Obtain appropriate authorisation before changed configuration items are entered into the CM system
3. Perform reviews to ensure that changes have not caused unintended effects on the baselines
4. Record changes to configuration items and the reasons for the changes as appropriate

The 5 activities that have to be performed in order to satisfy the CMMI practice to control configuration items are:

1. Control changes to configuration items throughout the life of the product
2. Obtain appropriate authorisation before changed configuration items are entered into the CM system
3. Check in and check out configuration items from the CM system for incorporation of changes in a manner that maintains the correctness and integrity of the configuration items
4. Perform reviews to ensure that changes have not caused unitended effects on the baselines
5. Record changes to configuration items and the reasons for the changes as appropriate

As the control of configuration items is very important in terms of ensuring the integrity of medical device software it is no surprise that 4 of the 5 activities required for this CMMI practice are necessary in order to adhere to the medical device standards.

The following list shows the mapping of the medical device standards against each of the activities required by the CMMI practice for controlling configuration items:

1. Control changes to configuration items throughout the life of the product
2. Obtain appropriate authorisation before changed configuration items are entered into the CM system
3. Perform reviews to ensure that changes have not caused unitended effects on the configuration baselines
4. Record changes to configuration items and the reasons for the changes as appropriate

**Summary of CMMED Goal 2.** Table 2, summarises goal 2 of the CMMED (*Track and Control Changes*). It may be observed that almost all of the activities of this CMMI goal will have to be performed in order to satisfy the medical device standards (in fact 8 of the 9 CMMI sub-practices will have to be performed). However, in order to satisfy the objectives of CMMED 1 additional sub-practice had to be added.

### 4.3 Goal 3: Establish Integrity

In order to fulfil CMMED goal 3: *Establish Integrity* the following specific practices have to be performed: *Establish CM Records* and *Perform Configuration Audits*.

**Table 2.** Summary of CMMED Goal 2

| Practice | CMMI activities | CMMI activities to meet medical device standards | Additional activities to meet medical device standards |
|---|---|---|---|
| Track change requests | 4 | 4 | 1 |
| Control Config items | 5 | 4 | 0 |
| Total | **9** | **8** | **1** |

**Establish CM Records.** The 3 activities that have to be performed in order to achieve regulatory compliance in relation to establishing CM records are:

1. Record CM actions in sufficient detail so the content and status of each configuration item is known and previous versions can be recovered
2. Identify the version of the configuration items that constitute a particular baseline.
3. Revise the status and history of the configuration item as necessary

The 6 activities that have to be performed in order to satisfy the CMMI practice for establishing CM records are:

1. Record CM actions in sufficient detail so the content and status of each configuration item is known and previous versions can be recovered
2. Ensure that relevant stakeholders have access to and knowledge of the configuration items
3. Specify the latest version of the baseline.
4. Identify the version of the configuration items that constitute a particular baseline.
5. Describe the differences between successive baselines
6. Revise the status and history of the configuration item as necessary

The process of establishing CM records is very important in terms of providing the traceability evidence that is required to meet the regulatory requirements associated with medical device software. Half of the activities (3 out of 6) required for this CMMI practice are necessary in order to adhere to the medical device standards and are therefore included in CMMED.

The CMMED activities for establishing CM records are:

1. Record CM actions in sufficient detail so the content and status of each configuration item is known and previous versions can be recovered
2. Identify the version of the configuration items that constitute a particular baseline.
3. Revise the status and history of the configuration item as necessary

**Perform Configuration Audits.** The medical device standards do not specify any activities that have to be performed in order to achieve regulatory compliance in relation to performing configuration audits. The list of the sub-activities that have to be performed in order to satisfy the CMMI practice for performing configuration audits are:

1. Assess the integrity of the baselines
2. Confirm configuration records correctly identify the configuration of the configuration items
3. Review the structure and integrity of the items in the CM system
4. Confirm the completeness and correctness of the items in the CM system
5. Confirm compliance with applicable CM standards and procedures
6. Track action items from the audit to closure

This practice in CMMI has no equivalent practice within the medical device regulations. The medical device regulations do not specify any need for auditing the CM processes and activities. Therefore CMMED contains no activities, as a result of mapping the regulatory medical device requirements for CM against each of the activities required for the CMMI practice relating to performing configuration audits.

**Summary of CMMED Goal 3.** Table 3 summaries goal 3 of the CMMED (*Establish Integrity*). It may now be determined that in order to satisfy medical device standards that not all of activities of this CMMI goal have to be performed (in fact only 3 of the 12 CMMI activities have to be performed. Additionally, no additional (medical device specific) activities have to be added in order to satisfy the objectives of CMMED.

**Table 3.** Summary of CMMED Goal 3

| Practice | CMMI activities | CMMI activities to meet medical device standards | Additional activities to meet medical device standards |
|---|---|---|---|
| Establish CM records | 6 | 3 | 0 |
| Perform configuration audits | 6 | 0 | 0 |
| **Total** | **12** | **3** | **0** |

## 5 Preliminary Feedback

In order to assist with preliminary feedback, the CM process outlined by this paper has been compared against the existing practices within an Irish medical device company. A high level summary of their comments are included below.

They liked the structure of the CMMED and in particular how it enabled them to create a list of all the CM practices that they should adopt in order to adhere to the medical device standards. They also made positive comments in relation to CMMED providing additional information in relation to how their existing CM practices could be improved by incorporating guidance from the CM CMMI process area in relation how mandatory medical device activities may be performed.

Upon further consultation with the authors it has also been decided that in order to assist with SPI within the company that a process diagram shall be created, this will provide a graphical representation of the logical flow of the practices within their CM process.

## 6 Summary and Conclusions

Table 4 provides a summary of the 3 goals within CMMED. There are 40 activities required by CMMED, consisting of 38 CMMI and 2 medical device specific activities. In order to satisfy the mandatory medical device CM requirements, 19 of these activities have to be adhered to (17 CMMI and 2 medical device specific activities).

It is clear that following the guidelines specified in the medical device regulations will at best, only partially meet the specific goals of this CMMI process area (this would only fulfil 17 of the 38 activities required by CMMI). Since failure to perform any specific practice implies failure to meet the specific goal, with respect to CMMI, it is clear, that the goals of CM cannot be obtained by satisfying medical device regulations and guidelines during software development. But is the opposite true, *can meeting the CMMI goals for CM successfully meet FDA and SW68 guidelines?* With the exception of 2 sub-practices, performing the CMMI specific practices for CM would in general more than meet the FDA and SW68 guidelines for this area.

If a medical device company follows the CMMI guidelines for CM (with the exception of 2 activities), this will more than fulfil the CM requirements specified in the medical device regulations. However, only a fraction of the CMMI guidelines for CM will be satisfied by adhering to the medical device regulations for CM

**Table 4.** Summary of CMMED Goals

| CMMED goal | CMMI activities | CMMI activities to meet medical device requirements | Additional activities to meet medical device requirements |
|---|---|---|---|
| Goal 1 | 17 | 6 | 1 |
| Goal 2 | 9 | 8 | 1 |
| Goal 3 | 12 | 3 | 0 |
| **Total** | **38** | **17** | **2** |

## Acknowledgements

This research is supported by the Science Foundation Ireland (SFI) funded project, Global Software Development in Small to Medium Sized Enterprises as part of Lero - the Irish Software Engineering Research Centre (http://www.lero.ie).

## References

1. Automotive SIG, The SPICE User Group Automotive Special Interest Group, Automotive SPICE Process Reference Model (2005)
2. Baskerville, R., Pries-Heje, J.: Knowledge Capability and Maturity in Software Management. The Data Base for Advances in Information Systems 30(2), 26–43 (Spring 1999)
3. Cass, A., Volcker, C.: SpiCE for SPACE: A method of Process Assessment for Space Projects, In: SPICE 2000 Conference Proceedings (2000), `http://www.synspace.com`
4. Chrissis, M.B., Konrad, M., Shrum, S.: CMMI: Guidelines for Process Integration and Product Improvement. Addison-Wesley, Reading (2003)
5. Cusumano, M., Yoffie, D.: Software Development on Internet Time. IEEE Computer 32(10), 60–69 (1999)
6. European Council Directive 93/42/EEC Concerning Medical Devices, (June 14, 1993)
7. Fayad, M., Laitinen, M.: Process Assessment Considered Wasteful. Communications of the ACM 40(11), 125–128 (1997)
8. FDA Regulations, Code of Federal Regulations 21 CFR Part 820, Food and Drug Administration (June 1997)
9. FDA/CDRH Guidance Document, General Principles of Software Validation, FDA (June 1997)
10. FDA/CDRH Guidance Document, Guidance for Off-the-Shelf Software Use in Medical Devices, FDA (September 1999)
11. FDA/CDRH Guidance Document, Guidance for the Content of Premarket Submissions for Software Contained in Medical Devices, FDA (May 2005)
12. GAMP, Guide for Validation of Automated Systems (GAMP 4), International Society for Pharmaceutical Engineering (December 2001)
13. Ghezzi, C., Jazayeri, M., Mandrioli, D.: Fundamentals of Software Engineering. Prentice-Hall, Englewood Cliffs (2003)
14. Humphrey, W.: Introduction to the Team Software Process. Addison-Wesley, Reading (2000)
15. ISO/IEC 15504, Information Technology – Process Assessment – Part 5: An exemplar Process Assessment Model, ISO/IEC JTC1/SC7 (October 2003)
16. ISO/IEC 12207, Information technology - Software lifecycle processes Amendment 2, International Standards Organisation (1995)
17. Jonassen-Hass, M.E.: Configuration Management Principles and Practice. Addison-Wesley, Reading (2002)
18. Medical device software life cycle processes, American National Standard / Association for the Advancement of Medical Instrumentation, SW68 (2001)
19. McCaffery, F., Donnelly, P., Dorling, A., Wilkie, G.: A Software Process Development, Assessment and Improvement Framework for the Medical Device Industry. In: Proceedings Fourth International SPICE Conference on Process Assessment and Improvement, Lisbon, Portugal (April 2004)

# A Systematic Review Measurement in Software Engineering: State-of-the-Art in Measures

Oswaldo Gómez[1], Hanna Oktaba[1], Mario Piattini[2], and Félix García[2]

[1] Institute of Investigations in Applied Mathematics and Systems
Autonomous National University of Mexico UNAM
Scholar Circuit University City, Coyoacán 04510, Mexico City, Mexico
oswaldog@uxmcc2.iimas.unam.mx, ho@hp.fciencias.unam.mx
[2] Alarcos Research Group, Department of Computer Science
University of Castilla-La Mancha, Paseo de la Universidad/4,13071, Ciudad Real, Spain
Mario.Piattini@uclm.es, Felix.Garcia@uclm.es

**Abstract.** The present work provides a summary of the state of art in software measures by means of a systematic review on the current literature. Nowadays, many companies need to answer the following questions: How to measure?, When to measure and What to measure?. There have been a lot of efforts made to attempt to answer these questions, and this has resulted in a large amount of data what is sometimes confusing and unclear information. This needs to be properly processed and classified in order to provide a better overview of the current situation. We have used a Measurement Software Ontology to classify and put the amount of *data* in this field in order. We have also analyzed the results of the *systematic review,* to show the trends in the software measurement field and the software process on which the measurement efforts have focused. It has allowed us to discover what parts of the process are not supported enough by measurements, to thus motivate future research in those areas.

**Keywords:** Software Measurement, Measure, Systematic Review.

## 1 Introduction

It is a well-known fact nowadays that software measurement helps us to better understand, evaluate, and control the products, processes, and software projects from the perspective of evaluating, tracking, forecasting, controlling and understanding [8]. On the one hand, software measurement allows organizations to know, compare and improve their software quality, performance, and processes. On the other hand, software measurement helps organizations to estimate and predict software characteristics to support better decisions [21]; [10]. As a consequence, software measures are proving to be very effective for understanding and improving software development and maintenance projects [4], showing problematic areas in system quality and institutionalizing software process improvement.

It should also be noted that there is a large amount of studies in software measurement, which makes it very easy to lose information and to get confused. For this reason, it is important to follow a specific, strict, and very well defined method for

J. Filipe, B. Shishkov, and M. Helfert (Eds.): ICSOFT 2006, CCIS 10, pp. 165–176, 2008.

searching in the current literature. If we take a look at software measurement, we realize that it is considered to be among the youngest disciplines, and it is currently in the phase in which terminology, principles, and methods are still being defined and consolidated [4]. This means that there is not a general agreement about the exact definitions of the main concepts related to measurement. In addition, no single standard contains a complete vision of software measurements [11].

With respect to the issues identified above, this article carries out a systematic review with a predefined search strategy, in order to summarize and classify the current and ongoing efforts in this field. The systematic review has been conducted according to the [16] proposal, which is very suitable for looking for information about measures on different sources in a disciplined and systematic way. Hence, Systematic review allows us to recognize, evaluate and do even more; it helps us to identify issues for planning future investigation and provides us with information about the consistency of our results [27]. We chose systematic review because of its scientific methodology that goes one step further than a simple overview.

The goal of this work is to find and clarify the answers to three different questions: What to measure, when to measure and how to measure. This is achieved by analyzing from the results of the literature review, the following issues: proportion of measured entities; measured attributes; validated measurement; measurement focus; and measurement in life cycle software process.

This paper is organized as follows. After this introduction; an overview of the systematic review process is given. In the third section, the way in which the systematic review has been carried out on the software measurement field is explained. Then, an analysis of the results is provided. Finally, the conclusions and future work are dealt with.

## 2 Systematic Reviews

It is often recognized in Software Engineering that different research studies are generally fragmented and limited, not properly integrated, and without agreed standards [16]. In order to avoid those problems we chose the systematic review to carry out this investigation on software measures. Systematic review aims to present a fair evaluation of a research topic by using trustworthy, rigorous and auditable methodology, along with a very well defined strategy that allows the completeness of the research to be executed (in this case on software measures). Furthermore, systematic literature review is a formal and methodological process that allows us to identify, evaluate, and interpret all existing studies that are related to our investigation on software measures based in this case on a research question, but it could be also based on topic area, or phenomenon of interest. This is done in such a way that it helps us to summarize the evidence that is currently available concerning a treatment or technology. It also serves to identify any gaps in the current research, and thus suggest areas for further investigations, and finally provide a framework/background to position new research activities appropriately.

The review provides us with the necessary information to properly address the software measures, by mapping the measure field, finding the relevant data, ideas, techniques and their correlation with our investigation. Besides, it can support the planning for a new piece of research. Moreover, with this systematic literature review we can integrate empirical investigation, in order to find out generalizations. We do this by establishing specific objectives to create critical analysis. An overview of the systematic review is provided in the next subsection.

## 2.1 The Systematic Review Process

In order to address and present a fair evaluation of a research topic, the systematic review is composed of the following phases:

*Review Planning Phase*: Here the investigation's goals are established. The *Review Protocol*, which is the most important item in this phase, is generated. First and foremost, this protocol defines the research question and the methods that will be executed in the review. In a broad manner, this phase involves the following, summarized, activities, defined by [27]:

*Question Formulization*: This activity is considered to be among the most important in the systematic review process. Here the investigation targets must be defined by focusing the question and by establishing its Quality and Amplitude.

*Source Selection*: Primary studies from sources are selected here, by defining a source selection criterion, setting the studies' languages, identifying and selecting the sources after an assessment of them and checking references.

*Study Selection*: It describes the process and criteria for the evaluation and selection of studies.

*Review Execution phase*: This phase involves identification, selection and evaluation of primary studies, based on the inclusion and exclusion criteria defined in the *Review Protocol*. It is composed of the following steps, in summary form:

*Selection Execution*: This section aims to register the selection process for primary studies by evaluating them with quality criteria.

*Information Extraction*: Once primary studies are selected, the relevant data must be extracted by following an *Information Inclusion and Exclusion Criteria Definition*, by defining *Data Extraction Forms*, and by resolving divergences among reviewers.

*Result Analysis*: In this phase all the information from the different studies is analyzed. This phase involves the next step: *Result Summarization*, which presents the data resulting from the collected studies by doing *Calculus Statistical*, *Results Tables*, *Sensitivity Analysis*, *Plotting*, which will lead to the *Conclusion* and *Final Comments*.

The whole process must be stored and the planning and the execution have to guarantee that the research can be done. It is worth mentioning here that the *Review Protocol* must be evaluated by experts. Finally, many of the activities of the review process involve iteration to refine the process, and therefore they are not necessarily sequential.

In the next section, we describe how the review process, which was designed as appropriate to our research goals, was performed

## 3 Systematic Review about Software Measures

First of all, it must be emphasized that this paper is an attempt to answer this fundamental question: What are the most current and useful measures in the literature? Since our whole protocol was produced around this question, this is the main step in our *Review Planning Phase*. Moreover, we hope that this work will be useful for project managers and software developers. The defined strategy was the following: first and foremost, the large collection of paper in current literature about software measurements was examined. Due to the great diversity of topics in this field, and with the aim of clarifying and summarizing them in the best way possible, we used the classifications of concepts defined in the Software Measurement Ontology proposed by [11]. This ontology aims at contributing to the harmonization of the different software measurement proposals and standards, by providing a coherent set of common concepts used in software measurement.

In order to do the research we built the following combinations of search strings:

"(measure OR metric OR quality OR quantitative) AND (process OR engineering OR maintenance OR management OR improvement OR Software testing OR development)".

All the possible combinations with these words were tested in the following web search engines: ACM Digital Library, Search IEEE magazines, Wiley Interscience, and Science@Direct.

The results obtained on the web engines are shown in Table 1.

**Table 1.** Total Search Results

| Sources | Search Results | Reviewed | Accepted |
|---|---|---|---|
| **Science@Direct** | 3569 | 78 | 10 |
| **ACM** | 950 | 85 | 28 |
| **IEEE** | 3740 | 111 | 32 |
| **Wiley** | 653 | 20 | 8 |
| **TOTAL** | 8912 | 294 | 78 |

As we can see in Table 1, search engines provided us with 8912 papers. Nevertheless, it should be pointed out that only 78 were accepted, which represents about 1 % of the total articles, hardly even that. It is apparent that many articles were rejected. This is so because if a more limited search had been carried out, it would certainly have been true that we would have started with fewer results from the search engines, but at the same time we would have lost important articles. Therefore, a very less restrictive search was defined: as a result of this, we obtained too many articles, of which very few were considered apt. Furthermore, we have discarded those measures that were outside the scope of our model. We have also discarded measures that did not provide any relevant information, as well as repeated measures proposed by more than one author so that each measure is included only once. Hence, our attention focused on papers where keywords and titles included the research strings. These strings were also searched for in the whole document by some search engines.

Regarding the execution phase of the systematic review, the selection and evaluation of information was initiated using the terms of the inclusion and exclusion criteria defined in the review protocol. These criteria established that selected studies were in English and that all of them showed current, useful software measurements, basically only studies about measures for software development, software project administration and maintenance were selected. All papers had to satisfy our quality criteria and in this sense it is important to point out that all the searched-for sources are serious and that the quality of their papers is guaranteed. Moreover the search engines were validated by experts. For this reason, our quality criteria also trusted in the quality of the sources.

Once the papers were selected, the information was extracted by means of an extraction template for objective results which includes study name, author, institution, journal, date, methodology, results, problems and subjective results which includes information through authors, general impressions and abstractions, according to the proposal provided by [27]; in particular, the aims of this template are to store the results of the execution phase process by extracting, not only the objective information, but also the subjective information from each article analyzed.

Finally, in the results analysis phase we analyzed the measures in order to show, among other aspects, the information about attributes, the entities measured and their characteristics, the amount of measures in a specific attribute or entity, etc. This phase is described in more detail in the following section.

## 4 Result Analysis

The measures extracted from the studies were summarized in terms of the Software Measurement Ontology, which helped us to find out what kinds of measures exist. More specifically, this ontology supported us in defining a template by categorizing the measures in the following three different ways: What to measure? How to measure? And When to measure?

Consequently, in order to summarize the existing measures, the ISO 15504, CMM, and CMMI establish a quality background for the improvement of maturity levels defining the Project, Process and Product as the kind of *entities* that can be measured. That is why we extracted *attribute* and *sub-attributes* [9] measured of these *entities*, from the articles reviewed and classified them into internal or external. With this part of the analysis we try to answer the question: What to measure? This is the first way in which we categorized the measurements. Table 2 shows these attributes.

Once the measurements were collected and stored in our template table, we analyzed the amount of measures which have been defined for the Process, Project and

**Table 2.** Definition of entities

| What? | | | | | | |
|---|---|---|---|---|---|---|
| Entibies | | | Attributes | Sub-attributes | Type of Attribute | |
| Project | Process | Product | | | Internal | External |

Product kind of entities. As we can see in Figure 1, the most measured kind of entity is the product, and the entities whose measurement has been less supported by the current literature are the project and process. The reason is that measuring product is easier than measuring process and project, in which we usually find ambiguous definition of attributes. For products, quality and technical attributes are very well defined because quality has been strongly focused on product. Finally, measurements on product entities help to measure process and project ones.

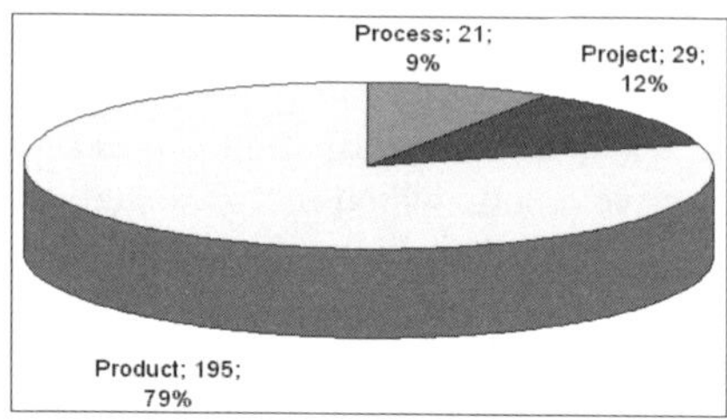


**Fig. 1.** Proportion of measured entities

Next, we shall look at another closely-related issue, which is the amount of measured attributes. Figure 2 shows the proportion of measure attributes according to our analysis of the accepted papers. As Figure 2 shows, size is one of the most measured attributes. The point is that the size is a base measure, not only needed in most of the derived measures, but the size measure is also easier to obtain because it focuses on one of the most "tangible" attributes which is the source code. Moreover, size has very well defined scales, units and methods of measurement like functions Points (FP) [14]; therefore it is very difficult to get confused with size measurements. Furthermore, cost estimation is derived from size and the overall productivity, and finally the schedule is based on the size and cost estimates [8]. Hence size is used on most of control measures in a software project. The arguments set out here lead to an explanation of why size has one of the highest values in Figure 2.

In order to show in a in a better way the information displayed in Figure 2, Table 3 show the attributes order by the most measured.

In connection with the most measured attributes, the complexity attribute is used in different contexts, for example: source code complexity, Design complexity, UML Diagrams complexity, Architecture complexity, etc. Hence it can be seen that complexity has gathered many measurements from its different applications. If we take a look at Figure 2 in greater detail, it should be pointed out that attributes like Activity, Role, Work products and Accuracy are the least measured. That is due to the fact that these attributes are mostly related with process and project kind of entities, for which there is not a well defined basic attribute.

Once the "What to Measure?" question was analyzed. The next step was to tackle the question: "How to measure?" To answer this question we gathered how the measurements of attributes in the selected papers were made and classified them in terms of the following characteristics: Representation, Description, Base or Derived Measurement, Scale [9], Empirically [30], [15], [2], [20] or Theoretically [28], [4], [29], [31], [23] validated. This analysis is summarized in Table 4.

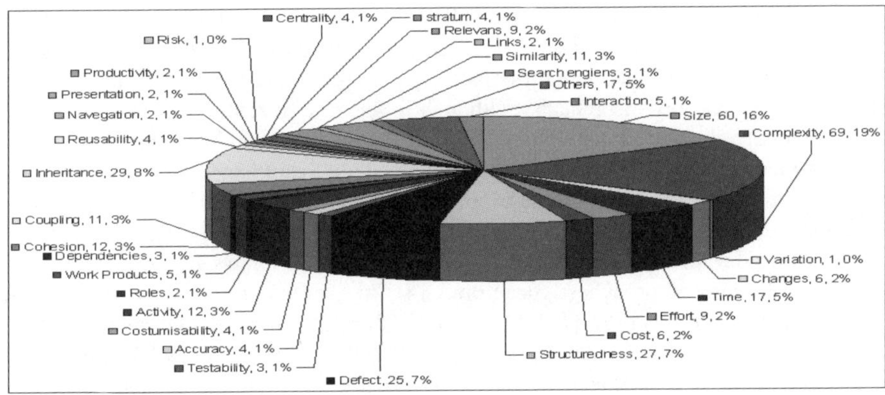


**Fig. 2.** Measured attributes

**Table 3.** Measures attributes

| | | | |
|---|---|---|---|
| Complexity | 19% | Productivity | 1% |
| Size | 16% | Testability | 1% |
| Inheritance | 8% | Costumisability | 1% |
| Defect | 7% | Roles | 1% |
| Structuredness | 7% | Work Products | 1% |
| Time | 5% | Dependencies | 1% |
| Others | 5% | Reusability | 1% |
| Activity | 3% | Navegation | 1% |
| Accuracy | 3% | Presentation | 1% |
| Cohesión | 3% | Centrality | 1% |
| Coupling | 3% | Stratum | 1% |
| Similarity | 3% | Links | 1% |
| Changes | 2% | Search engiens | 1% |
| Effort | 2% | Interaction | 1% |
| Cost | 2% | Variation | 0% |
| Relevans | 2% | Risk | 0% |

Let us have a look at the last characteristic, which has as its goal to discover if a measure has been validated empirically and/or theoretically. The aim of theoretical validation is to check whether the intuitive idea of the attribute being measured is considered in the defined measure. The main goal of empirical validation is to obtain objective information concerning the usefulness of the proposed metrics. Theoretical validation by itself is not enough to guarantee the usefulness of the measure, because it may occur that a measure is valid from a theoretical point of view, but it has no practical relevance in relation to a specific problem. As a consequence, a measure which has not been validated is not demonstrated to be useful. We therefore classified the measures in such a way as to know how many had been empirically and/or theoretically validated. This is shown in Figure 3.

As can be observed in Figure 3, about half of the measures found in the selected papers had been only empirically validated. This leads us to the conclusion that there is a great tendency to empirical validation. Furthermore, we can see that (24%) of the measurements had been validated only theoretically, although it was recognized in the papers that they need empirical validation. Finally only (20%) of the measurement had been both empirically and theoretically validated. It should be pointed out that it is necessary to get a common agreement to validated measures theoretically. Moreover empirically validation needs more data extracted from "real projects" in order to get practical conclusions.

Regarding the measurement focus found in the articles analysed, we have discovered the following approaches: Structured [4], measurement focussing in Process, Object Oriented (OO) [7] [5], [17], [19], [3], Quality [22]. Function Points [14], UML [19], Complexity [18], [12], *Project* [24] and OCL [26].Figure 4 shows the amount of measurement in each approach. It shows us that the most supported approaches by measure are Object Oriented (OO) ones. This is due to this kind of projects are currently the most popular in software development. Continue with this part of the analysis, there are efforts to get a universal WEB measures definition, with this review we found conceptual models and frameworks in order to classify WEB measures.

**Table 4.** Definition of measure attributes

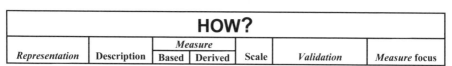

| HOW? | | | | | | |
|---|---|---|---|---|---|---|
| | | *Measure* | | | | |
| ***Representation*** | **Description** | **Based** | **Derived** | **Scale** | ***Validation*** | ***Measure* focus** |

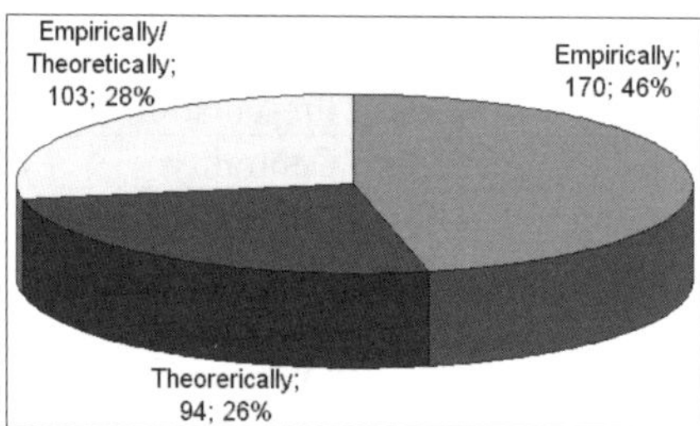


**Fig. 3.** Validated measures

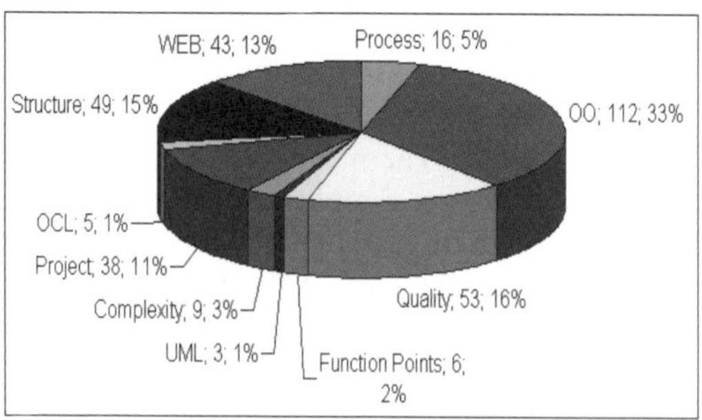


**Fig. 4.** Measure focus

Finally, we analyzed the third question: When to measure?, To classify in what parts of the lifecycle project the measure must be taken for projects and process entities, the PMBOK guide [1] was selected. In order to group when the measurements are taken for the product entity, the waterfall lifecycle model was applied. We chose these two models due to their wide acceptation and genericity. Figure 5 shows the proportion of product measurements in the different phases of the software life cycle:

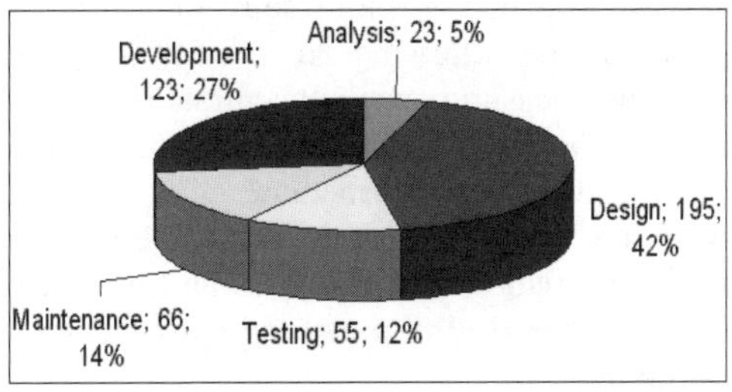


**Fig. 5.** Measure in life cycle software process

As we can see in Figure 5, most measurements are carried out during the Design, Testing and Development phases of the waterfall lifecycle software process. In the Design phase, products such as architecture, system designs, requirements analysis, etc. are generated. Hence it is necessary to support this phase with measurements, in order to know characteristics of these products when carrying out the design. Moreover, measurement in the Design phase can support the future products to be generated, which mean that this phase is one of the most measured. Continue with this analysis, it should be pointed out that the Development phase is one of the most measured, because most of the software products are created here, such as: manuals, source code and, among other products, the software itself. Therefore, it is possible to collect quantity information about these products here. According to PSP [13], measures about size, effort, time, faults, defects, LOC, etc. are commonly taken in this phase. Another factor to take into account is that once the software system is created, it is necessary to validate if this system fulfils the quality requirements. The counting faults and deriving the reliability is the most widely applied and accepted method

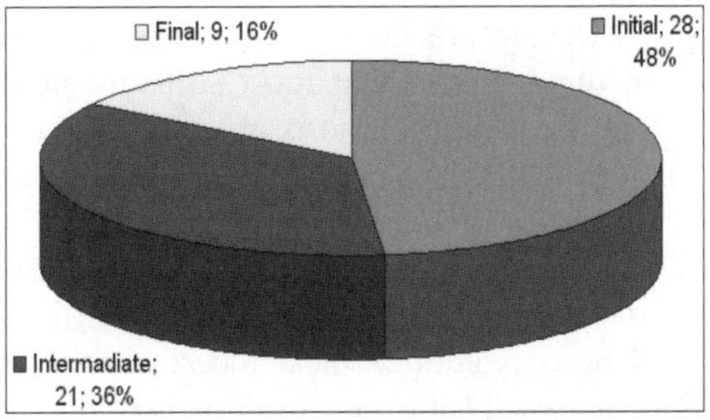


**Fig. 6.** Measure in life cycle projects

used to validate systems; most of this information focuses on the product and is commonly reported in terms of measurements. This is done in not only in the early phases but also especially in the testing phase, which is another of the most-measured phases in lifecycle software process.

In addition, the PMBOK guide defines the following general phases for project life: Initial, intermediate, and final phases. In Figure 6 we show the distributions of measures through these phases.It is worth mentioning here that in the initial phase there could be sub-phases with one or more deliverables, according to the kind of project. In these sub-phases the following are usually measured: size, complexity, level of risk, cash, etc. Most measurements concentrate on the Initial phase, as in this phase the planning for the whole project is executed- this in turn constitutes the main effort in project management. In the Intermediate phase, many control activities are carried out in order to ensure the success of the project. Periodical reports are thereby generated with quantity information about process and project measures and indicators. For these reasons this phase is also one of the most measured in project lifecycle for project and process entities.

## 5 Conclusions and Further Work

Software measurements are very important in software development process, because they help us, to control, estimate and improve process, projects and products, among other things. With that in mind, this article attempts to provide the state of art in software measurement, by carrying out a systematic review whose purpose is to summarize the most current and useful measures in the literature.

With this systematic review, we find out the following results:

(1) Measures are strongly aligned to product entity. Since this kind of entity has better attribute definition than project and product entities have, there are large amount of measures for the product. This leads to the conclusion that if an entity has a few measures, it is due to the fact that it doesn't have specific attribute.

(2) Complexity gathered a great amount of measures because this attribute is used in different contexts. While size is also one of the most measured attributes since it is used in cost and development schedule estimation

(3) There is a great tendency to obtain empirical validation. But it is necessary to get more data extracted from "real projects", in order to get practical conclusions and to improve software quality.

(4) Development and Design are the most measured phases in lifecycle software process because it is in these phases that most software products are generated.. It should be also noted that the testing phase is also one of the most measured phases. This is thanks to the fact that this phase involves quality activities for evaluating software quality characteristics, generally reported in terms of quantity values. But quality measures are considering in the early software development phases by counting faults which is the most widely applied method to determine software quality.

(5) For projects and process entities most measurements are concentrated in the Initial and Intermediate phases. That is because it is here that the project planning and control activities are developed.

(6) There are a large number of measures for OO projects. This is because these kinds of projects are currently the most popular in software development. Hence a lot of research has been done in this field.

(7) So many efforts had been made to get a universal WEB measures definition. In this review we found conceptual models and frameworks in order to classify WEB measures.

Finally, we need to relate the measurements found in this article to a specific software development process. The aim of this is to settle *when* a measure must be taken. To reach this goal, in our specific research, further work will take in the Process Model for the Software Industry (MoProSoft), which focuses on small companies and which is also the Mexican norm.

## Acknowledgements

This article was supported by the Process Improvement for Promoting Iberoamerican Software the Competitiveness of Small and Medium Enterprises (COMPETISOFT) and Science and Technology for Development (CYTED).

## References

1. ANSI/PMI. A Guide Project Management Body of Knowledge (PMBOK Guide) an American National Standard, ANSI/PMI 99-001-2004, 3rd edn., Project Management Institute, Inc, U.S.A (2004)
2. Basili, V., Shull, F., Lanubile, F.: Building knowledge through families of experiments. IEEE Transactions on Software Engineering 25(4), 435–437 (1999)
3. Bansiya, J., Davis, C.: A Hierarchical Model for Object-Oriented Design Quality Assessment. IEEE Transactions on Software Engineering 28(1), 4–17 (2002)
4. Briand, L., Morasca, S., y Basili, V.: Property-Based Software Engineering Measurement. IEEE Transactions on Software Engineering 22(1), 68–86 (1996)
5. Brito Abreu, F., Carapuça, R.: Object-Oriented Software Engineering: Measuring and controlling the development process. In: Proceedings of the 4th International Conference on Software Quality, McLean (USA) (1994)
6. Calero, C., Ruiz, J., Piattini, M.: Classifying web metrics using the web quality model. Online Information Review 29(3), 227–248 (2005)
7. Chidamber, S., Kemerer, C.: A Metrics Suite for Object Oriented Design. IEEE Transactions on Software Engineering 20(6), 476–493 (1994)
8. Ebert, C., Dumke, R., Bundschuh, M., Schmietendorf, A.: Best Practices in Software Measurement. How to use metrics to improve project and process performance, 1st edn. 295 Seiten-Springer, Berlin (2004)
9. Fenton, N., Pfleeger, S.L.: Software Metrics: A Rigorous & Practical Approach 2nd edn. PWS Publishing Company (1997)
10. Florac, W.A., Carleton, A.D.: Measuring the Software Process, 1st edn. Addison-Wesley, U.S.A (1999)
11. García, F., Bertoa, M.F., Calero, C., Vallecillo, A., Ruiz, F., Piattini, M., Genero, M.: Towards a consistent terminology for software measurement. Information and Software Technology, 1–14 (2005)

12. Henry, S., Kafura, S.: Software Structure Metrics Based on Information Flow. IEEE Transactions on Software Engineering 7(5), 510–518 (1981)
13. Humphrey, S.H.: PSP A Self-Improvement Process for Software Engineers, 1st edn. Addison-Wesley, U.S.A (2005)
14. IFPUG, IFPUG: Function Point Counting Practices Manual, Release 4.2. International Function Point Users Group, USA –IFPUG, Mequon, Wisconsin, USA (2004)
15. Juristo, N., Moreno, A.: Basics of Software Engineering Experimentation. Kluwer Academic Publishers, Dordrecht (2001)
16. Kitchenham, B.: Procedures for Performing Systematic Reviews. Joint Technical Report Software Engineering Group. Department of Computer Science Keele University, United King and Empirical Software Engineering, National ICT Australia Ltd, Australia, pp. 1–28 (2004)
17. Lorenz, M., Kidd, J.: Object-Oriented Software Metrics: A Practical Guide. Prentice Hall, Englewood Cliffs, Nueva Jersey (1994)
18. McCabe, T.: A Software Complexity Measure. IEEE Transactions on Software Engineering 2, 308–320 (1976)
19. Marchesi, M.: OOA Metrics for the Unified Modeling Language. 2nd Euromicro Conference on Software Maintenance and Reengineering 1998, 67–73 (1998)
20. Perry, D., Porte, A., Votta, L.: Empirical Studies of Software Engineering: A Roadmap. In: Finkelstein, A. (ed.), Future of Software Engineering. pp. 345–355, ACM (2000)
21. Pfleeger, S.L.: Assessing Software Measurement. IEEE Software, pp. 25–26 (March/April 1997)
22. Piattini, M., García, F.O.: Calidad en el desarrollo y mantenimiento de software, 1st edn., Ra-Ma. Spain (2003)
23. Poels, G., y Dedene, G.: Distance-based software measurement: necessary and sufficient properties for software measures. Information and Software Technology 42(1), 35–46 (2000)
24. Putnam, L.H., Myers, W.: Measures for Excellence - Reliable software on time, within budget. Prentice Hall, New Jersey (1992)
25. Raynus, J.: Software Process Improvement with CMM, 1st edn., Artech House, U.S.A (1999)
26. Reynoso, L., M., G., Piattini, M.: Measuring OCL Expressions: An Approach Based on Cognitive Techniques. In: Genero, M., Piattini, M., Calero, C. (eds.) Metrics for Software Conceptual Models ch. 5, Imperial College Press, UK (2004)
27. Travassos, G.H., Boilchi, J., Mian, P.G., Natali, A.C.C.: Systematic Review in Software Engineering. Technical Report Programa de Engenharia de Sistemas e Computaçâo PESC, Systems Engineering and Computer Science Department COPPE/UFRJ, Rio de Janeiro, pp. 1–30 (2005)
28. Weyuker, E.: Evaluating Software Complexity Measures. IEEE Transactions on Software Engineering 14(9), 1357–1365 (1988)
29. Whitmire, S.: Object Oriented Design Measurement. John Wiley, Chichester (1997)
30. Wohlin, C., Runeson, P., Höst, M., Ohlson, M., Regnell, B., Wesslén, A.: Experimentation in Software Engineering: An Introduction. Kluwer Academic Publishers, Dordrecht (2000)
31. Zuse, H.: A Framework of Software Measurement. Walter de Gruyter, Berlin (1998)

# Engineering a Component Language: CompJava

Hans Albrecht Schmid and Marco Pfeifer

University of Applied Sciences Konstanz, Brauneggerstr. 55, D - 78462 Konstanz, Germany
schmidha@htwg-konstanz.de, mpfeifer@htwg-konstanz.de

**Abstract.** After first great enthusiasm about the new generation of component languages, a closer inspection and use identified together with very strong points some disturbing drawbacks, which seem to have been an important impediment for a wider acceptance. A restricted acceptance of component languages would be harmful since the integration of architecture description with a programming language increases the quality of application architecture and applications, as our experience confirms. Therefore, we took an engineering approach to the construction of a new Java-based component language without these drawbacks. After deriving component language requirements, we designed a first language version meeting the requirements and developed a compiler. We used it in several projects; and re-iterated three times through the same cycle with improved language versions. The result, called CompJava, to be presented in the paper seems to be mature for use in an industrial environment.

**Keywords:** Components, Component language, Component composition, Component fragment, Connections.

## 1 Introduction

The new generation of component languages, like ArchJava [1] [2], ComponentJ [9], ACOEL [11], and to a smaller degree, KOALA [7] [8] made us enthusiastic about the new way of program construction without reference handling. The integration of architecture description with a programming language pushes the more abstract architecture-description language based approach (see ADL classification framework [6], [5]) forward towards a direct use during system development. Our experience confirms that this increases developer's awareness of application architecture, and even forces them emphasizing it. Consequently, it increases the quality of applications.

However, a closer inspection and use of component languages identified together with their strong points some small, but disturbing drawbacks.

For example, ArchJava is strong with regard to a clear composition of a component from subcomponents, except for some missing encapsulation and an overly complex connect-statement. But it shows substantial weakness with regard to using Java code for the construction of components, and filters for collaborating with subcomponents. Nor does it lend itself to the structuring of the implementation of a larger component into classes, which is practice in an industrial environment, neither regarding the ArchJava source nor the generated code. Another example is that ArchJava components behave practically like C++ classes with regard to types and inheritance (without explicit type

J. Filipe, B. Shishkov, and M. Helfert (Eds.): ICSOFT 2006, CCIS 10, pp. 177–191, 2008.

definition, implementation and derivation). Another weakness is that ArchJava re-defines constructs for concepts, like interfaces, which it shares with Java. More drawbacks and details are given in section 2.

It seems that these drawbacks may have been an impediment for a wider acceptance and broader use of component languages, which would be harmful. Therefore, we designed a new component language that does not have these drawbacks, following a sound engineering approach. We derived a list of component language requirements from the identified drawbacks. We constructed a component language that covers the requirements (the first version being available fall 2003). Then, we used the language in projects, and had three iterations with improved language definitions. Now, the language seems to be mature for industrial use.

Section 3 gives an overview about distinguishing structuring principles of CompJava, and section 4 introduces its type concept. Section 5 shows how components are composed in a structured way from component fragments, and section 6 shows how they are composed from subcomponents. Section 7 presents dynamic architectures using a Web server example.

## 2 Language Requirements

This section describes drawbacks identified in component languages, and derives specific requirements from them. These component language requirements complement general, but unlisted requirements, defined by a kind of intersection of the features of existing languages.

**Embedded OO-Programming Language.** A component language, like ArchJava, embeds a programming language, like Java, and uses its constructs to implement components. ArchJava has ports with both provided and required interfaces. It defines the interfaces of a port by listing, after the keyword provides or requires, operation specifications, or by listing method implementations. But you cannot define the interfaces of a port using Java interfaces. Thus, there are different constructs for the concept "interface" in the component language and the OO-language, which is a great drawback.

On the other hand, ArchJava allows deriving components from classes, like the worker component from the class Thread [1]. But how can a component, which is a first-class citizen of its own, "be" a class and inherit implementation from it?

Therefore, **requirement 1** is: a component language should not reinvent constructs for concepts it shares with its programming language. On the other hand, it should not intermingle differing concepts in the component language and programming language.

**Component Inheritance.** ArchJava transfers the type concept of class-based languages directly to components. It defines a component type implicitly as the type that is generated by a component, and it defines inheritance in such a way that a derived component inherits from a base component both the component type and its implementation.

This has two drawbacks. An independent definition of component types is required to define e.g. a product line architecture or a component framework. A product line architecture defines product component types which are implemented by different product components. Similarly, a component framework defines a set of collaborating component types which are implemented by different components. Second, a component should not inherit the implementation from another component, but should be composed with the other component in order to reuse its functionality. Therefore, **requirement 2** is that the definition of component types and inheritance among them should be provided, but implementation inheritance among components should be disallowed.

**Component Encapsulation.** ArchJava allows that a parent component invokes internal methods of a subcomponent which are not defined by a provided port. This breaks the encapsulation of the subcomponent. Further, a graceful evolution is inhibited since it is not possible that a sibling subcomponent invokes these methods instead of the parent component at a later point of the evolution. On the other hand, ACOEL allows that a parent component P exposes in a port interface a reference to a subcomponent S. When P passes that reference to a sibling component, the sibling may connect its own ports or those of its subcomponents to ports of S. This breaks component encapsulation since a component like S could be a subcomponent of two parent components.

**Requirement 3** is that a component should be completely encapsulated, i.e. it should collaborate only via its ports with external code. As a consequence, a subcomponent of a component must not collaborate with other components outside of its parent component. Therefore, the passing of component or port references should be restricted or prohibited.

**Interface Symmetry.** ArchJava has a complete symmetry among provided and required interfaces with regard to their definition and their use, since a port may comprise both of them. ACOEL [11] has a symmetry with regard to their definition, but not with regard to their use. A mix-in allows putting a filter between a provided port and the implementing class. But it does not allow putting a filter between the implementing class and a required port.

**Requirement 4** is that the definition and the handling of provided and required ports should be symmetrical.

**Ports and Connectors.** An ArchJava port may combine a provided and a required interface, like:

**port** *port1* **provides** *m1, m2* **requires** *m3, m4*;

As usual, a port with a required interface $I_1$ may be connected to a port with a provided interface $I_2$ when $I_2$ is a subtype of $I_1$. But an ArchJava connector may fork the calls from a required interface $I_1$ to several provided interfaces like $I_2$ and $I_3$ if each is a supertype of $I_1$, and their union is a subtype of $I_1$, and the intersection of $I_1$, $I_2$ and $I_3$ is empty. For example, with *port2* and *port3*:

**port** *port2* **provides** *m3, m6* **requires** *m1, m5*;
**port** *port3* **provides** *m4, m5, m6* **requires** *m2, m3*;

ArchJava allows to connect *port1*, *port2*, and *port3* by a connect statement. If *port1* would require additionally *m6* the connection would not be correct and rejected. This is not easy to check and understand for a programmer; it might be considered as a new kind of spaghetti problem (without dining philosophers!). Though it is easy for a compiler to check what happens, we should disallow it.

**Requirement 5** is that the definition of ports and connectors should be made in a way that is easily understandable to a programmer.

**Collaboration of Subcomponent Ports with Code.** ArchJava defines private ports in order to connect component code with a port of a subcomponent. However, a private port is a contradiction in itself since the ports of a component define its interfaces to the outside, i.e. the points of collaboration with external code: So what is the semantics of a private port? It is even more confusing that ArchJava allows connecting two private ports; what does that mean? Our conclusion is that the concept of private port is not meaningful. **Requirement 6** is that an adequate construct should connect component code with a port of a subcomponent.

**Implementation Isomorphism with OO-Based Approach.** ArchJava defines and generates a single component class which lists the provided methods of all public and private ports of the component. It does not lend itself to grouping together in a class the methods that implement the operations of the same port. Similarly, the required operations of all ports are always invoked from that list of methods. There is no direct way to group together in a class the methods that invoke the operations of the same port. This is in contrast to the usual OO-based implementation of a component where the provided methods of each port are implemented by a different class, and the required methods of each port are usually invoked by methods from different classes.

Therefore, **requirement 7** is that the source code of a component and the code generated from it should have some isomorphism with corresponding code written in class-based OO-languages.

**Implementation Efficiency.** The efficiency of the code generated from a component language may not be a primary concern when large architectural components with powerful operations are realized. But in many cases, the efficiency of a frequently performed operation invocation matters. Consider e.g. a scanner, used e.g. as a subcomponent of a compiler, which is certainly not a lightweight component. It fetches the next character from a source file over a required interface with a getCharacter-operation (compare section 6). The efficiency of that frequently performed operation invocation has a strong influence on the scanner overall performance.

**Requirement 8** is that the code generated from a component language should have about the same efficiency for basic constructs, like e.g. operation invocation over connected ports, as an equivalent (but not tricky) class-based implementation.

We state that requirement due to its importance for a wide acceptance of component languages, though we cannot cover it in this paper for space reasons.

## 3 CompJava Overview

Distinguishing features of CompJava are, besides the definition of component types and component type inheritance, its structuring facilities for component construction. CompJava allows not only, like the new generation of component languages, to compose components from subcomponents in a structured way. It allows composing them also in the same way from code building blocks, or from a combination of subcomponents and filters formed by code building blocks.

CompJava has code building blocks called component fragments. A component fragment might be considered as a simply structured light-weight component without ports: it provides exactly one interface, and it requires usually one interface. The provided interface of a component fragment is explicitly indicated in the form of a Java interface; the required interfaces of a component fragment are implicitly given by the visible ports and plugs of the enclosing component. There are three implementation variants of a component fragment: anonymous class, inner class and method block; from which a user may select the suitable one.

CompJava introduces plugs which are used mainly for connecting component fragments with subcomponent ports.

Ports of subcomponents are connected with the connect-statement to other ports or plugs. Component fragments are attached to the inside of the component ports or to plugs with an attach-statement. Thus, CompJava allows to compose in a clear, clean and structured way:

1. a low-level component from component fragments, as illustrated by Figure 1 a)
2. a high-level component from subcomponents, as illustrated by Figure 1 b)
3. a medium/high-level component from a combination of subcomponents and component fragments that are used as filters, as illustrated by Figure 1 c)

For a graphical depiction of the composition of a component, we have enriched UML 2 component diagrams with component fragments and plugs (depicted by a diamond). A component fragment is represented according to the selected implementation as an anonymous class, an inner class or as a method block (depicted like an anonymous class without class head).

The first version of the CompJava compiler has been available since winter 2003/2004, three more versions followed. The version available since fall 2006 is integrated in Eclipse. The CompJava Designer Eclipse-plugin is a graphical design tool that allows to draw enriched CompJava component diagrams and to generate component code skeletons from them.

The following sections introduce the CompJava language and show how their constructs satisfy the requirements. We use a compiler as a running example. The compiler component is composed from a scanner, parser and other subcomponents.

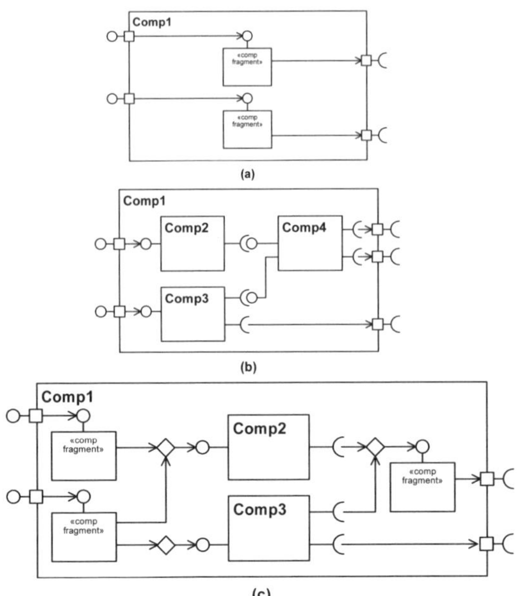


**Fig. 1.** Composition of a component from component fragments (a), from subcomponents (b), and from a combination of them (c)

## 4 Component Types

Let us consider first the scanner component. We define the provided interface of the scanner as a Java interface. It includes all scanner-related responsibilities, like setting the file name of the source file to be processed, and fetching the next token from it.

```
interface ScannerIF {
  Token getNext();
  void setSource( String sourceName);
}
```

Since the *ScannerIF* interface includes all source file processing related responsibilities, the component type *Scanner1Type* is defined with a single provided port.

```
component type Scanner1Type {
    port in provides ScannerIF;
}
```

A component type defines all interfaces of a component. That means components are completely encapsulated: all methods in a component, except for the main method, can be invoked from outside only via provided ports, and all methods can invoke an outside method only via required ports.

A port has either a provided, a required or an event interface. A port declaration gives the port name and after the corresponding keyword the associated interface. An event port is similar to a required port, but its operations must not have results, and several provided ports of event listeners may be connected to it. As we show in section 7, a component type may also define port arrays or port vectors.

A component type may extend another component type. It inherits all ports, and it may extend the interface of inherited provided ports or may add provided ports.

# 5 Low-Level Components

This section shows how low-level components are composed from component fragments.

## 5.1 Implementing Provided Ports

A component has a component type (indicated by the ofType-clause). It implements the provided ports, and may invoke operations from required ports, specified by its component type. In the Scanner1 component (see Figure 2), an attach-statement attaches the inside of the provided port *in* to a component fragment, an anonymous class implementing the *ScannerIF* interface.

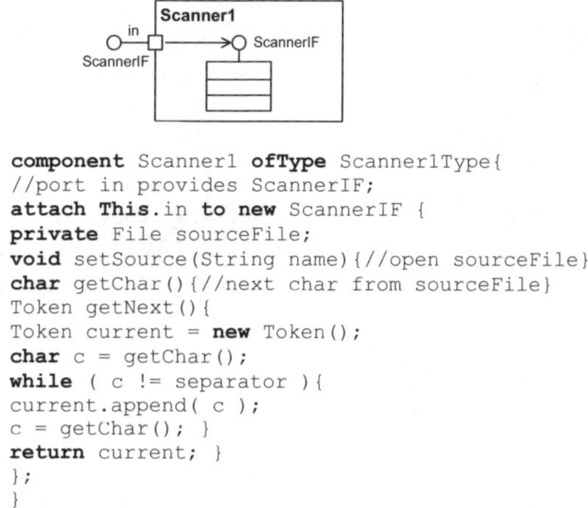


```
component Scanner1 ofType Scanner1Type{
//port in provides ScannerIF;
attach This.in to new ScannerIF {
private File sourceFile;
void setSource(String name){//open sourceFile}
char getChar(){//next char from sourceFile}
Token getNext(){
Token current = new Token();
char c = getChar();
while ( c != separator ){
current.append( c );
c = getChar(); }
return current; }
};
}
```

**Fig. 2.** *Scanner1* component with port *in* providing the *ScannerIF* implemented by a anonymous class

An attach-statement may be used to attach the inside of a provided port to a component fragment that implements an interface I. The condition is that I extends (including equals) the port interface; it is checked at compile time. A component fragment may be a Java construct: an instance of an anonymous class, as shown, or an instance of an inner class. The inside of a port is indicated by the keyword *This*, which stands for the component instance, followed by the port name. The declaration of inner and anonymous classes follows the Java standard; the only difference is that they are used inside of a component instead of a class.

When a component, like *Scanner1*, is quite small and not composed from other components, it might be a disadvantage that its implementation generates two object instances: one of the application-specific component fragment and another one of the component class. Therefore, CompJava allows that a component fragment is formed by a method block. A method block is a sequence of methods that implement a given interface (see Figure 3). A method block is not a Java construct, but an analogon to a Java block, which is a sequence of statements. When different provided ports are each attached to a method block, there is the restriction that their interfaces must have an empty intersection.

Consequently, CompJava provides component fragments which include method blocks, inner classes or anonymous classes, in order to structure the implementation of a component.

## 5.2 Accessing Required Ports

The Scanner mixes up two different concerns, scanning the program character stream, and handling of the source file to be parsed. Similarly, the *ScannerIF* interface mixes up two different concerns, accessing the tokens which the scanner creates, and determining the source file to be parsed. We should separate the different concerns, scanning and source file handling. To this purpose, we define two interfaces, *TokenIF* and *SourceAccess*:

```
interface TokenIF {
  Token getNext();
}
interface SourceAccess {
  char getChar();
}
```

The new scanner component does not include the source file handling but fetches the source file characters via a required interface. We define the component type *Scanner2Type* with a provided interface *TokenIF* and a required interface *SourceAccess*:

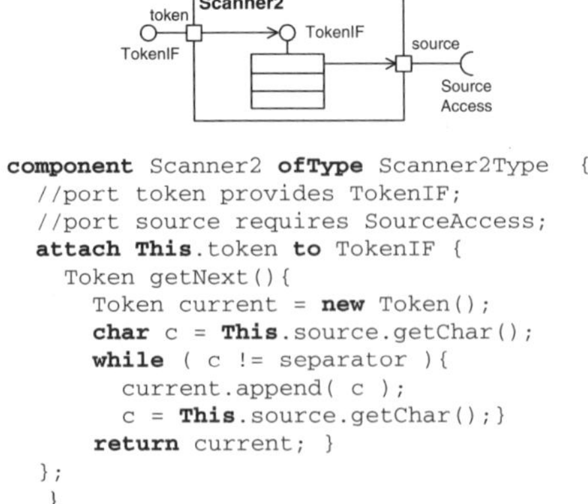


```
component Scanner2 ofType Scanner2Type  {
  //port token provides TokenIF;
  //port source requires SourceAccess;
  attach This.token to TokenIF {
    Token getNext(){
      Token current = new Token();
      char c = This.source.getChar();
      while ( c != separator ){
        current.append( c );
        c = This.source.getChar();}
      return current; }
  };
 }
```

**Fig. 3.** *Scanner2* component with port *token* providing the *TokenIF* implemented by a method block, and port *source* requiring the *SourceAccess* interface

```
component type Scanner2Type {
  port token provides TokenIF;
  port source requires SourceAccess;
}
```

The *Scanner2* component attaches the *token* port to a component fragment, a method block. It implements the *TokenIF* and scans the source file in order to determine the next token. When it needs the next character from the source file, it simply invokes the *getChar*-operation defined in the *SourceAccess* interface via the inside of the required port *source*.

# 6 Component Composition

A compiler is a top-level component that is composed from a scanner, a parser etc. For that reason, we declare its type without any ports. The type of the parser defines a required interface *TokenIF*, and other ones which we do not consider.

```
component type CompilerType {}
component type ParserType {
      port ...;
      port getToken requires TokenIF;
}
```

## 6.1 Subcomponents

A component may be composed from subcomponents. E.g. the *Compiler1* component (see Figure 4) is composed from a scanner, a parser, and other subcomponents like a code-generator which we disregard.

A component may contain subcomponent declarations and connect-statements that are processed with the initialization of the component.

A subcomponent declaration declares a subcomponent variable, like *myParser* and *myScanner*, of a component type; it may assign to it an instance of a matching component created with the *new* operator and the component constructor, like a *Parser* res. a *Scanner2* instance.

A connect-statement connects a required port of a subcomponent (instance), like *getToken* of *Parser*, to a provided port of a subcomponent (instance), like *token* of *Scanner2*, as Figure 4 shows. A constraint checked by the compiler is that a required port can be connected to only one provided port; but many required ports may be connected to the same provided port. An event port may be connected to many provided ports. The compilation of a connect-statement includes port-matching, i.e. checking if the provided port interface extends (incl. equals) the required port interface. We may use a connect-statement also to connect a port of a subcomponent directly with the inside of a matching port of the (parent) component.

## 6.2 Connecting Ports with Plugs

The *Compiler1* component contains a component fragment, an anonymous class implementing the interface *SourceHandling*, which the source port of the *Scanner2* should invoke. But a connect-statement does not allow connecting a subcomponent port with a component fragment. Therefore, we introduce plugs replacing private ports of ArchJava.

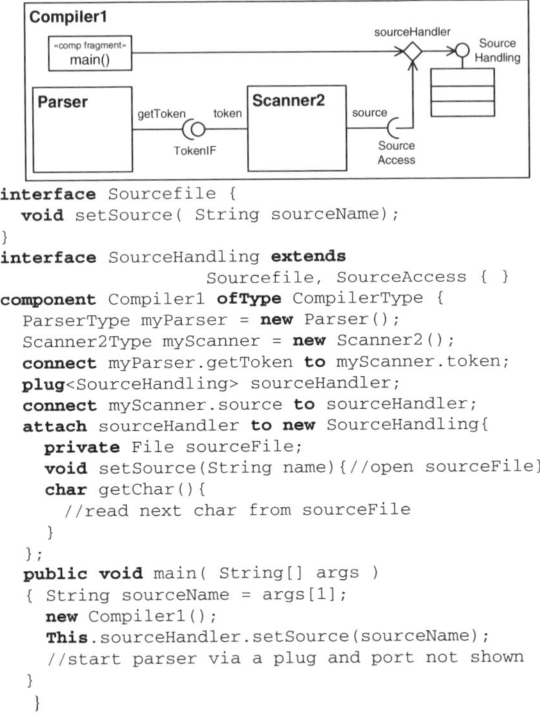


```
interface Sourcefile {
  void setSource( String sourceName);
}
interface SourceHandling extends
                   Sourcefile, SourceAccess { }
component Compiler1 ofType CompilerType {
  ParserType myParser = new Parser();
  Scanner2Type myScanner = new Scanner2();
  connect myParser.getToken to myScanner.token;
  plug<SourceHandling> sourceHandler;
  connect myScanner.source to sourceHandler;
  attach sourceHandler to new SourceHandling{
    private File sourceFile;
    void setSource(String name){//open sourceFile}
    char getChar(){
      //read next char from sourceFile
    }
  };
  public void main( String[] args )
  { String sourceName = args[1];
    new Compiler1();
    This.sourceHandler.setSource(sourceName);
    //start parser via a plug and port not shown
  }
   }
```

**Fig. 4.** Component *Compiler1* composed from subcomponents *Parser* and *Scanner2* and a component fragment implementing the interface *SourceHandling*

A plug is a generic construct that exceeds the generic possibilities provided by Java parametric interfaces or classes. The generic expression "plug<interface>" generates a plug of the interface type. It might be considered as a variable on which only a very limited set of operations may be executed: it may be used in connect- and attach-statements, or it may be used in a component fragment to invoke an operation defined in the plug interface.

The Compiler component (see Figure 4) declares a plug of the interface type *SourceHandling* named *sourceHandler*. The plug is used to pass operation invocations from the required port of the scanner subcomponent to a component fragment of the compiler component, which does all handling of the source file.

A connect-statement connects the required port *source* of the scanner with that plug, matching at compile time whether the plug interface extends the required port interface. The main method, which gets the filename of the source file passed as a parameter, invokes the *setSource*-operation via the same plug.

An attach-statement may attach a plug to a component fragment, as shown in Figure 4. It checks at compile time whether the interface of the component fragment extends the plug interface. The constraint is that the same plug may appear only once on the left-hand side of an attach- or connect-statement, but several times on their right-hand side and/or be used for operation invocations.

### 6.3 Factoring Out SourceHandling

Suppose that we want to reuse the anonymous source handling class with the interface *SourceHandling* shown in Figure 4. Then we should factor it out and transform it into a separate source file processing component with the component type *SourceType*.

```
component type SourceType {
  port source provides Sourcefile;
  port accessSource provides
                              SourceAccess;
}
```

The component Source (see Figure 5) contains a *SourceHandling* component fragment that is identical to the component fragment used by the *Compiler1* component (see Figure 4).

```
component Source ofType SourceType {
  plug<SourceHandling> sourceHandler;
  attach This.source to This.sourceHandler;
  attach This.accessSource to This.sourceHandler;
  attach This.sourceHandler to new SourceHandling{
    private File sourceFile;
    void setSource(String name){//open sourceFile}
    char getChar(){
      //read next char from sourceFile
}
  };
  }
```

**Fig. 5.** Component *Source* with the provided ports *source* and *accessSource* attached to plug *sourceHandler* attached to an anonymous class as component fragment

Since we want to attach both provided ports to the same component fragment, we declare the plug *sourceHandler* of type *SourceHandling*. It is attached to the component fragment with an attach-statement. The inside of each provided port is attached to the plug with each an attach-statement.

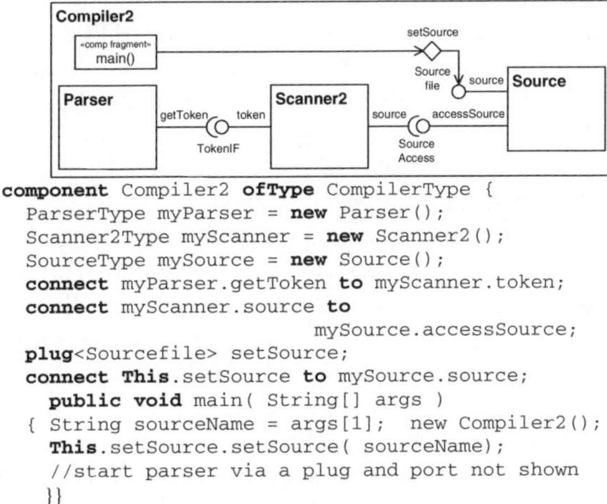


```
component Compiler2 ofType CompilerType {
  ParserType myParser = new Parser();
  Scanner2Type myScanner = new Scanner2();
  SourceType mySource = new Source();
  connect myParser.getToken to myScanner.token;
  connect myScanner.source to
                        mySource.accessSource;
  plug<Sourcefile> setSource;
  connect This.setSource to mySource.source;
    public void main( String[] args )
  { String sourceName = args[1];  new Compiler2();
    This.setSource.setSource( sourceName);
    //start parser via a plug and port not shown
    }}
```

**Fig. 6.** Component *Compiler2* composed from subcomponents *Parser*, *Scanner2* and *Source*

The component *Compiler2* (see Figure 6) is identical to *Compiler1*, except for replacing the *SourceHanding* component fragment by the *Source* component. It connects the port *source* of *Scanner2* with a connect-statement to the *accessSource* port of *Source*. The plug *setSource* is declared and connected to the *source* port of the *Source* component with the objective that the *main* method may invoke via that plug the *setSource*-operation of the *source* port.

## 7 Dynamic Architectures

The language constructs described so far allow to construct component systems with a static architecture, i.e. a static hierarchy of collaborating component instances. Though that is sufficient for a large class of systems, there are other systems which require a dynamic creation and connection of components.

A component instance may be created dynamically in a method of a component fragment with a new-operator and component constructor similarly as shown e.g. in Figure 4. Dynamically created components are connected at run-time with a reconnect-statement which is similar to a connect statement. A component should document explicitly all kinds of architectural interactions that are permitted between its subcomponents. To this purpose, a component uses connection patterns (as introduced by ArchJava [1] [2]) to describe the set of connections that can be made at run-time using *reconnect*-statements.

Since in a dynamic architecture, a component may have a variable number of subcomponents of the same type, we introduce component arrays and vectors (as a parametric Vector parameterized with a component type). Since it may also be required that a connection is made from the port of a component to a variable number of sibling components, we introduce port arrays or port vectors as arrays or parameterized vectors of an interface type. Though the primary emphasis of component and port arrays resp. vectors is on dynamic architectures, they may be of use also for static architectures with repetitive elements.

For example, consider a *WebServer* component. It has one *Router* and many *Worker* subcomponents. The *Router* receives incoming HTTP-requests and passes them through a required port of the port array *workers* to the connected *Worker* subcomponent that serves the request. The *WebServer* starts the *Router* via its provided port *start* and the plug *start*.

Figure 7 shows a shortened version of the *WebServer*. The running version with about three times the length of the presented version may be obtained from the authors. We present, in contrast to [1], an optimized solution that reuses idle *Worker* instances and their connections. A *Worker* contains a *WorkerThread* class. When an *httpRequest* is invoked via the *serve* port of a *Worker*, the *WorkerThread* is (re-) started by a notify-statement and takes up work with a call of its method *handleRequest*. When it has finished the processing of an HTTP-request, it goes into a wait state.

The *WebServer* has declared an array of *Worker* components. It connects the provided *serve* port of each *Worker* instance after its creation dynamically to the matching port of the required port array *workers* of the *Router* component.

The *WebServer* performs the administration of the *Worker* instances in the method block implementing the *WorkerAdministration* interface, which is attached to the *adminWorker* plug. It has a *setIdle*-operation which is invoked by a *Worker* after having finished the processing of an HTTP-request, and similar operations. The *requestWorker*-operation checks if an idle *Worker* is available, and returns its index. Otherwise, it creates a new *Worker* instance if the maximum worker number is not yet reached, and connects dynamically the new *Worker*'s *serve* port to the matching port of the *workers* port array of the *Router*, and its required *adminWorker* port to the *adminWorker* plug.

The *WebServer* has connected the required *request* port of the *Router* to the *adminWorker* plug. In that way, both the *Router* and all *Worker*'s may invoke operations of the worker administration, like *setIdle* or *requestWorker*.

The code of the *WebServer* component is easy to understand, in contrast to the code shown in [1].

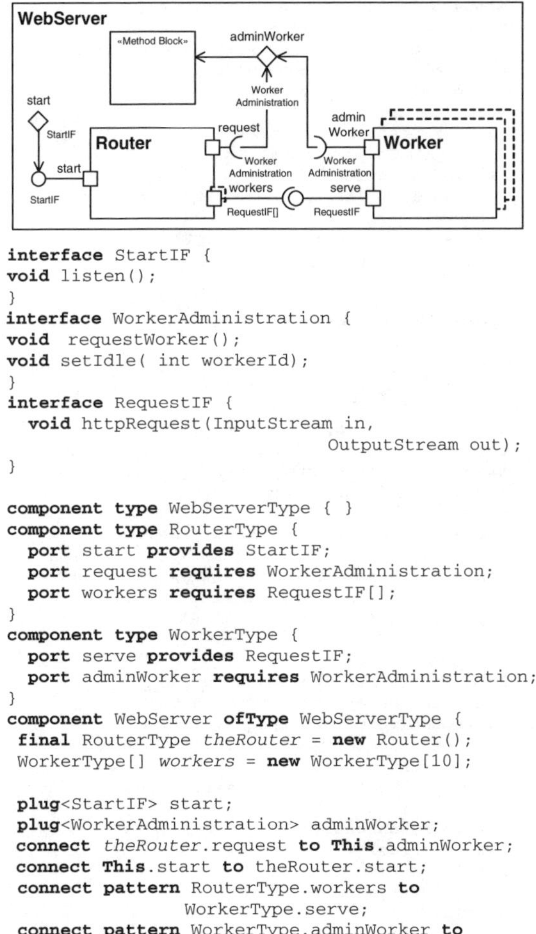


```
interface StartIF {
void listen();
}
interface WorkerAdministration {
void  requestWorker();
void setIdle( int workerId);
}
interface RequestIF {
  void httpRequest(InputStream in,
                              OutputStream out);
}

component type WebServerType { }
component type RouterType {
  port start provides StartIF;
  port request requires WorkerAdministration;
  port workers requires RequestIF[];
}
component type WorkerType {
  port serve provides RequestIF;
  port adminWorker requires WorkerAdministration;
}
component WebServer ofType WebServerType {
 final RouterType theRouter = new Router();
 WorkerType[] workers = new WorkerType[10];

 plug<StartIF> start;
 plug<WorkerAdministration> adminWorker;
 connect theRouter.request to This.adminWorker;
 connect This.start to theRouter.start;
 connect pattern RouterType.workers to
                WorkerType.serve;
 connect pattern WorkerType.adminWorker to
```

**Fig. 7.** Component *WebServer* composed from a worker administration component fragment together with one *Router* and a variable number of *Worker* subcomponents

```
                    plug<WorkerAdministration>;
 public static void main(String[] args) {
  new WebServer( ...).run();
 }
 void run() {
   This.start.listen();
 }
 attach This.adminWorker to WorkerAdministration {
  void setIdle( ...) { ...}
  int requestWorker(){
   if( no worker idle & workerID < maxWorkerID){
    workers[workerID] = new Worker(dir, workerID);
    reconnect workers[workerID].adminWorker to
                             This.adminWorker;
    reconnect theRouter.workers[workerID] to
                          workers[workerID].serve;
    return workerID; }
   //other methods...
  } };
}
component Router ofType RouterType {
 //port start provides StartIF;
 //port request requires WorkerAdministration;
 port workers = new RequestIF[10];
 attach This.start to StartIF {
  void listen() {
   ServerSocket server = new
            ServerSocket( This.request.getPort());
   while (true) {
    workerID = This.request.requestWorker();
    Socket sock = server.accept();
    This.workers[workerID].httpRequest(
    sock.getInputStream(),sock.getOutputStream());
   } }};
}
component Worker ofType WorkerType {
 //port serve provides RequestIF;
 //port adminWorker requires WorkerAdministration;
 WorkerThread myThread = new WorkerThread();
 myThread.start();
 BufferedReader in; // HTTP-request
 PrintWriter out; //   HTTP-response
 attach This.serve to RequestIF{
  synchronized void httpRequest(
           InputStream in, OutputStream out){
   this.in = new BufferedReader(new
                     InputStreamReader(in) );
   this.out = new PrintWriter(new BufferedWriter(
               new OutputStreamWriter(out)));
   myThread.notify();
  }
 };
 class WorkerThread extends Thread {
  //several data attributes and methods
  protected void handleRequest() {
    // open requested file and send answer ...
    out.println("HTTP/1.0 200 OK");
    // ... and file contents to Browser
  }
  public synchronized void run() {
   while (true) {
    this.wait();
    handleRequest();
    This.adminWorker.setIdle(this.workerNo);
   } }
 } //end WorkerThread
}
```

**Fig. 7.** (*Continued*)

## 8 Conclusions

CompJava, available for non-commercial use via www.compjava.org, composes components in a clear and simple way from two kinds of building blocks: component fragments and subcomponents.

We have introduced component fragments that may be considered as very simply structured lightweight components without ports. There are three implementation variants covering different performance and reusability requirements. Component fragments allow to structure low-level components in an adequate way, and they serve as filters for medium to high level components.

All these building blocks with well-defined and clear interfaces are attached/connected either directly or via plugs to themselves or to ports of the parent component.

Clean and efficient dynamic architectures are composed from dynamically instantiated and connected subcomponent instances together with component arrays and port arrays res. vectors.

The implementation of an instant messaging system [4] has proven that CompJava scales up to the construction of larger applications in an industrial environment.

CompJava has been extended for use as a distributed component language as described in [10].

## References


1. Aldrich, J., Chambers, C., Notkin, D.: ArchJava: Connecting Software Architecture to Implementation. In: Procs ICSE (May 2002)
2. Aldrich, J., Chambers, C., Notkin, D.: Architectural Reasoning in Archjava. In: Magnusson, B. (ed.) ECOOP 2002. LNCS, vol. 2374, pp. 185–193. Springer, Heidelberg (2002)
3. Gamma, E., Helm, R., Johnson, R., Vlissides, J.: Design Patterns: Elements of Reusable Object-Oriented Software. Addison-Wesley, Reading (1995)
4. Klenk, M.: Entwurf einer Chatapplikation mit der Komponentenprogrammiersprache CompJava. Diploma thesis, University of Applied Sciences Konstanz (2006)
5. Medvidovic, N., Rosenblum, D.S., Taylor, R.P.: A Language and Environment for Architecture-Based Software Development and Evolution. In: Procs ICSE 1999 (May 1999)
6. Medvidovic, N., Taylor, R.P.: A Classification and Comparison Framework for Software Architecture Description Languages (2000)
7. van Ommering, R., van der Linden, F., Kramer, J., Magee, J.: The KOALA Component Model for Consumer Electronics Software. IEEE Computer (2000)
8. van Ommering, R.: Building Product Populations with Software Components. In: Proc. ICSE 2002 (2002)
9. Seco, J.C., Caires, L.: A Basic Model of Typed Components. In: Bertino, E. (ed.) ECOOP 2000. LNCS, vol. 1850, Springer, Heidelberg (2000)
10. Schmid, H.A., Pfeifer, M., Schneider, T.: A Middleware-Independent Model and Language for Component Distribution. In: Proc. SEM 2005. ACM Press, New York (2005)
11. Sreedhar, V.C.: Mixin' Up Components. In: Procs ICSE (2002)
12. Szyperski, C.: Component Software, Beyond Object-Oriented Programming. Addison-Wesley, Reading (1997)

# PART III

# Distributed and Parallel Systems

# Towards a Quality Model for Grid Portals

Mª Ángeles Moraga[1], Coral Calero[1], Mario Piattini[1], and David Walker[2]

[1] Alarcos Research Group. UCLM-SOLUZIONA Research and Development Institute University of Castilla-La Mancha, Spain
{MariaAngeles.Moraga,Coral.Calero,Mario.Piattini}@uclm.es
[2] School of Computer Science. Cardiff University, United Kingdom
David.W.Walker@cs.cardiff.ac.uk

**Abstract.** Researchers require multiple computing resources when conducting their computational research; this makes necessary the use of distributed resources. In response to the need for dependable, consistent and pervasive access to distributed resources, the Grid came into existence. Grid portals subsequently appeared with the aim of facilitating the use and management of distributed resources. Nowadays, many Grid portals can be found. In addition, users can change from one Grid portal to another with only a click of a mouse. So, it is very important that users regularly return to the same Grid portal, since otherwise the Grid portal might disappear. However, the only mechanism that makes users return is high quality. Therefore, in this paper and with all the above considerations in mind, we have developed a Grid portal quality model from an existing portal quality model, namely, PQM. In addition, the model produced has been applied to two specific Grid portals.

**Keywords:** Quality models, Grid portals.

## 1 Introduction

Nowadays, many users have access to, and require, multiple computing resources to conduct their computational research [2]. This makes the use of distributed resources necessary. For this reason and with the aim of providing dependable, consistent and pervasive access to distributed resources, the Grid emerged [11]. The real and specific problem that underlies the Grid concept is coordinated resource sharing and problem solving in dynamic, multi-institutional virtual organizations [5].

Specifically, the Grid couples a wide variety of geographically distributed resources such as PCs, workstations and clusters, storage systems, data sources, databases and special purpose scientific instruments and presents them as a unified, integrated resource [11].

The main problem with the Grid, however, is the difficulty involved in using grid resources. That is due to its complex architecture. Therefore, in order for scientists to use grid resources effectively as a problem solving infrastructure, transparent and easy-of-use interfaces to the complex set of grid resources are necessary [8]. Nowadays, Grid Portals are coming into existence to resolve this problem. They can be considered as a mechanism for providing user-friendly access to grid resources, and

J. Filipe, B. Shishkov, and M. Helfert (Eds.): ICSOFT 2006, CCIS 10, pp. 195–203, 2008.

consistent access patterns, as well as easy usage of grid services. The original objective of this portal type was to create web-accessible problem-solving environments (PSEs) that allowed scientists to access distributed resources, and to monitor and execute distributed Grid applications from a Web browser [12]. Although at the beginning these portals were aimed at researchers, nowadays they can be used by any user who wants to use distributed resources.

Many Grid portals exist at the present time. An immediate effect of this widespread presence is the increasing range of resources available at the click of a mouse, that is, without the user wasting time and money by physically moving from one place to another [1], [16]. It is because of this that portals must offer a good level of quality, thus users are attracted to them and come back regularly.

Bearing this in mind, as well as the lack of quality models specifically for Grid portals, in this paper we present a Grid portal quality model (G-PQM) created from an existing portal quality model, namely, PQM (Portal Quality Model) [14].

The rest of the paper is organised as follows. In section 2 the quality model for Grid portals is shown while in section 3 this quality model is applied to two Grid portals. Finally, section 4 concludes and outlines further work.

## 2 Quality Model for Grid Portals

Grid portals appeared because of the need to make access by researchers to Grid resources easier. The developers of Grid portals seek to ensure that users return to their portal often. However, the only mechanism that makes users return is high quality [15]. Therefore, a quality model which is specifically for Grid portals, namely G-PQM (Grid Portal Quality Model), has been developed. The usefulness of this model is two-fold. On the one hand, this model helps users to evaluate the different Grid portals and to choose the one with the highest quality. And on the other hand, the model's dimensions can be used as indicators to help developers when building the portal.

To develop G-PQM a quality model for web portals, namely PQM (Portal Quality Model), was used as the basis. PQM is composed of six dimensions and seeks to determine the strong and weak points of a specific portal. We can also define corrective actions for the weaknesses, and improve the quality level of a portal [13]. In order to adapt this model to Grid portals, some definitions of the dimensions have been modified and, additionally, some dimensions have been inserted. In Figure 1, we can see the different phases used in developing the Grid portal quality model, G-PQM.

In our introduction, the first phase "Study of the Grid portals context" was presented.

### 2.1 Adaptation of the PQM Dimensions

We have adapted the following PQM dimensions:

- **Tangible:** This dimension indicates if "the Grid portal contains all the software and hardware infrastructures needed according to its functionality".
  - *Adaptability*: ability of the Grid portal to be adapted to different devices (for instance, PDA, PCs, mobile phone, etc.).

- o *Transparent access*: ability of the Grid portal to provide access to the Grid resources while isolating the user from their complexity.
- **Reliability:** It is the "ability of the portal to perform the specified services". In addition, this dimension will be affected by:
  - o *Fault tolerance*: capability of the Grid portal to maintain a specified level of performance in the event of software faults [9] (for example, a fault during the sending or the execution of a job).
  - o *Availability*: The portal must be always operative in order for users to be able to access it and use its Grid resources anywhere and anytime.
  - o *Search Quality*: The results that the portal provides when undertaking a search must be appropriate to the request made by the user.

*Quality in the use of resources*: the user can use Grid resources under specified conditions with the portal.

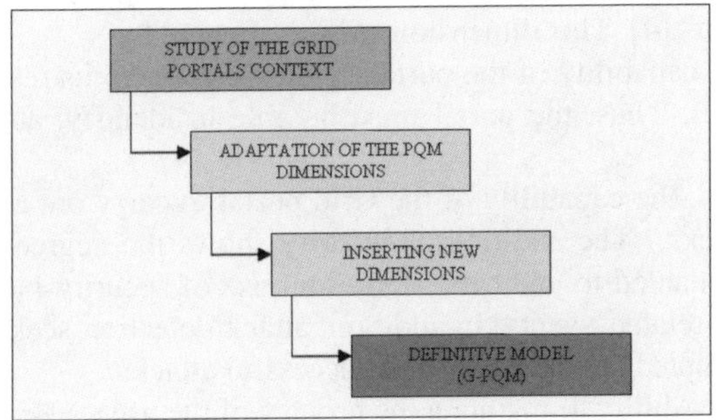


**Fig. 1.** Phases for the construction of the G-PQM model

- **Responsiveness:** It is the "willingness of the Grid portal to help and to provide its functionality in an immediate form to the users". In this dimension, we note the following sub-dimensions:
  - o *Scalability*: This refers to the ability of the portal to adapt smoothly to increasing workloads coming about as a result of additional users, an increase in traffic volume or the execution of more complex transactions [7].
  - o *Efficient access*: This relates to the response times experienced by portal users [7].
- **Empathy:** We define this dimension as the "ability of the Grid portal to provide caring and individual attention". In this dimension, we observe the following sub-dimensions:
  - o *Navigation*: The Grid portal must provide simple and intuitive navigation when being used.
  - o *Presentation*: The Grid portal must have a clear and uniform interface.
  - o *Integration*: All the components of the Grid portal must be integrated in a coherent form.
  - o *Personalization*: The portal must be capable of adapting to the user's priorities.
- **Data and Information Files Quality:** This dimension is defined as the "quality of the data contained in the portal and of the files which specify the available services

in the portal and the names of devices responsible for these services". According to Dedeke and Kahn, we can distinguish four different subdimensions [3]:

- o *Intrinsic*: this indicates what degree of care was taken in the creation and preparation of data/files.
- o *Representation*: this indicates what degree of care was taken in the presentation and organization of data/files for users.
- o *Contextual*: to what degree the data/files provided meet the needs of the users.
- o *Accessibility*: this indicates what degree of freedom users have to use data, define and/or refine the manner in which data/files are inputted, processed or presented to them.

## 2.2 Inserting New Dimensions

The following dimension has been added:

- **Security:** This is "the ability of the portal to prevent, reduce and properly respond to malicious harm" [4]. This dimension will be affected by:
  - o *Access control*: capability of the portal to allow access to its resources only to authorized persons. Thus, the portal must be able to identify, authenticate and authorize its users.
  - o *Security control*: the capability of the Grid portal to carry out auditing of security and detect attacks. The auditing of security shows the degree to which security personnel are enabled to audit the status and use of security mechanisms by analyzing security-related events. In addition, attack detection seeks to detect, record and notify attempted attacks as well as successful attacks.
  - o *Confidentiality*: Ability to maintain the privacy of the users.
  - o *Integrity*: the capability of the portal to protect components (of data, hardware, and software) from intentional or unauthorized modifications.

## 2.3 Definitive Model (G-PQM)

Taking into account the dimensions which have been adapted as well as the dimensions that have been introduced, the following model results (Figure 2):

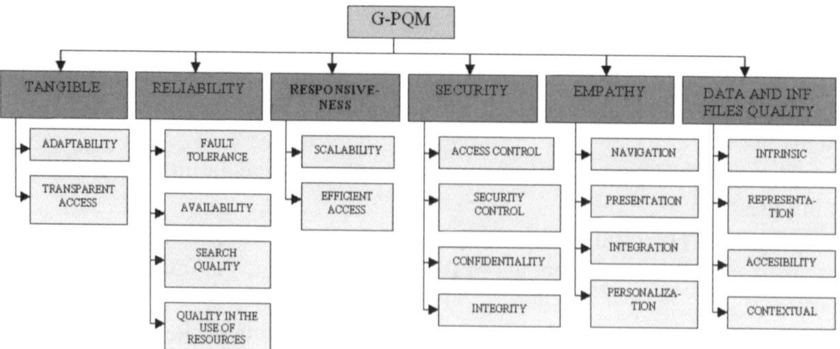


**Fig. 2.** Characteristics and subcharacteristics of G-PQM

# 3 Applying G-PQM

Having defined G-PQM, the next step is to apply it to some Grid portals with the objective of determining, on the one hand, the extent to which these portals satisfy the dimensions identified in the Grid portal quality model; and on the other hand, to identify possible improvements in the quality of these portals.

In our first approach, G-PQM has been applied to two Grid portals. It should be noted that we have applied G-PQM from the point of view of the users. G-PQM is, however, directed at portal developers. For this reason, some of the identified dimensions or sub-dimensions may not be measured (in this case, we will assign the value "not evaluable" to the (sub) dimension). In spite of this, we can obtain an overall assessment of the quality of these Grid portals.

## 3.1 GridPort Demo Portal

As a first step, the model has been applied to the GridPort demo portal which is a fully operational test portal that is intended to serve as a starting point for those interested in grid portal development (the reader can find more information about this portal at http://gridport.net/main/). This portal has been developed using the GridPort toolkit which enables the rapid development of highly functional grid portals that simplify the use of underlying grid services for the end-user [6]. The GridPort demo portal includes portlets that allow a user to do the following: view static and dynamic information about the resources in a grid, obtain short-term proxies from a myproxy server, submit batch jobs to resources on the grid, and browse and transfer files between resources on the grid [6].

The outcomes obtained are the following:

- Tangible:
  - Adaptability: The following software packages are prerequisites to using the GridPort Demo Portal: JDK 1.4.2, Jakarta Ant 1.6, TomCat, etc. These packages cannot be installed on all devices.
  - Transparent access: GridPort has Grid portlets whose aim is to provide transparent access to resources.
- Reliability:
  - Fault tolerance: Not evaluable.
  - Availability: During the testing, the portal was available anywhere and anytime.
  - Search Quality: Not applicable because the portal does not have a search engine.
  - Quality in the use of resources: Not evaluable.
- Responsiveness:
  - Scalability: The portal is not limited to a specific number of users.
  - Efficient access: During the testing, the time between the request for a page and obtaining it was found to be acceptable.
- Security:
  - Access control: The portal has mechanisms to identify (asking for username and password) and authenticate (has GridSphere authentication modules) users. Moreover, it has the capacity to authorize certain users to use certain resources.
  - Security control: Not evaluable.

  - Confidentiality: Not evaluable.
  - Integrity: users cannot carry out unauthorized actions.
- Empathy:
  - Navigation: The navigation is simple and intuitive.
  - Presentation: The interface is clear and uniform.
  - Integration: All the components of the Grid portal appear in a coherent, integrated form.
  - Personalization: The portal can adapt to the user's priorities.
- Data and information files quality:
  - Intrinsic:
    - From the point of view of data: Not evaluable.
    - From the point of view of information files: Not evaluable.
  - Representation:
    - From the point of view of data: During the testing, the data were presented in an organized form.
    - From the point of view of information files: Not evaluable.
  - Contextual:
    - From the point of view of data: the information obtained during the testing satisfied our needs.
    - From the point of view of information files: Not evaluable.
  - Accessibility:
    - From the point of view of data: users do not influence the manner in which data are inputted, processed or presented to them.
    - From the point of view of information files: Not evaluable.

We must take into account the fact that we have carried out the assessment from the point of view of the end user. That being so, we do not have all the necessary data, so the conclusions obtained from applying G-PQM are not as definitive as they should be. However, we can see that the main characteristics which must be improved are: adaptability (because the number of minimum requirements is excessive and this makes it impossible to adapt the portal to an arbitrary device) and data accessibility (because users cannot influence the way in which data are inputted, processed or presented to them). The rest of the characteristics which have been assessed, have given a favourable result. It would likewise be interesting to obtain more information related to the portal, for the purpose of detecting other weak points. We could thereby improve portal quality.

### 3.2 OGCE Portal

Secondly, we have applied the model to the OGCE portal, whose objective is to create an environment that facilitates the use of Grid resources. The results obtained from applying G-PQM are:

- Tangible:
  - Adaptability: The minimum requirements are: 500 MB free hard-disk space, Pentium III or higher (or a similarly capable processor)_and 128 MB free RAM.

- o Transparent access: OGCE Port (release 2) has Grid portlets which manage remote files, execute remote commands, etc. Furthermore, this portal has inter-portlet communication tools that allow portlets to share data.
- Reliability:
  - o Fault tolerance: Not evaluable.
  - o Availability: The portal was available anywhere and anytime.
  - o Search Quality: Not applicable because the portal does not have a search engine.
  - o Quality in the use of resources: Not evaluable.
- Responsiveness:
  - o Scalability: The portal is not limited to a specific number of users.
  - o Efficient access: The response time was very high in some testing, and the request was not even met in some instances.
- Security:
  - o Access control: The portal has mechanisms to identify (asking for username and password) and authenticate (has GridSphere authentication modules) users. Moreover, it has the capacity to authorize certain users to use certain resources.
  - o Security control: Not evaluable.
  - o Confidentiality: Not evaluable.
  - o Integrity: users cannot carry out unauthorized actions.
- Empathy:
  - o Navigation: The navigation is simple and intuitive.
  - o Presentation: The interface is clear and uniform.
  - o Integration: All the components of the OGCE portal are integrated in a coherent way.
  - o Personalization: The portal is capable of adapting itself to the user's priorities.
- Data and information files quality:
  - o Intrinsic:
    - ▪ From the point of view of data: Not evaluable.
    - ▪ From the point of view of information files: Not evaluable.
  - o Representation:
    - ▪ From the point of view of data: During the testing, the data were presented in an organized form.
    - ▪ From the point of view of information files: Not evaluable.
  - o Contextual:
    - ▪ From the point of view of data: the information obtained during the testing satisfied our needs.
    - ▪ From the point of view of information files: Not evaluable.
  - o Accessibility:
    - ▪ From the point of view of data: users do not influence the way in which data are inputted, processed or presented to them.
    - ▪ From the point of view of information files: Not evaluable.

As with the previous case, we have applied our model from the point of view of the end user, so there are some dimensions which cannot be assessed. However, taking into account the dimensions we *have* assessed, we can see that the following tasks to improve portal quality could be carried out: reduction of the number of minimum requirements, so as to allow the portal to adapt itself to any device; improvement of the efficiency of access; and above all, avoidance of a request not obtaining an answer and elimination of the appearance of a blank screen. On the other hand, we have obtained favourable results for the rest of the characteristics we have assessed. It will also be of interest to us to obtain information related to the dimensions which have not been assessed.

## 4 Conclusions and Future Work

Nowadays, many scientists require the use of the Grid to conduct their computational research. However, its use is not a trivial task. For this reason, and with the aim of allowing an easy access to Grid resources via a Web browser interface, Grid portals have come into existence.

Many different Grid portals can be found at the present time. Therefore, it is easy for users to move from one Grid portal to another, without the user wasting time and money. Thus, for users to be attracted to a particular Grid portal and come back regularly, the portal must offer a good level of quality.

Bearing all this in mind, a quality model for Grid portals, namely G-PQM, has been presented. This model can be used, on the one hand, to assess the quality level of a specific Grid portal, and on the other hand, to identify its weakness and define corrective actions which improve its level of quality. In addition, this model has been applied to two grid portals and some corrective actions have been defined in order to improve their level of quality.

Future work includes the validation of the model characteristics through surveys. In addition, measures for each one of the characteristics and sub-characteristics must be identified. Thereby, the G-PQM will be finished.

## Acknowledgements

This work was conducted when the first author was in stage at the University of Cardiff and is part of the CALIPO (TIC 2003-07804-C05-03) and DIMENSIONS (PBC-05-012-1) projects and the CALIPSO network (TIN2005-24055-E).

## References

1. Cox, J., Dale, B.G.: Service quality and e-commerce: an exploratory analysis. Managing Service Quality 11(2), 121–131 (2001)
2. Dahan, M., Thomas, M., Roberts, E., Seth, A., Urban, T., Walling, D., Boisseau, J.R.: Grid Portal Toolkit 3.0 (GridPort). In: 13th IEEE International Symposium on High Performance Distributed Computing (HPDC 2004), pp. 272–273 (2004)
3. Dedeke, A., Kahn, B.: Model-Based quality evaluation: a comparison of Internet classified operated by newspapers and non-newspaper firms. In: Proceedings of the Seventh International Conference on Information Quality, pp. 142–154 (2002)
4. Firesmith, D.: Specifying Reusable Security Requirements. Journal of Object Technology 3(1), 61–75 (2004)

5. Foster, I., Kesselman, C., Tuecke, S.: The Anatomy of the Grid. International J. Supercomputer Applications 15(3), 200–222 (2001)
6. GridPort. Retrieved 2006, from `http://gridport.net/main/`
7. Gurugé, A.: Corporate Portals Empowered with XML and Web Services. Digital Press, Amsterdam (2003)
8. He, G., Xu, Z.: Design and Implementation of a Web-based Computational Grid Portal. In: IEEE/WIC International Conference on Web Intelligence (WI 2003), pp. 478–481 (2003)
9. ISO, ISO/IEC 9126. Software Engineering-Product Quality. Parts1 to 4., International Organization for Standardization/International Electrotechnical Commission (2001)
10. Li, M., Baker, M.: The Grid: Core Technologies, John Willey & Sons England (2005)
11. Li, M., van Santen, P., Walker, D.W., Rana, O.F., Baker, M.A.: PortalLab: A Web Services Toolkit for Building Semantic Grid Portals. In: 3rd IEEE/ACM International Symposium on Cluster Computing and the Grid (CCGRID 2003), pp. 190–197 (2003)
12. Lin, M., Walker, D.W.: A Portlet Service Model for GECEM. In: Proceedings of the UK e-Science All Hands Meeting 2004, pp. 687–694 (2004)
13. Moraga, M.Á., Calero, C., Piattini, M.: Applying PQM to a Regional Portal. 5th Conference for Quality in Information and Communications Technology. In: Quatic 2004, Porto, Portugal, pp. 65–70. Instituto Português da Qualidade (2004a)
14. Moraga, M.Á., Calero, C., Piattini, M.: A first proposal of a portal quality model. In: IADIS International Conference. E-society 2004. International association for development of the information society (iadis), 'Avila, Spain, vol. 1(2), pp. 630–638 (2004b), ISBN: 972-98947-5-2
15. Offutt, A.J.: Quality attributes of web software applications. IEEE Software 19(2), 25–32 (2002)
16. Singh, M.: E-services and their role in B2C e-commerce. Managing Service Quality 12(6), 434–446 (2002)

# Algorithmic Skeletons for Branch and Bound

Michael Poldner and Herbert Kuchen

University of Münster, Department of Information Systems
Leonardo Campus 3, D-48149 Münster, Germany
poldner@wi.uni-muenster.de, kuchen@uni-muenster.de

**Abstract.** Algorithmic skeletons are predefined components for parallel programming. We will present a skeleton for branch & bound problems for MIMD machines with distributed memory. This skeleton is based on a distributed work pool. We discuss two variants, one with supply-driven work distribution and one with demand-driven work distribution. This approach is compared to a simple branch & bound skeleton with a centralized work pool, which has been used in a previous version of our skeleton library Muesli. Based on experimental results for two example applications, namely the $n$-puzzle and the traveling salesman problem, we show that the distributed work pool is clearly better and enables good runtimes and in particular scalability. Moreover, we discuss some implementation aspects such as termination detection as well as overlapping computation and communication.

**Keywords:** Parallel Computing, Algorithmic Skeletons, Branch & Bound, Load Distribution, Termination Detection.

## 1 Introduction

Today, parallel programming of MIMD machines with distributed memory is mostly based on message-passing libraries such as MPI [1, 2]. The resulting low programming level is error-prone and time consuming. Thus, many approaches have been suggested, which provide a higher level of abstraction and an easier program development. One such approach is based on so-called *algorithmic skeletons* [3, 4], i.e. typical patterns for parallel programming which are often offered to the user as higher-order functions. By providing application-specific parameters to these functions, the user can adapt an application independent skeleton to the considered parallel application. (S)he does not have to worry about low-level implementation details such as sending and receiving messages. Since the skeletons are efficiently implemented, the resulting parallel application can be almost as efficient as one based on low-level message passing.

Algorithmic skeletons can be roughly divided into data parallel and task parallel ones. Data-parallel skeletons (see e.g. [5, 6, 7, 8, 9, 10]) process a distributed data structure such as a distributed array or matrix as a whole, e.g. by applying a function to every element or by rotating or permuting its elements. Task-parallel skeletons [11, 12, 13, 14, 9, 10, 15] construct a system of processes communicating via streams of data. Such a system is mostly generated by nesting typical building blocks such as farms and pipelines. In the present paper, we will focus on a particular task-parallel skeleton, namely a branch & bound skeleton.

J. Filipe, B. Shishkov, and M. Helfert (Eds.): ICSOFT 2006, CCIS 10, pp. 204–219, 2008.

Branch & bound [16] is a well-known and frequently applied approach to solve certain optimization problems, among them integer and mixed-integer linear optimization problems [16] and the well-known traveling salesman problem [17]. Many practically important but NP-hard planning problems can be formulated as (mixed) integer optimization problems, e.g. production planning, crew scheduling, and vehicle routing. Branch & bound is often the only practically successful approach to solve these problems exactly. In the sequel we will assume without loss of generality that an optimization problem consists of finding a solution value which minimizes an objective function while observing a system of constraints. The main idea of branch & bound is the following. A problem is recursively divided into subproblems and lower bounds for the optimal solution of each subproblem are computed. If a solution of a (sub)problem is found, it is also a solution of the overall problem. Then, all other subproblems can be discarded, whose corresponding lower bounds are greater than the value of the solution. Subproblems with smaller lower bounds still have to considered recursively.

Only little related work on algorithmic skeletons for branch & bound can be found in the literature [18, 19, 20, 13]. However, in the corresponding literature there is no discussion of different designs. The MaLLBa implementation is based on a master/worker scheme and it uses a central queue (rather than a heap) for storing problems. The master distributes problems to workers and receives their solutions and generated subproblems. On a shared memory machine this approach can work well. We will show in the sequel that a master/worker approach is less suited to handle branch & bound problems on distributed memory machines. In a previous version of the Muesli skeleton library, a branch & bound skeleton with a centralized work pool has bee used, too [14]. Hofstedt outlines a B&B skeleton with a distributed work pool. Here, work is only shared, if a local work pool is empty. Thus, worthwhile problems are not propagated quickly and their investigation is concentrated on a few workers only.

The rest of this paper is structured as follows. In Section 2, we recall, how branch & bound algorithms can be used to solve optimization problems. In Section 3, we introduce different designs of branch & bound skeletons in the framework of the skeleton library Muesli [9, 10, 21]. After describing the simple centralized design considered in [14], we will focus on a design with a distributed work pool. Section 4 contains experimental results demonstrating the strengths and weaknesses of the different designs. In Section 5, we conclude and point out future work.

## 2 Branch and Bound

Branch & bound algorithms are general methods used for solving difficult combinatorial optimization problems. In this section, we illustrate the main principles of branch & bound algorithms using the 8-puzzle, a simplified version of the well-known 15-puzzle [22], as example. A branch & bound algorithm searches the complete solution space of a given problem for the best solution. Due to the exponentially increasing number of feasible solutions, their explicit enumeration is often impossible in practice. However, the knowledge about the currently best solution, which is called *incumbent*, and the use of *bounds* for the function to be optimized enables the algorithm to search parts of the solution space only implicitly. During the solution process, a pool of yet

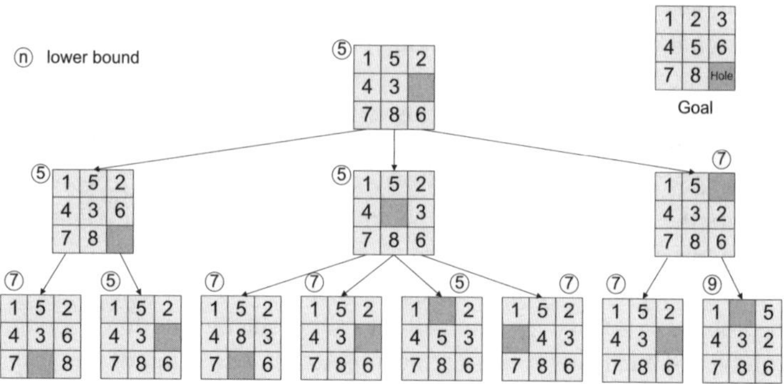


**Fig. 1.** Upper part of the state-space tree corresponding to an instance of the 8-puzzle and its goal board

unexplored subsets of the solution space, called the *work pool*, describes the current status of the search. Initially there is only one subset, namely the complete solution space, and the best solution found so far is infinity. The unexplored subsets are represented as nodes in a dynamically generated search tree, which initially only contains the root, and each iteration of the branch & bound algorithm processes one such node. This tree is called the *state-space tree*. Each node in the state-space tree has associated data, called its *description*, which can be used to determine, whether it represents a *solution* and whether it has any successors. A branch & bound problem is solved by applying a small set of basic rules. While the signature of these rules is always the same, the concrete formulation of the rules is problem dependent. Starting from a given initial problem, subproblems with pairwise disjoint state spaces are generated using an appropriate *branching rule*. A generated subproblem can be estimated applying a *bounding rule*. Using a *selection rule*, the subproblem to be branched from next is chosen from the work pool. Last but not least subproblems with non-optimal or inadmissible solutions can be eliminated during the computation using an *elimination rule*. The sequence of the application of these rules may vary according to the strategy chosen for selecting the next node to process [23]. As an example of the branch and bound technique, consider the 8-puzzle [22]. Figure 1 illustrates the goal state of the 8-puzzle and the first three levels of the state-space tree.

The 8-puzzle consists of eight tiles, numbered 1 through 8, arranged on a $3 \times 3$ board. Eight positions on the board contain exactly one tile and the remaining position is empty. The objective of the puzzle is to repeatedly fill the hole with a tile adjacent to it in horizontal or vertical direction, until the tiles are in row major order. The aim is to solve the puzzle in the least number of moves.

The branching rule describes, how to split a problem represented by a given initial board into subproblems represented by the boards resulting after all valid moves. A minimum number of tile moves needed to solve the puzzle can be estimated by adding the number of tile moves made so far to the Manhattan distance between the current position of each tile and its goal position. The computation of this lower bound is described by the bounding rule.

The state-space tree represents all possible boards that can be reached from the initial board. One way to solve this puzzle is to pursue a breadth first search or a depth first search of the state-space tree until the sorted board is discovered. However, we can often reach the goal faster by selecting the node with the best lower bound to branch from. This selection rule corresponds to a best-first search strategy. Other selection rules such as a variant of depth-first search are discussed in [23, 24, 25].

Branch & bound algorithms can be parallelized at a low or at a high level. In case of a low-level parallelization, the sequential algorithm is taken as a starting point and just the computation of the lower bound, the selection of the subproblem to branch from next, and/or the application of the elimination rule are performed by several processes in a data parallel way. The overall behavior of such a parallel algorithm resembles of the sequential algorithm.

In case of a high-level parallelization, the effects and consequences of the parallelism are not restricted to a particular part of the algorithm, but influence the algorithm as a whole. Several iterations of the main loop are performed in a task-parallel way, such that the state-space tree is explored in a different (non-deterministic!) order than in the sequential algorithm.

## 3 Branch and Bound Skeletons

In this section, we will consider different implementation and design issues of branch & bound skeletons. For the most interesting distributed design, several work distribution strategies are discussed and compared with respect to scalability, overhead, and performance. Moreover, a corresponding termination detection algorithm is presented.

A B&B skeleton is based on one or more branch & bound algorithms and offers them to the user as predefined parallel components. Parallel branch & bound algorithms can be classified depending on the organization of the work pool. A central, distributed, and hybrid organization can be distinguished. In the MaLLBa project, a central work pool is used [19, 20]. Hofstedt [13] sketches a distributed scheme, where work is only delegated, if a local work pool is empty. Shinano et al. [24, 25] and Xu et al. [26] describe hybrid approaches. A more detailed classification can be found in [27], where also complete and partial knowledge bases, different strategies for the use of knowledge and the division of work as well as the chosen synchronicity of processes are distinguished.

Moreover, different selection rules can be fixed. Here, we use the classical best-first strategy. Let us mention that this can be used to simulate other strategies such as the depth-first approach suggested by Clausen and Perregaard [23]. The bounding function just has to depend on the depth in the state-space tree.

We will consider the skeletons in the context of the skeleton library Muesli [9,10,21]. Muesli is based on MPI [1, 2] internally in order to inherit its platform independence.

### 3.1 Design with a Centralized Work Pool Manager

The simplest approach is a kind of the master/worker design as depicted in Figure 2. The work pool is maintained by the master, which distributes problems to the workers and receives solutions and subproblems from them. The approach taken in a previous

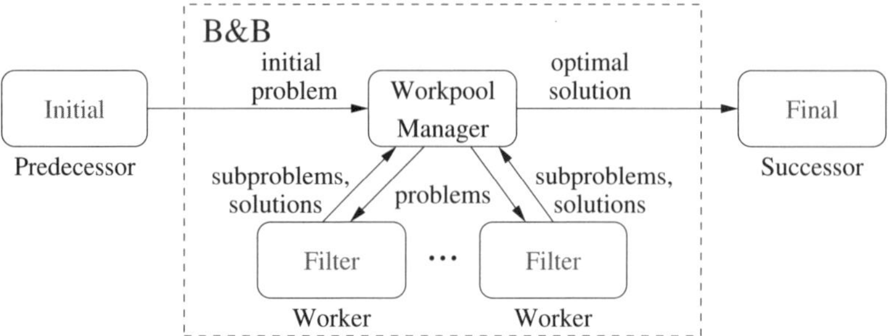


**Fig. 2.** Branch & bound skeleton with centralized work pool manager

version of the skeleton library Muesli is based on this centralized design. When a worker receives a problem, it either solves it or decomposes it into subproblems and computes a lower bound for each of the subproblems. The work pool is organized as a heap, and the subproblem with the best lower bound at the time is stored in its root. Idle workers are served with new problems taken from the root. This selection rule implicitly implements a best-first search strategy. Subproblems are discarded, if their bounds indicate that they cannot produce better solutions than the best known solution. An optimal solution is found, if the master has received a solution, which is better than all the bounds of all the problems in its work pool and no worker currently processes a subproblem. If at least one worker is processing, it can lead to a new incumbent. When the execution is finished, the optimal solution is sent to the master's successor in the overall process topology [1] and the skeleton is ready to accept and solve the next optimization problem. The code fragment in Fig. 3 illustrates the application of our skeleton in the context of the Muesli library. It constructs the process topology shown in Fig. 2.

```
int main(int argc, char* argv[]) {
   InitSkeletons(argc,argv);
   // step 1: create a process topology
   Initial<Problem> initial(generateProblem);
   Filter<Problem,Problem> filter(generateCases,1);
   BranchAndBound<Problem> bnb(filter,n,
                               betterThan,isSolution);
   Final<Problem> final(fin);
   Pipe pipe(initial,bnb,final);
   // step 2: start process topology
   pipe.start();
   TerminateSkeletons();
}
```

**Fig. 3.** Example application using a branch and bound skeleton with centralized work pool manager

[1] Remember that task-parallel skeletons can be nested.

In a first step the process topology is created using C++ constructors. The process topology consists of an initial process, a branch & bound process, and a final process connected by a pipeline skeleton. The initial process is parameterized with a `generateProblem` method returning the initial optimization problem that is to be solved. The filter process represents a worker. The passed function `generateCases` describes, how to branch & bound subproblems. The constructor `BranchAndBound` produces $n$ copies of the worker and connects them to the internal work pool manager (which is not visible to the user). `bool betterThan(Problem x1, Problem x2)` has to deliver `true`, iff the lower (upper) bound for the best solution of problem `x1` is better than the lower (upper) bound for the best solution of problem `x2` in case of a minimization (maximization) problem. This function is used internally for the work pool organization. The function `bool isSolution(Problem x)` can be used to discover, whether its argument `x` is a solution or not. The final process receives and processes the optimal solution. Problems and solutions are encoded by the same type `Problem`.

The advantage of a single central work pool maintained by the master is that it provides a good overall picture of the work still to be done. This makes it easy to provide each worker with a good subproblem to branch from and to prune the work pool. Moreover, the termination of the workers is easy to implement, because the master knows about all idle workers at any time, and the best solution can be detected easily. The disadvantage is that accessing the work pool tends to be a bottleneck, as the work pool can only be accessed by one worker at a time. This may result in high idle times on the workers' site. Another disadvantage is that the master/worker approach incurs high communication costs, since each subproblem is sent from its producer to the master and propagated to its processing worker. If the master decides to eliminate a received subproblem, time is wasted for its transmission. Moreover, the communication time required to send a problem to a worker and to receive in return some subproblems may be greater than the time needed to do the computation locally. The master's limited memory capacity for maintaining the work pool is another disadvantage of this architecture.

As we will see in the next subsection, these disadvantages can be avoided by a distributed maintenance of the work pool. However, this design requires a suitable scheme for distributing subproblems and some distributed termination detection.

### 3.2 Distributed Work Pool

Figure 5 illustrates the design of the distributed branch and bound (DBB) skeleton provided by the Muesli skeleton library. It consists of a set of peer solvers, which exchange problems, solutions, and (possibly) load information. Several topologies for connecting the solvers are possible. For small numbers of processors, a ring topology can be used, since it enables an easy termination detection. For larger numbers of processors, topologies like torus or hypercube may lead to a faster propagation of work from hot spots to idle processors. For simplicity, we will assume a ring topology in the sequel. Compared to more complicated topologies the ring also simplifies the dynamic adaption of the number of workers in case that more or less computation capacity has to be devoted to the branch & bound skeleton within the overall computation. This (not yet implemented) feature will enable a well-balanced overall computation.

In our example, $n = 5$ solvers are used. Each solver maintains its own local work pool and has one entrance and one exit. Exactly one of the solvers, called the *master solver*, serves as an entrance to the DBB-skeleton and receives new optimization problems from the predecessor. Any of the $n$ solvers may deliver the detected optimal solution to the successor of the branch & bound skeleton in the overall process topology. All solvers know each other for a fast distribution of newly detected best solutions[2]. If the skeleton only consists of a single solver neither communication nor distributed termination detection are necessary. In this case all communication parts as well as the distributed termination detection algorithm are bypassed to speed up the computation.

The code fragment in Fig. 4 shows an example application of our distributed B&B skeleton. It constructs the process topology depicted in Fig. 5. Work request messages are only sent when using a demand-driven work distribution.

```
int main(int argc, char* argv) {
   InitSkeletons(argc,argv);
   // step 1: create a process topology
   Initial<Problem> initial(generateProblem);
   BBSolver<Problem> solver("ring",branch,bound,
                            betterThan,isSolution);
   DistributedBB<Problem> bnb =
                  DistributedBB<Problem>(solver,n);
   Final<Problem> final(fin);
   Pipe pipe(initial,bnb,final);
   // step 2: start process topology
   pipe.start();
   TerminateSkeletons();
}
```

**Fig. 4.** Task parallel example application of a fully distributed Branch and Bound skeleton

The construction of the process topology resembles that in the previous example. Instead of a filter a `BBSolver` process is used as a worker. In addition to the `better Than` and `isSolution` function two other argument functions are passed to the constructor, namely a `branch` and a `bound` function. The constructor `DistributedBB` produces $n$ copies of the solver. One of the solvers is automatically chosen as the master solver.

As described in the previous section, a task-parallel skeleton consumes a stream of input values and produces a stream of output values. If the master solver receives a new optimization problem, the communication with the predecessor is blocked until the received problem is solved. This ensures that the skeleton processes only one optimization problem at a time. There are different variants for the initialization of parallel branch & bound algorithms with the objective of providing each worker with a certain amount of work within the start-up phase. Ideally, the work load is distributed equally to all workers. However, the work load is hard to predict without any domain knowledge. For this reason the skeleton uses the most common approach, namely *root initialization*, i.e. the

[2] Thus, the topology is in fact a kind of wheel with spokes rather than a ring.

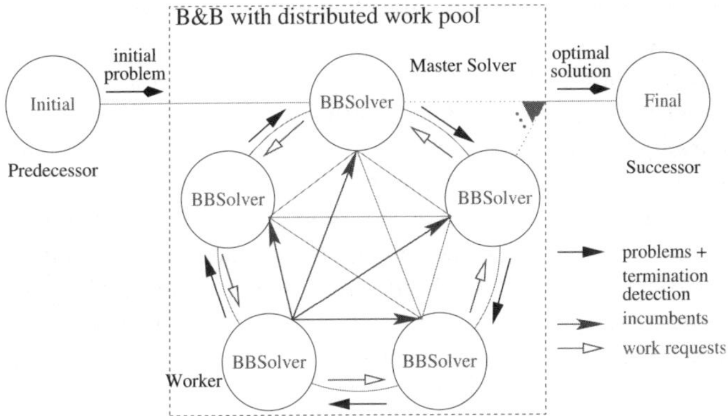


**Fig. 5.** Branch & bound skeleton with distributed work pool

root of the state space tree is inserted into the local work pool of the master solver. Subproblems are distributed according to the load balancing scheme applied by the solvers. This initialization has the advantage that it is very easy to implement and no additional code is necessary. Other initialization strategies are discussed in the literature. A good survey can be found in [28].

Each worker repeatedly executes two phases: a communication phase and a solution phase. Let us first consider the communication phase. In order to avoid that computation time is wasted with the solution of irrelevant subproblems, it is essential to spread and process new best solutions as quickly as possible. For this reason, we distinguish problem messages and incumbent messages. Each solver first checks for arriving incumbents with MPI_Testsome. If it has received new incumbents, the solver stores the best and discards the others. Moreover, it removes subproblems whose lower bound is worse than the incumbent from the work pool. Then, it checks for arriving subproblems and stores them in the work pool, if their lower bounds are better than the incumbent.

The solution phase starts with selecting an unexamined subproblem from the work pool. As in the master/worker design, the work pool is organized as a heap and the selection rule implements a best-first search strategy. The selected problem is decomposed into $m$ subproblems by applying `branch`. For each of the subproblems, we proceed as follows. First, we check, whether it is solved. If a new best solution is detected, we update the local incumbent and broadcast it. A worse solution is discarded. Finally, if the subproblem is not yet solved, the `bound` function is applied and the subproblem is stored in the work pool (see Fig. 5).

### 3.3 Load Distribution and Knowledge Sharing

Since the work pools of the different solvers, grow and shrink differently, some load balancing mechanism is required. Many global and local load distribution schemes have been studied in the literature [29, 30, 31, 32, 33, 34] and many of them are suited in the context of a distributed branch & bound skeleton. Here, we will focus on two local load balancing schemes, a supply- and a demand-driven one. The local schemes avoid the larger overhead of a global scheme. On the other hand, they need more time to distribute work over long distances.

With the simple supply-driven scheme, each worker sends in each $i$th iteration its second best problem to its right neighbor in the ring topology. It always processes the best problem itself, in order to avoid communication overhead compared to the sequential algorithm. The supply driven approach has the advantage that it distributes work slightly more quickly than a demand driven approach, since there is no need for work requests. This may be beneficial in the beginning of the computation. A major disadvantage of this approach is that many subproblems are transmitted in vain, since they will be sooner or later discarded at their destination due to better incumbents, in particular for small $i$. Thus, high communication costs are caused.

The demand-driven approach distributes load only in case that a neighbor requests it. In our case, a neighbor sends the lower bound of the best problem in its work pool (see Fig. 5). If this value is worse than the lower bound of the second best problem of the worker receiving this information, it is interpreted as a work request and a problem is transmitted to the neighbor. In case that the work pool of the neighbor is empty, the information message indicates this fact rather than transmitting a lower bound. An information message is sent every $i$th iteration of the main loop. In order to avoid flooding the network with "empty work pool" messages, such messages are never sent twice. If the receiver of an "empty work pool message" is idle, too, it stores this request and serves it as soon as possible. The advantage of this algorithm is that distributing load only occurs, if it is necessary and beneficial. The overhead of sending load information messages is very low due to their small sizes. For small $i$ the overhead is bigger, but idle processors get work more quickly.

### 3.4 Termination Detection

In the distributed setting, it is harder to detect that the computation has finished and the optimal solution has been found. The termination detection algorithm used in the DBB-skeleton is a variant of Dijkstra's algorithm outlined in [22]. Our implementation utilizes the specific property of MPI that the order in which messages are received from a sender $S$ is always equal to the order in which they were sent by $S$. This characteristic can be used for the purpose of termination detection in connection with local load distribution strategies as described above.

As mentioned, we arrange the workers in a ring topology, since this renders the termination detection particularly easy and simplifies the dynamic addition and removal of workers. For a small number of processors (as in our system), the large diameter of the ring topology is no serious problem for the distribution of work.

Let $n$ be the number of solvers of the DBB-skeleton. When the master solver receives a new optimization problem, it initializes the termination detection by sending a token along the ring in the same direction as the load is distributed. The token only consists of an `int` value. Initially, the token has the value $n$. If a solver receives a new subproblem, this event is noted by setting a flag to `true`. On arrival of a token the solver uses the rules stated by the following pseudo code:

```
IF (workpool is empty AND flag == false)
    token := token - 1;
IF (workpool is not empty OR flag == true) {
    token := n; flag := false; }
IF (token > 0) send token to successor;
IF (token == 0) computation is finished;
```

Only if all workers are idle, the token is decremented by every worker and the computation is finished. No more problems can be in the network, since the token cannot overtake other messages on its way. Note that this algorithm only works for load balancing strategies which send load in the same direction as the token.

## 4 Experimental Results

We have tested the different versions of the branch & bound skeleton experimentally on a IBM workstation cluster [35] using up to 16 Intel Xeon EM64T processors with 3.6 GHz, 1 MB L2 Cache, and 4 GB memory, connected by a Myrinet [36]. As example applications we have considered the $n$-puzzle as explained in section 2 as well as a parallel version of the traveling salesman problem (TSP) algorithm by Little et al. [17]. Both differ w.r.t. the quality of their bounding functions and hence in the number of considered irrelevant subproblems.

The presented B&B algorithm for the $n$-puzzle has a rather bad bounding function based on the Manhattan distance of each tile to its destination. It is bad, since the computed lower bounds are often much below the value of the best solution. As a consequence, the best-first search strategy is not very effective and the number of problems considered by the parallel skeleton differs enormously over several runs with the same inputs. This number largely depends on the fact whether a subproblem leading to the optimal solution is picked up early or late. Note that the parallel algorithm behaves non-deterministically in the way the search-space tree is explored. In order to get reliable results, we have repeated each run 100 times and computed the average runtimes.

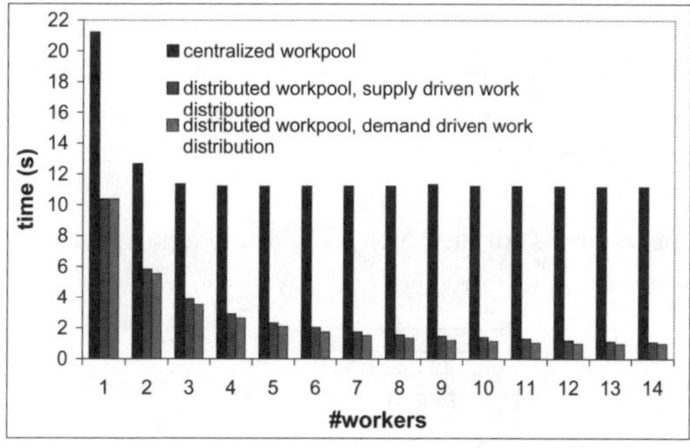


**Fig. 6.** Runtimes for the 16 city TSP using the central work pool manager and the distributed work pool with supply- and demand-driven work distribution depending on the number of workers

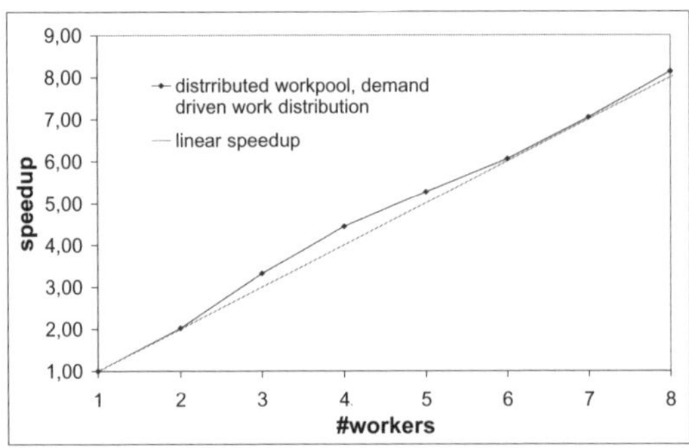


**Fig. 7.** Speedups for the 30 city TSP using the distributed work pool with demand-driven work distribution depending on the number of workers. The speedups are the averages taken from 300 runs with different, randomly generated maps.

The goal of the TSP is to find the shortest round trip through $n$ cities. Little's algorithm represents each problem by its residual adjacency matrix, a set of chosen edges representing a partially completed tour, and a lower bound on the length of any full tour, which can be generated by extending the given partial tour. New problems are produced by selecting a *key edge* and generating two new problems, in which the chosen edge is included and excluded from the emerging tour, respectively. The key edge is selected based on the impact that the exclusion of the edge will have on the lower bound. The lower bounds are computed based on the fact that each city has to be entered and left once and that consequently one value in every row and column of the adjacency matrix has to be picked. The processing of a problem mainly requires three passes through the adjacency matrix.

The TSP algorithm computes rather precise lower bounds. Thus, the best-first strategy works fine, and the parallel implementation based on Quinn's formulation of the algorithm [22] considers only very few problems more than the sequential algorithm, as explained below.

Consequently, the runtimes were relatively similar over several runs with the same parameters. For the TSP, we have used a real world 16 city map taken and adapted from [37] and 300 randomly generated 30 city maps. The real world map has much more sub-tours with similar lengths. Thus, proportionally more subproblems are processed which do not lead to the optimal solution than for the artificial map, where the best solution is found more easily.

**Table 1.** Distribution of problems for the 16 city TSP using a distributed work pool and demand driven work distribution

| #workers | runtime (s) | # considered problems | | | | | | | | |
|---|---|---|---|---|---|---|---|---|---|---|
| | | | worker | | | | | | | |
| | | total | 1 | 2 | 3 | 4 | 5 | 6 | 7 | 8 |
| 1 | 10.38 | 263019 | 263019 | | | | | | | |
| 2 | 5.54 | 274002 | 139922 | 134080 | | | | | | |
| 3 | 3.52 | 263583 | 90783 | 86039 | 86761 | | | | | |
| 4 | 2.64 | 262794 | 66536 | 65141 | 65475 | 65642 | | | | |
| 5 | 2.10 | 273175 | 55863 | 52386 | 55993 | 53878 | 55055 | | | |
| 6 | 1.74 | 270525 | 45916 | 45150 | 44938 | 45574 | 42638 | 46309 | | |
| 7 | 1.52 | 263180 | 39495 | 38749 | 37492 | 37197 | 37134 | 36273 | 36840 | |
| 8 | 1.35 | 265698 | 34196 | 33763 | 33525 | 32793 | 32424 | 32466 | 32231 | 34300 |

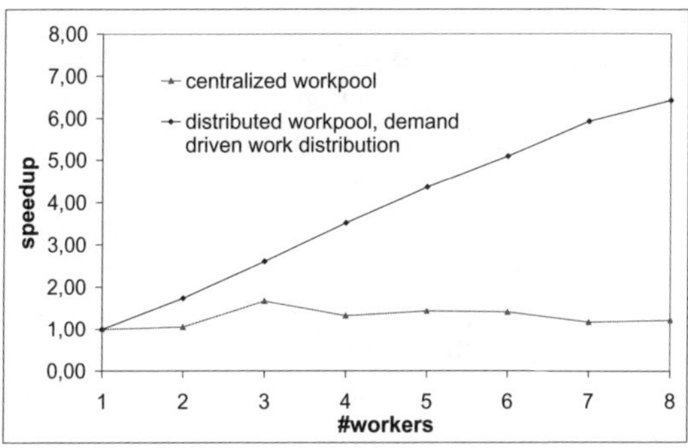


**Fig. 8.** Speedups for 24-puzzle using the central work pool manager and the distributed work pool with demand-driven work distribution depending on the number of workers

**Table 2.** Distribution of problems for the 16 city TSP using a distributed work pool and supply driven work distribution

| #workers | runtime (s) | # considered problems | | | | | | | | |
|---|---|---|---|---|---|---|---|---|---|---|
| | | total | worker | | | | | | | |
| | | | 1 | 2 | 3 | 4 | 5 | 6 | 7 | 8 |
| 1 | 10.38 | 263019 | 263019 | | | | | | | |
| 2 | 5.84 | 262522 | 162536 | 99986 | | | | | | |
| 3 | 3.92 | 271179 | 93060 | 89886 | 88233 | | | | | |
| 4 | 2.91 | 269021 | 66004 | 67709 | 66572 | 68736 | | | | |
| 5 | 2.32 | 271717 | 53161 | 54569 | 55420 | 55074 | 53493 | | | |
| 6 | 2.03 | 265100 | 43227 | 47420 | 47739 | 42342 | 42185 | 42187 | | |
| 7 | 1.75 | 265862 | 34390 | 34701 | 35100 | 36693 | 37917 | 51595 | 35466 | |
| 8 | 1.35 | 264509 | 44379 | 32157 | 29789 | 29228 | 30017 | 31704 | 33124 | 34111 |

When comparing the supply- and the demand-driven approach (see Figure 6 and the 3rd columns of Tables 1, 2), we notice that, as expected, the demand driven scheme is better, since it produces less communication overhead. The fact that the problems are distributed slightly slower causes no serious performance penalty.

For the supply driven scheme, we have used an optimal number $i$ for the amount of iterations that a worker waits before delegating a problem to a neighbor. If $i$ is chosen too large, important problems will not spread out fast enough. If $i$ is too small, the communication overhead will be too large. We found that the optimal value for $i$ depends on the application problem and on the number of workers. If the number of workers increases, $i$ has to be increased as well. In our experiments, the optimal values for $i$ were ranging between 2 and 20 for up to 8 workers.

**Table 3.** Distribution of problems for the 16 city TSP using a central work pool manager

| #workers | runtime (s) | # considered problems | | | | | | | | |
|---|---|---|---|---|---|---|---|---|---|---|
| | | total | worker | | | | | | | |
| | | | 1 | 2 | 3 | 4 | 5 | 6 | 7 | 8 |
| 1 | 22.01 | 263018 | 263018 | | | | | | | |
| 2 | 12.74 | 267057 | 133808 | 133249 | | | | | | |
| 3 | 11.41 | 267019 | 116339 | 104349 | 46331 | | | | | |
| 4 | 11.26 | 267030 | 115396 | 103522 | 45945 | 2167 | | | | |
| 5 | 11.26 | 267039 | 116064 | 103735 | 45406 | 1712 | 106 | | | |
| 6 | 11.25 | 267199 | 116082 | 103679 | 45470 | 1791 | 123 | 54 | | |
| 7 | 11.26 | 267050 | 114111 | 103167 | 46558 | 2767 | 319 | 89 | 39 | |
| 8 | 11.25 | 267024 | 115671 | 103675 | 45227 | 2071 | 226 | 81 | 45 | 28 |

As expected, we see that for the centralized B&B skeleton the work pool manager quickly becomes a bottleneck and it has difficulties to keep more than 2 workers busy (see Figures 6, 8 and Table 3). This is due to the fact that the amount of computations done for a problem is linear in the size of the problem, just as the communication complexity for sending and receiving a problem. Thus, relatively little is gained by delegating a problem to a worker. The work pool manager has to spend only little work less for transmitting the problem than its processing would require. This property is typical for virtually all practically relevant branch & bound problems we are aware of. It has the important consequence that a centralized work pool manager does not work well for branch & bound on distributed memory machines. Also note that the centralized scheme needs one more processor, the work pool manager, than the distributed one rendering this approach even less attractive.

Both variants of the design with a distributed work pool do not have these drawbacks (see Figures 6, 7, 8 and Tables 1, 2). Here, the communication overhead is much smaller. Each worker fetches most problems from its own work pool, such that they require no communication. This is particularly true for the demand driven approach. This scheme has the advantage that after some start-up phase, in which all workers are supplied with problems, there is relatively little communication and the workers mainly process locally available problems. This is essential for achieving good runtimes and speedups. We anticipate that this insight not only applies to branch & bound but also to other skeletons with a similar characteristic such as divide & conquer and other search skeletons. We are currently working on experimental results supporting this claim.

Interestingly we could even observe slightly superlinear speedups for the 30 city TSPs. They can be explained by the fact that a parallel B&B algorithm may tackle important subproblems earlier than the sequential one, since it processes the state-space tree in a different order [38].

It is clear that a parallel B&B algorithm will typically consider more problems than a corresponding sequential one, since it eagerly processes several problems in parallel, which would be discarded in the sequential case, since their lower bounds are worse than a detected solution. Interestingly for both considered example applications, TSP and $n$-puzzle, the corresponding overhead was very small and only few additional problems have been processed by the parallel implementation (see the 3rd columns of Tables 1, 2, 3). For instance, for the 16 city TSP no more than $274002 - 263019 = 10983$ additional problems are processed by the parallel algorithm; this is less than 4.2 %. This is essential for achieving reasonable speedups.

As an implementation detail of the centralized approach let us mention that it is important that the work pool manager receives in each iteration all available subproblems and solutions from the workers rather than just one of them. The reason is that MPI_Waitany (used internally) is unfair and that an overloaded work pool manager will hence almost exclusively communicate with a small number of workers. If a starving worker has an important subproblem (one that leads to the optimal solution) or a good solution, which it is not able to deliver to the work pool manager, this will cause very bad runtimes.

Another implementation detail of the centralized approach concerns the amount of buffering. In order to be able to overlap computation and communication, it is a good idea that the work pool manager not only sends one problem to each worker and then waits for the results, but that it sends $m$ problems such that the worker can directly tackle the next problem after finishing the previous one. Here it turned out that one has to be careful not to choose $m$ too large, since then problems which would otherwise be discarded due to appearing better incumbents will be processed (in vain). In our experiments, $m = 2$ was a good choice.

## 5 Conclusions

We have considered two different implementation schemes for the branch & bound skeleton. Besides a simple approach with a central work pool manager, we have investigated a scheme with a distributed work pool. As our analysis and experimental results show, the communication overhead is high for the centralized approach and the work pool manager quickly becomes a bottleneck, in particular, if the number of computation steps for each problem grows linearly with the problem size, as it is the case for virtually all practically relevant branch & bound problems. Thus, the centralized scheme does not work well in practice.

On the other hand, our scheme with a distributed work pool works fine and provides good runtimes and scalability. The latter is not trivial, as discussed e.g. in the book of Quinn [22], since parallel B&B algorithms tend to process an increasing number of irrelevant problems the more processors are employed. In particular, the demand-driven design works well due to its low communication overhead.

For the supply-driven approach, we have investigated, how often a problem should be propagated to a neighbor. Depending on the application and the number of workers, we have observed the best runtimes, if a problem was delegated between every 2nd and every 20th iteration.

We are not aware of any previous comparison of different implementation schemes of branch & bound skeletons for MIMD machines with distributed memory in the literature. In the MaLLBa project [18, 19], a branch & bound skeleton based on a master/worker approach and a queue for storing subproblems has been developed. But as we pointed out above, this scheme is more suitable for shared memory machines than for distributed memory machines. Hofstedt [13] sketches a B&B skeleton with a distributed work pool. Here, work is only delegated, if a local work pool is empty. A quick propagation of "interesting" subproblems are missing. According to our experience, this leads to a suboptimal behavior. Moreover, Hofstedt gives only few experimental results based on reduction steps in a functional programming setting rather than actual runtimes and speedups.

As future work, we intend to investigate alternative implementation schemes of skeletons for other search algorithms and for divide & conquer.

## References

1. Gropp, W., Lusk, E., Skjellum, A.: Using MPI. MIT Press, Cambridge (1999)
2. MPI: Message passing interface forum, mpi. In: MPI: A Message-Passing Interface Standard (2006), `http://www.mpi-forum.org/docs/mpi-11-html/mpi-report.html`

3. Cole, M.: Algorithmic Skeletons: Structured Management of Parallel Computation. MIT Press, Cambridge (1989)
4. Cole, M.: The skeletal parallelism web page (2006), `http://homepages.inf.ed.ac.uk/mic/Skeletons/`
5. Bisseling, I.F.R.: Mondriaan sparse matrix partitioning for attacking cryptosystems – a case study. In: Proceedings of ParCo 2005, Malaga (to appear, 2005)
6. Botorog, G.H., Kuchen, H.: Efficient parallel programming with algorithmic skeletons. In: Fraigniaud, P., Mignotte, A., Bougé, L., Robert, Y. (eds.) Euro-Par 1996. LNCS, vol. 1123, pp. 718–731. Springer, Heidelberg (1996)
7. Botorog, G.H., Kuchen, H.: Efficient high-level parallel programming. Theoretical Computer Science 196, 71–107 (1998)
8. Kuchen, H., Plasmeijer, R., Stoltze, H.: Efficient distributed memory implementation of a data parallel functional language. In: Halatsis, C., Philokyprou, G., Maritsas, D., Theodoridis, S. (eds.) PARLE 1994. LNCS, vol. 817, Springer, Heidelberg (1994)
9. Kuchen, H.: A skeleton library. In: Monien, B., Feldmann, R.L. (eds.) Euro-Par 2002. LNCS, vol. 2400, pp. 620–629. Springer, Heidelberg (2002)
10. Kuchen, H.: Optimizing sequences of skeleton calls. In: Lengauer, C., Batory, D., Consel, C., Odersky, M. (eds.) Domain-Specific Program Generation. LNCS, vol. 3016, pp. 254–273. Springer, Heidelberg (2004)
11. Benoit, A., Cole, M., Hillston, J., Gilmore, S.: Flexible skeletal programming with eskel. In: Cunha, J.C., Medeiros, P.D. (eds.) Euro-Par 2005. LNCS, vol. 3648, pp. 761–770. Springer, Heidelberg (2005)
12. Cole, M.: Bringing skeletons out of the closet: A pragmatic manifesto for skeletal parallel programming. Parallel Computing 30(3), 389–406 (2004)
13. Hofstedt, P.: Task parallel skeletons for irregularly structured problems. In: Pritchard, D., Reeve, J.S. (eds.) Euro-Par 1998. LNCS, vol. 1470, pp. 676–681. Springer, Heidelberg (1998)
14. Kuchen, H., Cole, M.: The integration of task and data parallel skeletons. Parallel Processing Letters 12(2), 141–155 (2002)
15. Pelagatti, S.: Task and data parallelism in p3l. In: Rabhi, F.A., Gorlatch, S. (eds.) Patterns and Skeletons for Parallel and Distributed Computing, pp. 155–186. Springer, Heidelberg (2003)
16. Nemhauser, G.L., Wolsey, L.A.: Integer and combinatorial optimization. Wiley, Chichester (1999)
17. Little, J.D.C., Murty, K.G., Sweeny, D.W., Karel, C.: An algorithm for the traveling salesman problem. Operations Research 11, 972–989 (1963)
18. Alba, E., Almeida, F., et al.: Mallba: A library of skeletons for combinatorial search. In: Monien, B., Feldmann, R.L. (eds.) Euro-Par 2002. LNCS, vol. 2400, pp. 927–932. Springer, Heidelberg (2002)
19. Almeida, F., Dorta, I., et al.: Mallba: Branch and bound paradigm. In: Technical Report DT-01-2, University of La Laguna, Spain, Dpto. Estadistica, I.O. y Computacion (2001)
20. Dorta, I., Leon, C., Rodriguez, C., Rojas, A.: Parallel skeletons for divide and conquer and branch and bound techniques. In: Proc. 11th Euromicro Conference on Parallel, Distributed and Network-based Processing (PDP2003) (2003)
21. Kuchen, H.: The skeleton library web pages (2006), `http://www.wi.uni-muenster.de/PI/forschung/Skeletons/index.php`
22. Quinn, M.J.: Parallel Computing: Theory and Practice. McGraw-Hill, New York (1994)
23. Clausen, J., Perregaard, M.: On the best search strategy in parallel branch-and-bound: Best-first search versus lazy depth-first search search. Annals of Operations Research 90, 1–17 (1999)

24. Shinano, Y., Higaki, M., Hirabayashi, R.: A generalized utility for parallel branch and bound algorithms. In: Proc. 7th IEEE Symposium on Parallel and Distributed Processing, pp. 392–401. IEEE Computer Society Press, Los Alamitos (1995)
25. Shinano, Y., Higaki, M., Hirabayashi, R.: Control schemes in a generalized utility for parallel branch and bound algorithms. In: Proc. 11th International Parallel Processing Symposium, pp. 621–627. IEEE, Los Alamitos (1997)
26. Xu, Y., Ralphs, T., Ladyi, L., Salzman, M.: Alps: A framework for implementing parallel tree search algorithms. In: Proc. 9th INFORMS Computing Society Conference (2005)
27. Trienekens, H.: Parallel branch & bound algorithms. PhD Thesis, University of Rotterdam (1990)
28. Henrich, D.: Initialization of parallel branch-and-bound algorithms. In: Proc. 2nd International Workshop on Parallel Processing for Artificial Intelligence (PPAI-1993). Elsevier, Amsterdam (1994)
29. Henrich, D.: Local load balancing for data-parallel branch-and-bound. In: Proc. Massively Parallel Processing Applications and Development, pp. 227–234 (1994)
30. Henrich, D.: Lastverteilung fuer feinkoernig parallelisiertes branch-and-bound. PhD Thesis, TH Karlsruhe (1995)
31. Lüling, R., Monien, B.: Load balancing for distributed branch and bound algorithms. In: Proc. 6th International Parallel Processing Symposium (IPPS 1992), pp. 543–549. IEEE, Los Alamitos (1992)
32. Mahapatra, N., Dutt, S.: Adaptive quality equalizing: High-performance load balancing for parallel branch-and-bound across applications and computing systems. In: Proc. International Parallel Processing and Distributed Processing Symposium (IPDPS 1998) (1998)
33. Sanders, P.: Tree shaped computations as a model for parallel applications. In: Proc. Workshop on Application Based Load Balancing (ALV 1998), TU Munich (1998)
34. Shina, A., Kalé, L.: A load balancing strategy for prioritized execution of tasks. In: Lehrmann Madsen, O. (ed.) ECOOP 1992. LNCS, vol. 615, Springer, Heidelberg (1992)
35. ZIV: Ziv-cluster (2006), `http://zivcluster.uni-muenster.de/`
36. Myricom: The myricom homepage (2006), `http://www.myri.com/`
37. Reinelt, G.: Tsplib – a traveling salesman problem library. ORSA Journal on Computing 3, 376–384 (1991), `http://www.iwr.uni-heidelberg.de/groups/comopt/software/TSPLIB95/`
38. Lai, T., S.S.: Anomalies in parallel branch-and-bound algorithms. Communications of the ACM 27, 594–602 (1984)

# A Hybrid Topology Architecture for P2P File Sharing Systems

J.P. Muñoz-Gea, J. Malgosa-Sanahuja, P. Manzanares-Lopez, J.C. Sanchez-Aarnoutse, and A.M. Guirado-Puerta

Department of Information Technologies and Communications
Polytechnic University of Cartagena, Campus Muralla del Mar, 30202, Cartagena, Spain
{juanp.gea,josem.malgosa,pilar.manzanares,juanc.sanchez antonio.guirado}@upct.es

**Abstract.** Over the Internet today, there has been much interest in emerging Peer-to-Peer (P2P) networks because they provide a good substrate for creating data sharing, content distribution, and application layer multicast applications. There are two classes of P2P overlay networks: structured and unstructured. Structured networks can efficiently locate items, but the searching process is not user friendly. Conversely, unstructured networks have efficient mechanisms to search for a content, but the lookup process does not take advantage of the distributed system nature. In this paper, we propose a hybrid structured and unstructured topology in order to take advantages of both kind of networks. In addition, our proposal guarantees that if a content is at any place in the network, it will be reachable with probability one. Simulation results show that the behaviour of the network is stable and that the network distributes the contents efficiently to avoid network congestion.

**Keywords:** Peer-to-peer, structured networks, unstructured networks, application layer multicast.

## 1 Introduction

The main characteristic of an overlay network is that all the computer terminals that shape it are organized defining a new network structure overlayed to the existent one. They are purely distributed systems, and can be used in a lot of interesting fields: for example, to transmit multicast traffic in a unicast network (like Internet), technique known as Application Layer Multicast (ALM). However, the most popular overlay networks are peer-to-peer (P2P) networks, commonly used to efficiently download large amounts of information. In this last scenario there are two types of P2P overlay networks: structured and unstructured.

The technical meaning of structured is that the P2P overlay network topology is tightly controlled. Such structured P2P systems have a property that consistently assigns uniform random *NodeIDs* to the set of peers into a large space of identifiers. With this identifier, the overlay network places the terminal in a specific position into a graph. On the other hand, in unstructured P2P networks the terminals are located in the overlay network by one (or several) *rendez-vous* terminals with network management functions.

J. Filipe, B. Shishkov, and M. Helfert (Eds.): ICSOFT 2006, CCIS 10, pp. 220–229, 2008.

Although unstructured P2P networks require the presence of one controller (*rendez-vous*) at least, they have the advantage that the information searching process supports complex queries (it is a similar methodology to that used to search for information in Google and supports keyword and phrases searching). That does not happen when the P2P network is structured. In this case the advantages are that it enables efficient discovery of data items and it doesn't require any central controller. In addition, it is also much easier to reorganize when changes occur (registering and leaving terminals) and, consequently, the overlay network is more scalable and robust. Section 2 describes in depth the searching and location process of structured and unstructured networks.

In this work we try to design a file-sharing system that shares the advantages of both types of P2P networks. The users locate the contents in an unstructured way. If this search fails, the system will use an application layer multicast service (given by a structured P2P network), to locate the terminal that owns the searched information. It is necessary to remark that, with the system proposed in this work, the location of any existing content always success.

There are several proposals that try to support sophisticated search requirements, like [1][2][3]. These proposals organize P2P overlays into a hierarchy, and they have a high degree of complexity.

The remainder of the paper is organized as follows: Section 2 describes the main characteristics of both, unstructured and structured networks. Section 3 describes the system proposed in this paper in detail. Section 4 summarizes the more relevant contributions of the proposed solution. Section 5 shows the simulation results and finally, Section 6 concludes the paper.

## 2 P2P Overlay Networks

The topology in a P2P structured overlay network is algorithmly fixed. Both, the nodes and the contents, have assigned an identifier (*NodeID* and *Key* respectively) belonging to the same scope. These P2P systems use a *hash* function applied to a MAC or IP terminal address and to the data content respectively, to generate these identifiers. The overlay network organizes its peers into a graph that maps each data *Key* to a peer, so that content is placed not at random peers but at specified locations. This structured graph enables efficient discovery of data items using the given *Keys*: a lookup algorithm is defined and it is responsible for locating the content, knowing its identifier only. However, in its simple form, this class of systems does not support complex queries. They only support exact-match lookups: one needs to know the exact *Key* of a data item to locate the node(s) responsible for storing that item. In practice, however, P2P users often have only partial information for identifying these items and tend to submit broad queries (e.g., all the articles written by "John Smith") [4]. Some examples of P2P structured networks are: CAN, Chord, Tapestry, Kademlia and Viceroy [5], [6], [7].

Unstructured P2P networks are composed of nodes that are linked to the network without any previous knowledge of the topology. The terminals need to know beforehand the location of a central controller, also denoted *rendez-vous* point, responsible for including them within the overlay network and for storing their contents list. The overlay networks organize peers in a random graph in a flat or hierarchical manner (e.g., Super-Peers layer). The search requests are sent to the *rendez-vous* node, and this evaluates the query locally on its own content, and supports complex queries. If the content is not located in the *rendez-vous*, most of the available networks use flooding or random walks or expanding-ring Time-To-Live (TTL) search on the graph to query content stored by overlay peers. This is inefficient because queries for content that are not widely replicated must be sent to a large fraction of peers, and there is no coupling between topology and data items'location [8]. Some examples of P2P unstructured networks are: Gnutella, FastTrack/Kazaa, BitTorrent and eDonkey 2000 [8].

In sum, for a human being, the searching process is easier in an unstructured network, since this is made using patterns of very high level (like in Google, for example). Nevertheless, there exists much inefficiency in the location process of the content. In structured networks, exactly the opposite happens: the location is quasi-immediate, but the searching process is more tedious.

## 3 Description of the System

Our proposal tries to define a hybrid system. Therefore, the user can search contents using more or less general fy the searching criterion that content which he wishes to download, like in unstructured networks. Nevertheless, the network will be organized in a structured way, which will facilitate the location of the contents.

All the nodes are immersed in a structured overlay network (anyone of the previously mentioned types). In addition, the nodes divide automatically into different sub-groups, in a more or less uniform way, surrounding a *rendez-vous* node. This node has the best peformances in terms of CPU, bandwidth and reliability (see Section 3.2). When searching for a content, the user will send the search parameters to its *rendez-vous*, and this will return information about who has the contents in this sub-group.

All the *rendez-vous* nodes of the network are going to be members of a multicast group defined within the same structured network. This way, if the search fails, the *rendez-vous* node will send the request to the rest of *rendez-vous* nodes in a multicast way. Fig. 1 describes the general architecture of the system.

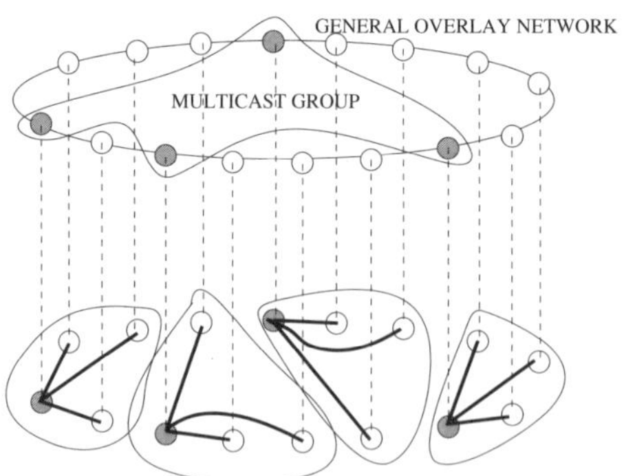


**Fig. 1.** General architecture of the system

### 3.1 Obtaining the Identifiers and Joining the General Network

Every node needs to obtain a *NodeID*. In this work, this identifier is obtained applying a hash function (MD5 or SHA-1) to its MAC or IP address. In the same way each node also needs to obtain a *SubgroupID* that identifies the sub-group to which the node is going to belong. We propose to use a previously well-known server to obtain this identifier. Each sub-group will have a maximum number of nodes, and the nodes will be assigned by order to each one of the sub-groups until completing their maximum capacity. When the existing sub-groups are completed new sub-groups will be created.

As is usual in any structured network a node needs to know at least one address of another node in the overlay network. The previous server can also provide this information. Finally, the node will have to link to the P2P overlay network, using the mechanism imposed by the structured network.

### 3.2 Joining the Sub-group

Each node of the sub-group will be able to establish a TCP connection with its *rendez-vous* node, and they will send their content list to it. Each sub-group is identified by a *SubgroupID*. Initially, a node looks for its *rendez-vous*. To do this, it uses the structured network to locate the node which *NodeID* fits with its *SubgroupID*. This node knows the IP address of the *rendez-vous* node of its sub-group. The last step consists of transmitting this information to the requester node. Note that in this way the system builds an unstructured network by using an underground structured network. In addition, this last property allows us to define the *rendez-vous* nodes dynamically and to guarantee the stability of the network throughout time.

### 3.3 Management of the Hierarchy

When the new node finds its *rendez-vous*, it notifies its resources of bandwidth and CPU. The *rendez-vous* nodes control the nodes that are linked to their sub-group and they form an ordered list of future *rendez-vous* candidates: the longer a node remains connected (and the better resources it has), the better candidate it becomes. This list is transmitted to all the members of the sub-group, and when the *rendez-vous* fails, the first node in the list becomes its successor. Later, it must inform all sub-group members that this node is now the new *rendez-vous*. Also, it must to modify this information in the node which *NodeID* fits with its *SubgroupID*.

### 3.4 Management of the *Rendez-Vous* Nodes

All the *rendez-vous* nodes are members of a multicast group defined at application level. When a node becomes *rendez-vous*, it must be linked to this multicast group, in order to spread the unsuccessful searches to the rest of sub-groups. Structured P2P networks can be used to implement an application layer multicast service, for example CAN-Multicast [9], Chord-Multicast [10] and Scribe [11]. Each one uses a different P2P overlay and it can implement the multicast service using flooding (CAN-Multicast, Chord-Multicast) or the construction of a tree (Scribe). Anyone of the previous methods provides an efficient mechanism to identify and to send messages to all the members of a group.

Our proposal uses Chord-Multicast. It is not necessary that the multicast process reaches all the group members before sending the searching results to the requester node. When one node responds affirmatively to a request it sends to the requester's *rendez-vous* the coincidences of the search in its database. Next, this *rendez-vous* gives back immediately the IP address and the corresponding metadata to the requester node. Therefore, the requester node obtains the searching results as soon as possible.

### 3.5 Registering the Shared Files

In a similar way to KaZaA, when a node establishes connection with its *rendez-vous* it sends the metadata of those files that it wants to share. This allows the *rendez-vous* to maintain a data base including the identifiers of the files that all the nodes of the sub-group are sharing and the corresponding IP address of the node that contains them. The information sent by the node includes the name of the file, its size and its description.

### 3.6 Search

When a user wishes to make the search of certain content, his node sends a request on the TCP connection established with its *rendez-vous*. For each coincidence of the search in the data base, the *rendez-vous* gives back the IP address and the corresponding metadata.

If the search fails, the user has the possibility of asking for to its *rendez-vous* node that tries to contact with other *rendez-vous*. The identification of those nodes is simple, since all belong to the same application layer multicast group.

## 4 Advantages of the System

Next we are going to describe some of the contributions of the system proposed in this work. First, it is necessary to emphasize that all the nodes are assigned to a sub-group and not to a server. The nodes are able to automatically find the *rendez-vous* responsible for their sub-group.

It is also necessary to emphasize that this system is able to manage the heterogeneity of the network too. The most stable nodes and those with better benefits will become *rendez-vous* nodes, which will increase the network performances.

On the other hand, the application layer multicast service provides an effective way to share information among *rendez-vous* nodes. In this way the maintenance of multicast group is practically made in an automatic mode. In addition, this guarantees that any content in the network can be located by any user.

Finally, the searches will be made in a simple way, similar to those made in current unstructured file-sharing applications.

## 5 Simulations

One of the advantages of this system, commented previously, is that any content present in the network could be located by any user. Nevertheless, the searches of contents present in the same sub-group will be faster and more efficient than when the searches need to use other *rendez-vous* nodes.

There are several interesting parameters that is necessary to quantify. First, the probability that the requested content is registered in the *rendez-vous* of the node's sub-group. Second, the evolution of the previous parameter throughout time. Since the users are making successive searches of contents in other sub-groups of the network, these automatically will be registered in their own *rendez-vous* node, increasing the value of this probability. Finally, it is also interesting to find out the average number of *rendez-vous* nodes that will be consulted in order to locate a content.

In order to quantify the previous parameters a simulator in C language has been programmed. The contents are classified in three classes based on the degree of interest that they can motivate in the users ("very interesting", "interesting" and "of little interest"). At the beginning, the available contents are distributed in a random way among all the nodes of the network. As has been mentioned before, the *rendez-vous* share information using a Chord-Multicast procedure.

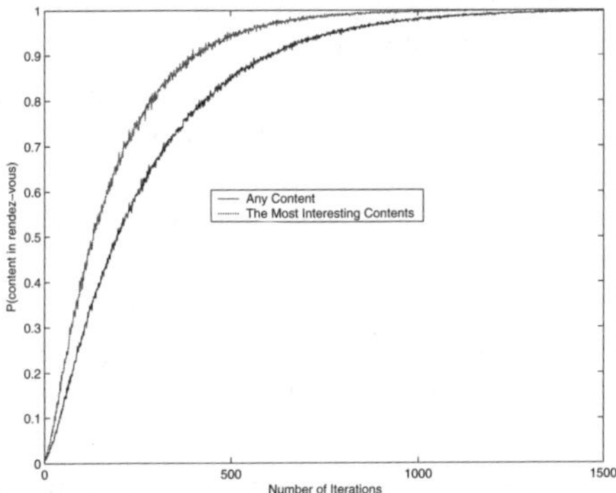


**Fig. 2.** Probability that a content is located in the same sub-group as the requester node

The simulation results show the probability that a content is located in the same sub-group as the requester node, as well as the average and maximum number of *rendez-vous* consulted until content location. All these results are obtained based on the number of simulation iterations. In each one, all the nodes of the network ask for a content that they do not have.

Figures 2, 3 and 4 present the simulation results corresponding to a network with 12,800 different contents and 6,400 nodes, with 128 *rendez-vous* nodes.

Fig. 2 shows the probability that the content is in the same requester's sub-group, for both the most interesting contents and for any content. It is observed that this probability grows as the number of iterations increases, but converging to a value of one, which assures that our system is stable. This also indicates that our architecture assures that, in a few steps, the contents will be equally distributed among all the sub-groups. It is also possible to observe that in the transitory, the probability of finding an interestig content in the *rendez-vous* increases more quickly than the probability of finding any content.

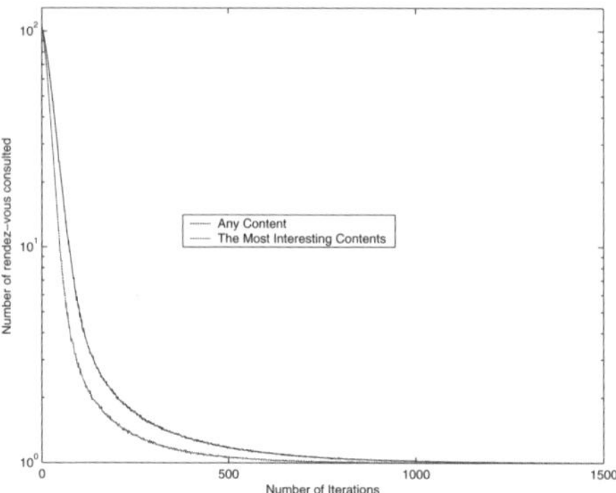


**Fig. 3.** Average number of *rendez-vous.* consulted until content location (in semilogarithmic scale).

Fig. 3 shows the average number of *rendez-vous* consulted to find a content. It is observed that the number of consulted *rendez-vous* quickly decreases, and when the number of iterations reaches 500 this value converges to one, which indicates that the content is in the same sub-group as the requester node. This shows us that the load coming from other sub-groups is minimal. It is also observed that this parameter decreases more quickly in the case of the more interesting contents than in the case of other contents.

Finally, Fig. 4 shows, in linear scale, the maximum number of *rendez-vous* consulted to locate any content. This parameter oscillates a lot in the initial transitory, but when it finishes it converges to values near the unit, agreeing practically with the average number.

Next, we are going to check the effect that both the number of contents and the number of *rendez-vous* nodes have on the probability that a content is located in the same requester's sub-group. Figure 5 shows the previous probability but with 6,400 and 19,200 contents. It can be observed that when the number of contents in the network

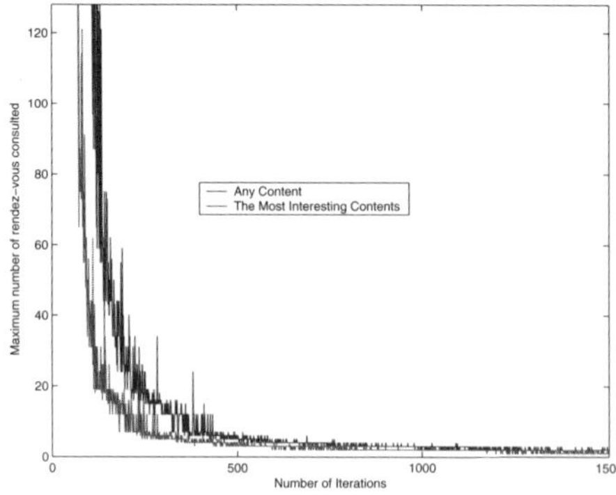


**Fig. 4.** Maximum number of *rendez-vous.* consulted until content location (in linear scale).

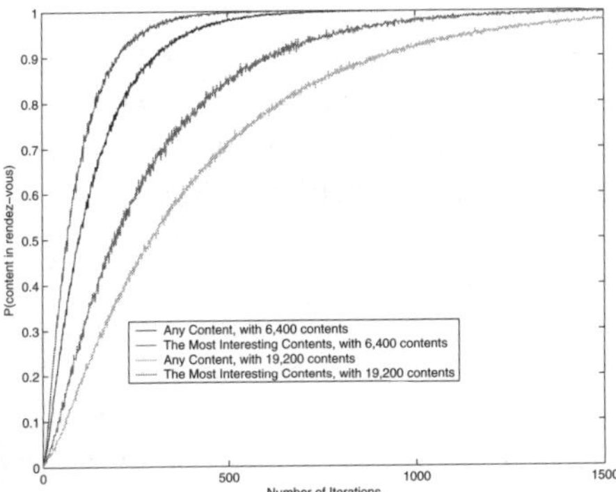


**Fig. 5.** Probability that a content is located in the same sub-group as the requester node, with 6,400 and 19,200 contents

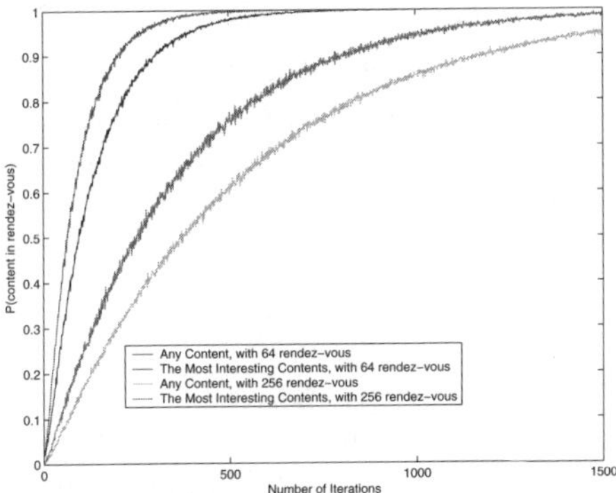


**Fig. 6.** Probability that a content is located in the same sub-group as the requester node, with 64 and 256 *rendez-vous*

diminishes the probability of finding it in the same requester's sub-group increases more quickly. On the other hand, when the number of contents in the network increases, a greater number of iterations is needed for the previous probability to reach the value of one.

Besides, Figure 6 shows the previous probability in a similiar network but with 64 and 256 *rendez-vous* nodes. It can be observed that the effect of the number of *rendez-vous* on this probability is quite similar to the effect of the number of contents. When the number of *rendez-vous* diminishes, the probability of finding a content in the same requester's sub-group increases more quickly. On the other hand, when the number of *rendez-vous* increases, a greater number of iterations is needed to obtain a probability close to one.

Next, we are going to compare the presented approach with the existing ones. In [1], peers are organized into groups, and each group has its autonomous intra-group structured overlay network and lookup service. Groups are organized in a top-level structured overlay network. To find a peer that it is responsible for a key, the top-level overlay first determines the group responsible for the key; the responsible group then uses its intra-group overlay to determine the specific peer that is responsible for the key. However, due to the use of structured networks this system does not support complex queries. The main advantage of this system is to reduce the expected number of hops that are required for a lookup, but in any case this is bigger than in our system, since in steady state only one hop is required.

In [2], they call an instance of a structured overlay as a *organizational ring*. A *multi-ring* protocol stitches together the *organizational rings* and implements a *global ring*. Each ring has a globally unique *ringID*, which is known by all the members of the ring. Every search message carries, in addition to a target key, the *ringID* in which the key is stored. Then, the node forwards the message in the global ring to the group that corresponds to the desired *ringId*. When a key is inserted into a organizational ring, it is necessary that a special indirection record is inserted into the global ring that associates the key with the *ringID* of the organizational ring where key is stored. However, the expected number of hops that are required for a lookup is similar to the previous work.

Finally, in [3], it is proposed the use of a universal ring, but it provides only bootstrap functionality while each service runs in a separate P2P overlay. The universal ring provides: an indexing service that enables users to find services of interest, a multicast service used to distributed software updates, a persistent store and distribution network that allows users to obtain the code needed to participate in a service's overlay and a service to provide users with a contact node to join a service overlay.

## 6 Conclusions

This paper presents a hybrid P2P overlay network that makes easier for the user both the searching process and the content location. The simulation results show that in this type of networks the contents are distributed in a way that minimizes the overload on the *rendez-vous* nodes.

We have also verified that an increase of both the number of *rendez-vous* and of contents increases the number of necessary iterations to guarantee that the content is located in the same requester's sub-group.

## Acknowledgements

This work has been supported by the Spanish Researh Council under project TEC2005-08068-C04-01/TCM and with funds of DG Technological Innovation and Information Society of Industry and Environment Council of the Regional Government of Murcia and with funds ERDF of the European Union.

## References

1. Garcés-Erice, L., Biersack, E.W., Ross, K.W., Felber, P.A., Urvoy-Keller, G.: Hierarchical peer-to-peer systems. In: Kosch, H., Böszörményi, L., Hellwagner, H. (eds.) Euro-Par 2003. LNCS, vol. 2790, pp. 1230–1239. Springer, Heidelberg (2003)

2. Mislove, A., Druschel, P.: Providing Administrative Control and Autonomy in Structured Peer-to-Peer Overlays. In: Voelker, G.M., Shenker, S. (eds.) IPTPS 2004. LNCS, vol. 3279, pp. 162–172. Springer, Heidelberg (2005)
3. Castro, M., Druschel, P., Kermarrec, A.M., Rowstron, A.: One ring to rule them all: Service discovery and binding in structured peer-to-peer overlay networks. In: Proceedings of the SIGOPS European Workshop, Saint-Emilion, France (2002)
4. Garcés-Erice, L., Felber, P.A., Biersack, E.W., Urvoy-Keller, G., Ross, K.W.: Data indexing in peer-to-peer dht networks. In: Proceedings of the 24th International Conference on Distributed Computing Systems, Tokyo, Japan, pp. 200–208 (2004)
5. Stoica, I., Morris, R., Liben-Nowell, D., Karger, D.R., Kaashoek, M.F., Dabek, F., Balakirshnan, H.: Chord: a scalable peer-to-peer lookup protocol for internet applications. IEEE/ACM Transactions on Networking 11(1), 17–32 (2003)
6. Zhao, B.Y., Huang, L., Stribling, J., Rhea, S.C., Joseph, A.D., Kubiatowicz, J.D.: Tapestry: A resilient global-scale overlay for service deployment. IEEE Journal on Selected Areas in Communications (JSAC) 22, 41–53 (2004)
7. Maymounkov, P., Mazières, D.: Kademlia: A Peer-to-Peer Information System Based on the XOR Metric. In: Druschel, P., Kaashoek, M.F., Rowstron, A. (eds.) IPTPS 2002. LNCS, vol. 2429, pp. 53–65. Springer, Heidelberg (2002)
8. Lua, K., Crowcroft, J., Pias, M., Sharma, R., Lim, S.: A survey and comparison of peer-to-peer overlay networks schemes. IEEE Communications Surveys & Tutorials 7, 72–93 (2005)
9. Ratsanamy, S., Handley, M., Karp, R., Shenker, S.: Application-level multicast using content-addressable networks. In: Proceedings of the Third International Workshop on Networked Group Communication, London, UK, pp. 14–29 (2001)
10. El-Ansary, S., Alima, L.O., Brand, P., Haridi, S.: Efficient broadcast in structured p2p networks. In: Kaashoek, M.F., Stoica, I. (eds.) IPTPS 2003. LNCS, vol. 2735, pp. 304–314. Springer, Heidelberg (2003)
11. Castro, M., Druschel, P., Kermarrec, A.-M., Rowstron, A.: Scribe: A large-scale and decentralized application-level multicast infrastructure. IEEE Journal on Selected Areas in Communications (JSAC) 20, 100–110 (2002)

# Parallel Processing of "Group-By Join" Queries on Shared Nothing Machines

M. Al Hajj Hassan and M. Bamha

LIFO, Université d'Orléans
B.P. 6759, 45067 Orléans Cedex 2, France
{mohamad.alhajjhassan,mostafa.bamha}@univ-orleans.fr

**Abstract.** SQL queries involving join and group-by operations are frequently used in many decision support applications. In these applications, the size of the input relations is usually very large, so the parallelization of these queries is highly recommended in order to obtain a desirable response time. The main drawbacks of the presented parallel algorithms that treat this kind of queries are that they are very sensitive to data skew and involve expansive communication and Input/Output costs in the evaluation of the join operation. In this paper, we present an algorithm that minimizes the communication cost by performing the group-by operation before redistribution where only tuples that will be present in the join result are redistributed. In addition, it evaluates the query without the need of materializing the result of the join operation and thus reducing the Input/Output cost of join intermediate results. The performance of this algorithm is analyzed using the scalable and portable BSP (Bulk Synchronous Parallel) cost model which predicts a near-linear speed-up even for highly skewed data.

**Keywords:** PDBMS, Parallel joins, Data skew, Join product skew, GroupBy-Join queries, BSP cost model.

## 1 Introduction

Data warehousing, On-Line Analytical Processing (OLAP) and other multidimensional analysis technologies have been employed by data analysts to extract interesting information from large database systems in order to improve the business performance and help the organisations in decision making. In these applications, aggregate queries are widely used to summarize large volume of data which may be the result of the join of several tables containing billions of records [1,2]. The main difficulty in such applications is that the result of these analytical queries must be obtained interactively [1,3] despite the huge volume of data in warehouses and their rapid growth especially in OLAP systems [1]. For this reason, parallel processing of these queries is highly recommended in order to obtain acceptable response time [4]. Research has shown that join, which is one of the most expansive operations in DBMS, is parallelizable with near-linear speed-up only in ideal cases [5]. However, data skew degrades the performance of parallel systems [6,5,7,8,9,10]. Thus, effective parallel algorithms that evenly distribute the load among processors and minimizes the inter-site communication must be employed in parallel and distributed systems in order to obtain acceptable performance.

J. Filipe, B. Shishkov, and M. Helfert (Eds.): ICSOFT 2006, CCIS 10, pp. 230–241, 2008.

In traditional algorithms that treat "GroupBy-Join" queries[1], join operations are performed in the first step and then the group-by operation [2,11]. But the response time of these queries is significantly reduced if the group-by operation is performed before the join [2], because group-by reduces the size of the relations thus minimizing the join and data redistribution costs. Several algorithms that perform the group-by operation before the join operation were presented in the literature [12,13,14,11].

In the "Early Distribution Schema" algorithm presented in [14], all the tuples of the tables are redistributed before applying the join or the group-by operations, thus the communication cost in this algorithm is very high. However, the cost of its join operation is reduced because the group-by is performed before the expansive join operation. In the second algorithm, "Early GroupBy Scheme" [14], the group-by operation is performed before the distribution and the join operations thus reducing the volume of data. But in this algorithm, all the tuples of the group-by results are redistributed even if they do not contribute in the join result. This is a drawback, because in some cases only few tuples of relations formed of million of tuples contribute in the join operation, thus the distribution of all these tuples is useless.

These algorithms fully materialize the intermediate results of the join operations. This is a significant drawback because the size of the result of this operation is generally large with respect to the size of the input relations. In addition, the Input/Output cost in these algorithms is very high where it is reasonable to assume that the output relation cannot fit in the main memory of each processor, so it must be reread in order to evaluate the aggregate function.

In this paper, we present a new parallel algorithm used to evaluate "GroupBy-Join" queries on Shared Nothing machines (a multiprocessors machine where each processor has its own memory and disks [15]). In this algorithm, we do not materialize the join operation as in the traditional algorithms where the join operation is evaluated first and then the group-by and aggregate functions [11]. So the Input/Output cost is minimal because we do not need to save the huge volume of data that results from the join operation.

We also use the histograms of both relations in order to find the tuples which will be present in the join result. After finding these tuples, we apply on them the grouping and aggregate function, in each processor, before performing the join. Using our approach, we reduce the size of data and communication costs to minimum. It is proved in [5,6], using the BSP model, that histogram management has a negligible cost when compared to the gain it provides in reducing the communication cost. In addition, Our algorithm avoids the problem of data skew because the hashing functions are only applied on histograms and not on input relations.

The performance of this algorithm is analyzed using the scalable and portable BSP cost model [16] which predicts for our algorithm a near-linear speed-up even for highly skewed data.

The rest of the paper is organized as follows. In section 2, we present the BSP cost model used to evaluate the processing time of the different phases of the algorithm. In section 3, we give an overview of different computation methods of "GroupBy-Join" queries. In section 4, we describe our algorithm. We then conclude in section 5.

[1] GroupBy-Join queries are queries involving group-by and join operations.

## 2 The BSP Cost Model

*Bulk-Synchronous Parallel* (BSP) cost model is a programming model introduced by L. Valiant [17] to offer a high degree of abstraction like PRAM models and yet allow portable and predictable performance on a wide variety of multi-processor architectures [18,16]. A BSP computer contains a set of processor-memory pairs, a communication network allowing inter-processor delivery of messages and a global synchronization unit which executes collective requests for a synchronization barrier. Its performance is characterized by 3 parameters expressed as multiples of the local processing speed:

- the number of processor-memory pairs $p$,
- the time $l$ required for a global synchronization,
- the time $g$ for collectively delivering a 1-relation (communication phase where each processor receives/sends at most one word). The network is assumed to deliver an $h$-relation in time $g * h$ for any arity $h$.

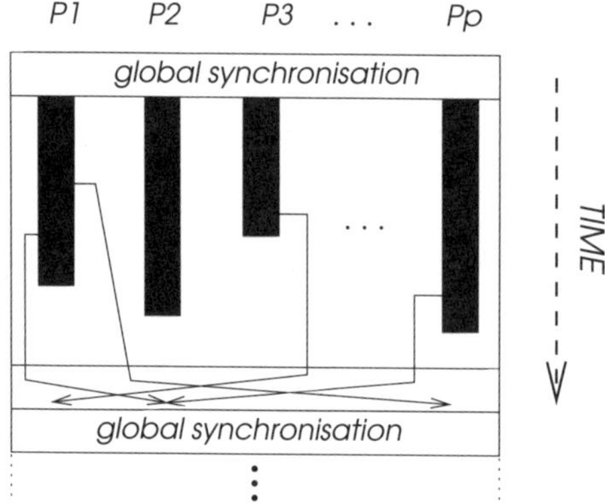


**Fig. 1.** A BSP superstep

A BSP program is executed as a sequence of *supersteps*, each one divided into (at most) three successive and logically disjoint phases. In the first phase each processor uses only its local data to perform sequential computations and to request data transfers to/from other nodes. In the second phase the network delivers the requested data transfers and in the third phase a global synchronization barrier occurs, making the transferred data available for the next superstep. The execution time of a superstep $s$ is thus the sum of the maximal local processing time, of the data delivery time and of the global synchronization time:

$$\text{Time}(s) = \max_{i:processor} w_i^{(s)} + \max_{i:processor} h_i^{(s)} * g + l$$

where $w_i^{(s)}$ is the local processing time on processor $i$ during superstep $s$ and $h_i^{(s)} = \max\{h_{i+}^{(s)}, h_{i-}^{(s)}\}$ where $h_{i+}^{(s)}$ (resp. $h_{i-}^{(s)}$) is the number of words transmitted (resp. received) by processor $i$ during superstep $s$. The execution time, $\sum_s \text{Time}(s)$, of a BSP program composed of $S$ supersteps is therefore a sum of 3 terms: $W + H*g + S*l$ where $W = \sum_s \max_i w_i^{(s)}$ and $H = \sum_s \max_i h_i^{(s)}$. In general $W$, $H$ and $S$ are functions of $p$ and of the size of data $n$, or (as in the present application) of more complex parameters like data skew and histogram sizes. To minimize execution time of a BSP algorithm, design must jointly minimize the number $S$ of supersteps and the total volume $h$ (resp. $W$) and imbalance $h^{(s)}$ (resp. $W^{(s)}$) of communication (resp. local computation).

## 3 Computation of "Group-By Join" Queries

In DBMS, the aggregate functions can be applied on the tuples of a single table, but in most SQL queries, they are applied on the output of the join of multiple relations. In the later case, we can distinguish two types of "GroupBy-Join" queries. We will illustrate these two types using the following example.

In this example, we have three relations that represent respectively Suppliers, Products and the quantity of a product shipped by a supplier in a specific date.

```
SUPPLIER (Sid, Sname, City)
PRODUCT (Pid, Pname, Category)
SHIPMENT (Sid, Pid, Date, Quantity)
```

**Query 1**

```
Select p.Pid, SUM (Quantity)
From PRODUCT as p, SHIPMENT as s
Where p.Pid = s.Pid
Group By p.Pid
```

**Query 2**

```
Select p.Category, SUM (Quantity)
From PRODUCT as p, SHIPMENT as s
Where p.Pid = s.Pid
Group By p.Category
```

The purpose of $Query1$ is to find the total quantity of every product shipped by all the suppliers, while that of $Query2$ is to find the total amount of every category of product shipped by all the suppliers.

The difference between $Query1$ and $Query2$ lies in the group-by and join attributes. In $Query1$, the join attribute ($Pid$) and the group-by attribute are the same. In this case, it is preferable to carry out the group-by operation first and then the join operation [13,14], because the group-by operation reduces the size of the relations to be joined. As a consequence, applying the group-by operation before the join operation in PDBMS[2] results in a huge gain in the communication cost and the execution time of the "GroupBy-Join" queries.

In the contrary, this can not be applied on Query 2, because the join attribute ($Pid$) is different from the group-by attribute ($category$).

[2] PDBMS : Parallel DataBase Management Systems.

In this paper, we focus on "GroupBy-Join" queries when the join attributes are part of the group-by attributes. In our algorithm, we succeeded to redistribute only tuples that will be present in the join result after applying the aggregate function. Therefore, the communication cost is reduced to minimum.

## 4 Presented Algorithm

In this section, we present a detailed description of our parallel algorithm used to evaluate "GroupBy-Join" queries when the join attributes are part of the group-by attributes. We assume that the relation $R$ (resp. $S$) is partitioned among processors by horizontal fragmentation and the fragments $R_i$ for $i = 1, ..., p$ are almost of the same size on each processor, i.e. $|R_i| \simeq \frac{|R|}{p}$ where $p$ is the number of processors.

For simplicity of description and without loss of generality, we consider that the query has only one join attribute $x$ and that the group-by attribute set consists of $x$, an attribute $y$ of $R$ and another attribute $z$ of $S$. We also assume that the aggregate function $f$ is applied on the values of the attribute $u$ of $S$. So the treated query is the following:

Select $R.x, R.y, S.z, f(S.u)$
From $R, S$
Where $R.x = S.x$
Group By $R.x, R.y, S.z$

In the rest of this paper, we use the following notation for each relation $T \in \{R, S\}$:

- $T_i$ denotes the fragment of relation $T$ placed on processor $i$,
- $Hist^w(T)$ denotes the histogram[3] of relation $T$ with respect to the attribute $w$, i.e. a list of pairs $(v, n_v)$ where $n_v \neq 0$ is the number of tuples of relation $T$ having the value $v$ for the attribute $w$. The histogram is often much smaller and never larger than the relation it describes,
- $Hist^w(T_i)$ denotes the histogram of fragment $T_i$,
- $Hist^w_i(T)$ is processor $i$'s fragment of the histogram of $T$,
- $Hist^w(T)(v)$ is the frequency $(n_v)$ of value $v$ in relation $T$,
- $Hist^w(T_i)(v)$ is the frequency of value $v$ in sub-relation $T_i$,
- $AGGR^w_{f,u}(T)$ [4] is the result of applying the aggregate function $f$ on the values of the aggregate attribute $u$ of every group of tuples of $T$ having identical values of the group-by attribute $w$. $AGGR^w_{f,u}(T)$ is formed of a list of tuples having the form $(v, f_v)$ where $f_v$ is the result of applying the aggregate function on the group of tuples having value $v$ for the attribute $w$ ($w$ may be formed of more than one attribute),
- $AGGR^w_{f,u}(T_i)$ denotes the result of applying the aggregate function on the attribute $u$ of the fragment $T_i$,

[3] Histograms are implemented as a balanced tree (B-tree): a data structure that maintains an ordered set of data to allow efficient search and insert operations.

[4] $AGGR^w_{f,u}(T)$ is implemented as a balanced tree (B-tree).

- $AGGR^{w}_{f,u,i}(T)$ is processor $i$'s fragment of the result of applying the aggregate function on $T$,
- $AGGR^{w}_{f,u}(T)(v)$ is the result $f_v$ of the aggregate function of the group of tuples having value $v$ for the group-by attribute $w$ in relation $T$,
- $AGGR^{w}_{f,u}(T_i)(v)$ is the result $f_v$ of the aggregate function of the group of tuples having value $v$ for the group-by attribute $w$ in sub-relation $T_i$,
- $\|T\|$ denotes the number of tuples of relation $T$, and
- $|T|$ denotes the size (expressed in bytes or number of pages) of relation $T$.

The algorithm proceeds in four phases. We will give an upper bound of the execution time of each superstep using BSP cost model. The notation $O(...)$ hides only small constant factors: *they depend only on the program implementation but neither on data nor on the BSP machine parameters*.

**Phase 1: Creating Local Histograms**
In this phase, the local histograms $Hist^{x}(R_i)_{i=1,...,p}$ (resp. $Hist^{x}(S_i)_{i=1,...,p}$) of blocks $R_i$ (resp. $S_i$) are created in parallel by a scan of the fragment $R_i$ (resp. $S_i$), on processor $i$, in time $c_{i/o} * \max_{i=1,...,p} |R_i|$ (resp. $c_{i/o} * \max_{i=1,...,p} |S_i|$) where $c_{i/o}$ is the cost of writing/reading a page of data from disk.

In addition, the local fragments $AGGR^{x,z}_{f,u}(S_i)_{i=1,...,p}$ of blocks $S_i$ are created on the fly while scanning relation $S_i$ in parallel, on each processor $i$, by applying the aggregate function $f$ on every group of tuples having identical values of the couple of attributes $(x, z)$. At the same time, the local histograms $Hist^{x,y}(R_i)_{i=1,...,p}$ are also created.
(In this algorithm the aggregate function may be $MAX, MIN, SUM$ or $COUNT$. For the aggregate $AVG$ a similar algorithm that merges the $COUNT$ and the $SUM$ algorithms is applied).

In principle, this phase costs:

$$Time_{phase1} = O\big(c_{i/o} * \max_{i=1,...,p} (|R_i| + |S_i|)\big).$$

**Phase 2: Creating the Histogram of $R \bowtie S$**
The first step in this phase is to create the histograms $Hist^{x}_{i}(R)$ and $Hist^{x}_{i}(S)$ by a parallel hashing of the histograms $Hist^{x}(R_i)$ and $Hist^{x}(S_i)$. After hashing, each processor $i$ merges the messages it received to constitute $Hist^{x}_{i}(R)$ and $Hist^{x}_{i}(S)$.

While merging, processor $i$ also retains a trace of the network layout of the values $d$ of the attribute $x$ in its $Hist^{x}_{i}(R)$ (resp. $Hist^{x}_{i}(S)$): this is nothing but the collection of messages it has just received. This information will help in forming the communication templates in phase 3.

The cost of redistribution and merging step is (cf. to proposition 1 in [19]):

$$\begin{aligned} &Time_{phase2.a} = \\ &O\Big(min\big(g * |Hist^{x}(R)| + \|Hist^{x}(R)\|, g * \frac{|R|}{p} + \frac{\|R\|}{p}\big) \\ &+ min\big(g * |Hist^{x}(S)| + \|Hist^{x}(S)\|, g * \frac{|S|}{p} + \frac{\|S\|}{p}\big) \\ &+ l\Big), \end{aligned}$$

where $g$ is the BSP communication parameter and $l$ the cost of a barrier of synchronisation.

We recall that, in the above equation, for a relation $T \in \{R, S\}$, the term $min(g * |Hist^x(T)| + ||Hist^x(T)||, g * \frac{|T|}{p} + \frac{||T||}{p})$ is the necessary time to compute $Hist^x_{i=1,\dots,p}(T)$ starting from the local histograms $Hist^x(T_i)_{i=1,\dots,p}$.

The histogram[5] $Hist^x_i(R \bowtie S)$ is then computed on each processor $i$ by intersecting $Hist^x_i(R)$ and $Hist^x_i(S)$ in time:

$$Time_{phase2.b} = O\Big(\max_{i=1,\dots,p} \big(min(||Hist^x_i(R)||, ||Hist^x_i(S)||)\big)\Big).$$

The total cost of this phase is:

$$\begin{aligned} &Time_{phase2} = Time_{phase2.a} + Time_{phase2.b} \\ &O\Big(min\big(g * |Hist^x(R)| + ||Hist^x(R)||, g * \frac{|R|}{p} + \frac{||R||}{p}\big) \\ &+ min\big(g * |Hist^x(S)| + ||Hist^x(S)||, g * \frac{|S|}{p} + \frac{||S||}{p}\big) \\ &+ \max_{i=1,\dots,p} \big(min(||Hist^x_i(R)||, ||Hist^x_i(S)||)\big) + l\Big). \end{aligned}$$

**Phase 3: Data Redistribution**

In order to reduce the communication cost, only tuples of $Hist^{x,y}(R)$ and $AGGR^{x,z}_{f,u}(S)$ that will be present in the join result will be redistributed.

To this end, we first compute on each processor $j$ the intersections $\overline{Hist}^{(j)x}(R_i) = Hist^{(j)x}(R_i) \cap Hist_j(R \bowtie S)$ and $\overline{Hist}^{(j)x}(S_i) = Hist^{(j)x}(S_i) \cap Hist_j(R \bowtie S)$ for $i = 1, ..., p$ where $Hist^{(j)x}(R_i)$ (resp. $Hist^{(j)x}(S_i)$) is the fragment of $Hist^x(R_i)$ (resp. $Hist^x(S_i)$) which was sent by processor $i$ to processor $j$ in the second phase.

The cost of this step is:

$$O(\sum_i ||Hist^{(j)x}(R_i)|| + \sum_i ||Hist^{(j)x}(S_i)||).$$

We recall that,
$\sum_i ||Hist^{(j)x}(R_i)|| = || \cup_i Hist^{(j)x}(R_i)|| \leq min(||Hist^x(R)||, \frac{||R||}{p})$
and
$\sum_i ||Hist^{(j)x}(S_i)|| = || \cup_i Hist^{(j)x}(S_i)|| \leq min(||Hist^x(S)||, \frac{||S||}{p})$,
thus the total cost of this step is:

$$Time_{phase3.a} = O\Big(min\big(||Hist^x(R)||, \frac{||R||}{p}\big) + min\big(||Hist^x(S)||, \frac{||S||}{p}\big)\Big).$$

Now each processor $j$ sends each fragment $\overline{Hist}^{(j)x}(R_i)$ (resp. $\overline{Hist}^{(j)x}(S_i)$) to processor $i$. Thus, each processor $i$ receives $\sum_j |\overline{Hist}^{(j)x}(R_i)| + \sum_j |\overline{Hist}^{(j)x}(S_i)|$ pages of data from the other processors.

[5] The size of $Hist(R \bowtie S) \equiv Hist(R) \cap Hist(S)$ is generally very small compared to $|Hist(R)|$ and $|Hist(S)|$ because $Hist(R \bowtie S)$ contains only values that appears in both relations $R$ and $S$.

In fact,$Hist^x(R_i)=\cup_j Hist^{(j)x}(R_i)$ and $|Hist^x(R_i)|=\sum_j |Hist^{(j)x}(R_i)| \geq \sum_j |Hist^{(j)x}(R_i) \cap Hist^x(R \bowtie S)|$, thus $|Hist^x(R_i)| \geq \sum_j |\overline{Hist}^{(j)x}(R_i)|$ (this also applies to $Hist^x(S_i)$).

Therefore, the total cost of this stage of communication is at most:

$$Time_{phase3.b} = O\Big(g * (|\overline{Hist}^x(R_i)| + |\overline{Hist}^x(S_i)|) + l\Big).$$

**Remark 1.** $\cup_j \overline{Hist}^{(j)x}(R_i)$ *is simply the intersection of* $Hist^x(R_i)$ *and the histogram* $Hist^x(R \bowtie S)$ *which will be noted:*

$$\overline{Hist}^x(R_i) = \cup_j \overline{Hist}^{(j)x}(R_i) = Hist^x(R_i) \cap Hist^x(R \bowtie S).$$

*Hence* $\overline{Hist}^x(R_i)$ *is only the restriction of the fragment of* $Hist^x(R_i)$ *to values which will be present in the join of the relations R and S. (this also applies to* $\overline{Hist}^x(S_i)$*).*

Now, each processor obeys all the distributing orders it has received, so only tuples of $\overline{Hist}^{x,y}(R_i) = Hist^{x,y}(R_i) \cap \overline{Hist}^x(R_i)$ and $\overline{AGGR}^{x,z}_{f,u}(S_i) = AGGR^{x,z}_{f,u}(S_i) \cap \overline{Hist}^x(S_i)$ are redistributed.

To this end, we first evaluate $\overline{Hist}^{x,y}(R_i)$ and $\overline{AGGR}^{x,z}_{f,u}(S_i)$. The cost of this step is of order:

$$Time_{phase3.c} = O\Big(\max_{i=1,\ldots,p} (||Hist^{x,y}(R_i)|| + ||AGGR^{x,z}_{f,u}(S_i)||)\Big),$$

which is the necessary time to traverse all the tuples of $Hist^{x,y}(R_i)$ and $AGGR^{x,z}_{f,u}(S_i)$ and access $\overline{Hist}^x(R_i)$ and $\overline{Hist}^x(S_i)$ respectively on each processor $i$.

Now, each processor $i$ distributes the tuples of $\overline{Hist}^{x,y}(R_i)$ and $\overline{AGGR}^{x,z}_{f,u}(S_i)$. After distribution, all the tuples of $\overline{Hist}^{x,y}(R_i)$ and $\overline{AGGR}^{x,z}_{f,u}(S_i)$ having the same values of the join attribute $x$ are stored on the same processor. So, each processor $i$ merges the blocks of data received from all the other processors in order to create $\overline{Hist}^{x,y}_i(R)$ and $\overline{AGGR}^{x,z}_{f,u,i}(S)$.

The cost of distributing and merging the tuples is of order (cf. to proposition 1 in [19]):

$$\begin{aligned} &Time_{phase3.d} = \\ &O\Big(min(g * |\overline{Hist}^{x,y}(R)| + ||\overline{Hist}^{x,y}(R)||, \\ &\qquad g * \frac{|R|}{p} + \frac{||R||}{p}) \\ &\quad + min(g * |\overline{AGGR}^{x,z}_{f,u}(S)| + ||\overline{AGGR}^{x,z}_{f,u}(S)||, \\ &\qquad g * \frac{|S|}{p} + \frac{||S||}{p}) + l\Big), \end{aligned}$$

where the terms:

$$min(g * |\overline{Hist}^{x,y}(R)| + ||\overline{Hist}^{x,y}(R)||, g * \frac{|R|}{p} + \frac{||R||}{p})$$

and

$$min(g * |\overline{AGGR}^{x,z}_{f,u}(S)| + ||\overline{AGGR}^{x,z}_{f,u}(S)||, g * \frac{|S|}{p} + \frac{||S||}{p})$$

represent the necessary time to compute $\overline{Hist_i}^{x,y}(R)$ and $\overline{AGGR}_{f,u,i}^{x,z}(S)$ starting from $\overline{Hist}^{x,y}(R_i)$ and $\overline{AGGR}_{f,u}^{x,z}(S_i)$ respectively.

The total time of the redistribution phase is:

$$
\begin{aligned}
&Time_{phase3} = \\
&O\Big(min\big(g * |\overline{Hist}^{x,y}(R)| + ||\overline{Hist}^{x,y}(R)||, \\
&\qquad g * \frac{|R|}{p} + \frac{||R||}{p}\big) + min\big(||Hist^{x}(R)||, \frac{||R||}{p}\big) \\
&+ min\big(g * |\overline{AGGR}_{f,u}^{x,z}(S)| + ||\overline{AGGR}_{f,u}^{x,z}(S)||, \\
&\qquad g * \frac{|S|}{p} + \frac{||S||}{p}\big) + min\big(||Hist^{x}(S)||, \frac{||S||}{p}\big) \\
&+ \max_{i=1,\dots,p} \big(||Hist^{x,y}(R_i)|| + ||AGGR_{f,u}^{x,z}(S_i)||\big) + l\Big).
\end{aligned}
$$

We mention that we only redistribute $\overline{Hist}^{x,y}(R_i)$ and $\overline{AGGR}_{f,u}^{x,z}(S_i)$ and their sizes are generally very small compared to $|R_i|$ and $|S_i|$ respectively. In addition, the size of $|Hist^{x}(R \bowtie S)|$ is generally very small compared to $|Hist^{x}(R)|$ and $|Hist^{x}(S)|$. Thus, we reduce the communication cost to minimum.

**Phase 4: Global Computation of the Aggregate Function**

In this phase, we compute the global aggregate function on each processor. We use the following algorithm where $AGGR_{f,u,i}^{x,y,z}(R \bowtie S)$ holds the final result on each processor $i$. The tuples of $AGGR_{f,u,i}^{x,y,z}(R \bowtie S)$ have the form $(x, y, z, v)$ where $v$ is the result of the aggregate function.

**Par** `(on each node in parallel)` $i = 1, ..., p$
  $AGGR_{f,u,i}^{x,y,z}(R \bowtie S)$ `= NULL` [6]
  **For every tuple** $t$ **of relation** $\overline{Hist_i}^{x,y}(R)$ **do**
    $freq = \overline{Hist_i}^{x,y}(R)(t.x, t.y)$
    **For every entry** $v_1$ = $\overline{AGGR}_{f,u,i}^{x,z}(S)(t.x, z)$ **do**
      `Insert a new tuple` $(t.x, t.y, z, f(v_1, freq))$
      `into` $AGGR_{f,u,i}^{x,y,z}(R \bowtie S)$`;`
    **EndFor**
  **EndFor**
**EndPar**

The time of this phase is: $O\big(\max_{i=1,\dots,p} ||AGGR_{f,u,i}^{x,y,z}(R \bowtie S)||\big)$, because the combination of the tuples of $\overline{Hist_i}^{x,y}(R)$ and $\overline{AGGR}_{f,u,i}^{x,z}(S)$ is performed to generate all the tuples of $AGGR_{f,u,i}^{x,y,z}(R \bowtie S)$.

**Remark 2.** *In practice, the imbalance of the data related to the use of the hash functions can be due to:*

- *a bad choice of the hash function used. This imbalance can be avoided by using the hashing techniques presented in the literature making it possible to distribute evenly the values of the join attribute with a very high probability [20],*

[6] This instruction creates a B-tree to store histogram's entries.

– *an intrinsic data imbalance which appears when some values of the join attribute appear more frequently than others. By definition a hash function maps tuples having the same join attribute values to the same processor. These is no way for a clever hash function to avoid load imbalance that result from these repeated values [10]. But this case cannot arise here owing to the fact that histograms contains only distinct values of the join attribute and the hashing functions we use are always applied to histograms.*

The global cost of evaluating the "GroupBy-Join" queries is of order:

$$
\begin{aligned}
&Time_{total} = O\Big(c_{i/o} * \max_{i=1,\ldots,p} (|R_i| + |S_i|) \\
&+ \max_{i=1,\ldots,p} ||AGGR^{x,y,z}_{f,u,i}(R \bowtie S)|| \\
&+ min(g * |Hist^{x}(R)| + ||Hist^{x}(R)||, g * \frac{|R|}{p} + \frac{||R||}{p}) \\
&+ min(g * |Hist^{x}(S)| + ||Hist^{x}(S)||, g * \frac{|S|}{p} + \frac{||S||}{p}) \\
&+ min\big(g * |\overline{Hist}^{x,y}(R)| + ||\overline{Hist}^{x,y}(R)||, \\
&\qquad g * \frac{|R|}{p} + \frac{||R||}{p}\big) \\
&+ min\big(g * |\overline{AGGR}^{x,z}_{f,u}(S)| + ||\overline{AGGR}^{x,z}_{f,u}(S)||, \\
&\qquad g * \frac{|S|}{p} + \frac{||S||}{p}\big) \\
&+ \max_{i=1,\ldots,p} (||Hist^{x,y}(R_i)|| + ||AGGR^{x,z}_{f,u}(S_i)||) + l\Big).
\end{aligned}
$$

**Remark 3.** *In the traditional algorithms, the aggregate function is applied on the output of the join operation. The sequential evaluation of the "groupBy-Join" queries requires at least the following lower bound:* $bound_{inf_1} = \Omega\big(c_{i/o} * (|R| + |S| + |R \bowtie S|)\big)$. *Parallel processing with* $p$ *processors requires therefore:* $bound_{inf_p} = \frac{1}{p} * bound_{inf_1}$.

*Using our approach in the evaluation of the "GroupBy-Join" queries, we only redistribute tuples that will be effectively present in the "groupBy-Join" result, which reduces the communication cost to minimum. This algorithm has an asymptotic optimal complexity because all the terms in* $Time_{total}$ *are bounded by those of* $bound_{inf_p}$.

## 5 Conclusions

The algorithm presented in this paper is used to evaluate the "GroupBy-Join" queries on Shared Nothing machines when the join attributes are part of the group-by attributes. Our main contribution in this algorithm is that we do not need to materialize the costly join operation which is necessary in all the other algorithms presented in the literature, thus we reduce its Input/Output cost. It also helps us to avoid the effect of data skew which may result from computing the intermediate join results and from redistributing all the tuples if AVS (Attribute Value Skew) exists in the relation. In addition, we partially evaluate the aggregate function before redistributing the data between processors or evaluating the join operation, because group-by and aggregate functions reduce the volume of data. To reduce the communication cost to minimum, we use the histograms to distribute only the tuples of the grouping result that will effectively be present in the output of the join operation. This algorithm is proved to have a near-linear speed-up, using the BSP cost model, even for highly skewed data. Our experience with the join operation [5,6,19] is evidence that the above theoretical analysis is accurate in practice.

## References

1. Datta, A., Moon, B., Thomas, H.: A case for parallelism in datawarehousing and OLAP. In: Ninth International Workshop on Database and Expert Systems Applications, DEXA 1998, pp. 226–231. IEEE Computer Society, Vienna (1998)
2. Chaudhuri, S., Shim, K.: Including Group-By in Query Optimization. In: Proceedings of the Twentieth International Conference on Very Large Databases, Santiago, Chile, pp. 354–366 (1994)
3. Tsois, A., Sellis, T.K.: The generalized pre-grouping transformation: Aggregate-query optimization in the presence of dependencies. In: VLDB, pp. 644–655 (2003)
4. Bamha, M.: An Optimal Skew-insensitive Join and Multi-join Algorithm for Distributed Architectures. In: Andersen, K.V., Debenham, J., Wagner, R. (eds.) DEXA 2005. LNCS, vol. 3588, pp. 616–625. Springer, Heidelberg (2005)
5. Bamha, M., Hains, G.: A Skew-Insensitive Algorithm for Join and Multi-join Operations on Shared Nothing Machines. In: Ibrahim, M., Küng, J., Revell, N. (eds.) DEXA 2000. LNCS, vol. 1873. Springer, Heidelberg (2000)
6. Bamha, M., Hains, G.: A frequency adaptive join algorithm for Shared Nothing machines. Journal of Parallel and Distributed Computing Practices (PDCP), 3(3), 333–345 (1999); appears also In: Columbus, F. (ed.) Progress in Computer Research, II, Nova Science Publishers (2001)
7. Seetha, M., Yu, P.S.: Effectiveness of parallel joins. IEEE, Transactions on Knowledge and Data Enginneerings 2, 410–424 (1990)
8. Hua, K.A., Lee, C.: Handling data skew in multiprocessor database computers using partition tuning. In: Lohman, G.M., Sernadas, A., Camps, R. (eds.) Proc. of the 17th International Conference on Very Large Data Bases, Barcelona, Catalonia, Spain, pp. 525–535. Morgan Kaufmann, San Francisco (1991)
9. Wolf, J.L., Dias, D.M., Yu, P.S., Turek, J.: New algorithms for parallelizing relational database joins in the presence of data skew. IEEE Transactions on Knowledge and Data Engineering 6, 990–997 (1994)
10. DeWitt, D.J., Naughton, J.F., Schneider, D.A., Seshadri, S.: Practical Skew Handling in Parallel Joins. In: Proceedings of the 18th VLDB Conference, Vancouver, British Columbia, Canada, pp. 27–40 (1992)
11. Yan, W.P., Larson, P.K.: Performing group-by before join. In: Proceedings of the 10th IEEE International Conference on Data Engineering, pp. 89–100. IEEE Computer Society Press, Los Alamitos (1994)
12. Shatdal, A., Naughton, J.F.: Adaptive parallel aggregation algorithms. SIGMOD Record (ACM Special Interest Group on Management of Data) 24, 104–114 (1995)
13. Taniar, D., Jiang, Y., Liu, K., Leung, C.: Aggregate-join query processing in parallel database systems, In: Proceedings of The Fourth International Conference/Exhibition on High Performance Computing in Asia-Pacific Region HPC-Asia 2000, vol. 2, pp. 824–829. IEEE Computer Society Press, Los Alamitos (2000)

14. Taniar, D., Rahayu, J.W.: Parallel processing of ’groupby-before-join’ queries in cluster architecture. In: Proceedings of the 1st International Symposium on Cluster Computing and the Grid, Brisbane, Qld, Australia, pp. 178–185. IEEE Computer Society Press, Los Alamitos (2001)
15. DeWitt, D.J., Gray, J.: Parallel database systems: The future of high performance database systems. Communications of the ACM 35, 85–98 (1992)
16. Skillicorn, D.B., Hill, J.M.D., McColl, W.F.: Questions and Answers about BSP. Scientific Programming 6, 249–274 (1997)
17. Valiant, L.G.: A bridging model for parallel computation. Communications of the ACM 33, 103–111 (1990)
18. Bisseling, R.H.: Parallel Scientific Computation: A Structured Approach using BSP and MPI. Oxford University Press, Oxford (2004)
19. Bamha, M., Hains, G.: An Efficient Equi-semi-join Algorithm for Distributed Architectures. In: Sunderam, V.S., van Albada, G.D., Sloot, P.M.A., Dongarra, J. (eds.) ICCS 2005. LNCS, vol. 3515, pp. 755–763. Springer, Heidelberg (2005)
20. Carter, J.L., Wegman, M.N.: Universal classes of hash functions. Journal of Computer and System Sciences 18, 143–154 (1979)

# Impact of Wrapped System Call Mechanism on Commodity Processors

Satoshi Yamada[1], Shigeru Kusakabe[1], and Hideo Taniguchi[2]

[1] Grad. School of Information Sci. & Electrical Eng., Kyushu University
6-10-1 Hakozaki, Higashi-ku, Fukuoka, 812-8581, Japan
satoshi@ale.csce.kyushu-u.ac.jp, kusakabe@csce.kyushu-u.ac.jp
[2] Faculty of Engineering, Okayama University
3-1-1 Tsushima-naka, Okayama, 700-8530, Japan
tani@cs.okayama-u.ac.jp

**Abstract.** Split-phase style transactions separate issuing a request and receiving the result of an operation in different threads. We apply this style to system call mechanism so that a system call is split into several threads in order to cut off the mode changes from system call execution inside the kernel. This style of system call mechanism improves throughput, and is also useful in enhancing locality of reference. In this paper, we call this mechanism as Wrapped System Call (WSC) mechanism, and we evaluate the effectiveness of WSC on commodity processors. WSC mechanism can be effective even on commodity platforms which do not have explicit multithread support. We evaluate WSC mechanism based on a performance evaluation model by using a simplified benchmark. We also apply WSC mechanism to variants of `cp` program to observe the effect on the enhancement of locality of reference. When we apply WSC mechanism to `cp` program, the combination of our split-phase style system calls and our scheduling mechanism is effective in improving throughput by reducing mode changes and exploiting locality of reference.

**Keywords:** System call, mode change, locality of reference.

## 1 Introduction

Although recent commodity processors are built based on a procedural sequential computation model, we believe some dataflow-like multithreading models are effective not only in supporting non-sequential programming models but also in achieving high throughput even on commodity processors. Based on this assumption, we are developing a programming environment, which is based on a dataflow-like fine-grain multithreading model[1]. Our work also includes a dataflow-like multithread programming language and an operating system, CEFOS(Communication and Execution Fusion OS)[2].

In our dataflow-like multithreading model, we use a split-phase style system call mechanism in which a request of a system call and the receipt of the system call result are separated in different threads. Split-phase style transactions are useful in hiding latencies of unpredictably long operations in several situations. We apply this style to

J. Filipe, B. Shishkov, and M. Helfert (Eds.): ICSOFT 2006, CCIS 10, pp. 242–253, 2008.

system calls and call as Wrapped System Call (WSC) mechanism. WSC mechanism is useful both in reducing overhead caused by system call mechanisms on commodity processors and in enhancing locality of reference.

In this paper, we evaluate the effectiveness of WSC mechanism on commodity processors. Section 2 introduces our operating system, CEFOS, and some of its features including WSC mechanism. Section 3 discusses the performance estimation and experimental benchmark results of WSC mechanism from the view point of system call overhead. Section 4 evaluates WSC mechanism for variants of `cp` program from the view point of locality of reference. We conclude WSC mechanism can reduce system call overhead and enhance locality of reference even on commodity platforms, which have no explicit support to dataflow-like multithreading.

## 2 Scheduling Mechanisms in CEFOS

### 2.1 CEFOS for Fine-Grained Multithreading

While running user programs under the control of an operating system like Unix, frequent context switches and communications between user processes and the kernel are performed behind the scenes. A system call requests a service of the kernel, and then voluntarily causes mode change. Activities involving operating system level operations are rather expensive on commodity platforms.

Table 1 shows the result of a micro-benchmark LMbench [3] on platforms with commodity processors and Linux. The row "null call" shows the overhead of a system call and the row "2p/0K" shows that of a process switch when we have two processes of zero KB context. Thus, the row "*x* p/*y* K" shows the overhead of a process switch for the pair of *x* and *y* which represent the number and the size of processes, respectively. The rows "L1$", "L2$" and "MainMem" show the access latency for L1 cache, L2 cache and main memory, respectively. As seen from Table 1, activities involving operating system level operations such as system calls and context switches are rather expensive on commodity platforms.

Therefore, one of the key issues to improve system throughput is to reduce the frequency of context switches and communications between user processes and the kernel. In order to address this issue, we employ mechanisms for efficient cooperation between the operating system kernel and user processes based on a dataflow-like multithreading model in CEFOS.

Figure 1 shows the outline of the architecture of CEFOS consisting of two layers: the external kernel in user mode and the internal kernel in supervisor mode. Internal kernel

**Table 1.** Results of LMbench (Clock Cycles)

| processor | null call | 2p/0K | 2p/16K | L1$ | L2$ | MainMem |
|---|---|---|---|---|---|---|
| Celeron 500MHz | 315 | 675 | 3235 | 3 | 11 | 93 |
| Pentium4 2.53 GHz | 1090 | 3298 | 5798 | 2 | 18 | 261 |
| Intel Core Duo 1.6GHz | 464 | 1327 | 2820 | 3 | 14 | 152 |
| PowerPC G4 1GHz | 200 | 788 | 2167 | 4 | 10 | 127 |

corresponds to the kernel of conventional operating systems. A process in CEFOS has a thread scheduler to schedule its ready threads.

A program in CEFOS consists of one or more partially ordered threads which may be fine-grained compared to conventional threads such as Pthreads. A thread in our system does not have a sleep state and we separate threads in a split-phase style at the points where we anticipate long latencies. Each thread is non-preemptive and runs to its completion without going through sleep states like Pthreads. While operations within a thread are executed based on a sequential model, threads can be flexibly scheduled as long as dependencies among threads are not violated.

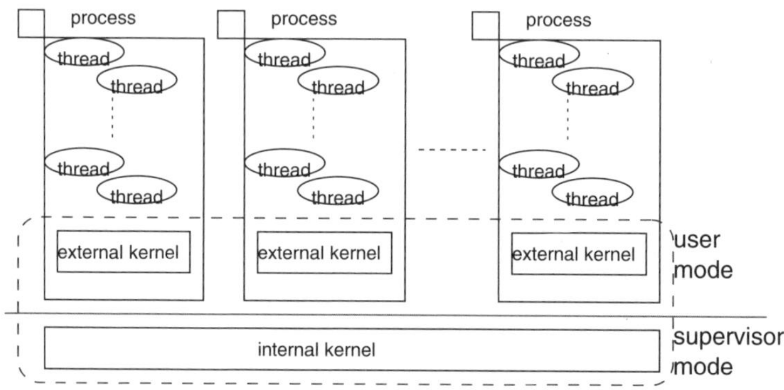


**Fig. 1.** Overview of CEFOS

A process in CEFOS has a thread scheduler and schedules its ready threads basically in the user-space. Since threads in CEFOS are a kind of user-level thread, we can control threads with small overhead. The external-kernel mechanism in CEFOS intermediates interaction between the kernel and thread schedulers in user processes. Although there exist some works on user level thread scheduling such as Capriccio [4], our research differs in that we use fine-grain thread scheduling. In order to simplify control structures, process control is only allowed at the points of thread switching. Threads in a process are not totally-ordered but partially-ordered, and we can introduce various scheduling mechanisms as long as the partial order relations among threads are not violated. Thus, CEFOS has scheduling mechanisms such as WSC mechanisms and Semi-Preemption mechanism.

### 2.2 Display Requests and Data (DRD) Mechanism

Operating systems use system calls or upcalls [5] for interactions between user programs and operating system kernel. System calls issue the demands of user processes through SVC and Trap instructions, and upcalls invoke specific functions of processes. The problem in these methods is overhead of context switches [6]. We employ Display Requests and Data (DRD) mechanisms [7] for cooperation between user processes and the kernel in CEFOS as we show below:

1. Each process and the kernel share a common memory area (CA).
2. Each process and the kernel display requests and necessary information on CA.
3. At some appropriate occasions, each process and the kernel check the requests and information displayed on CA, and change the control of its execution if necessary.

This DRD mechanism assists cooperation between processes and the kernel with small overhead. A sender or receiver of the request does not directly trigger the execution of request at the instance the request is generated. If the sender triggers directly the execution of receiver's side, the system may suffer from large overhead to switch. On the other hand, the system handles the request at its convenience with small overhead if we use DRD mechanism. For an extreme example, all requests from a process to the kernel are buffered and the kernel is called only when the process exhausted its ready threads.

The external kernel mechanism in CEFOS intermediates interaction between the internal kernel and thread schedulers in user processes by using this DRD mechanism. Thus, CEFOS realizes scheduling mechanisms such as WSC mechanism and Semi-Preemption mechanism by using DRD mechanism.

### 2.3 WSC Mechanism

WSC mechanism buffers system call requests from user programs until the number of the requests satisfies some threshold and then transfers the control to the internal kernel with a bucket of the buffered system call requests. Each system call request consists of four kinds of elements listed below.

- type of the system call
- arguments of the system call
- the address where the system call stores its result
- ID of the thread which the system call syncs after the execution

The buffered system calls are executed like a single large system call and each result of the original system calls is returned to the appropriate thread in the user process. Figure 2 illustrates the control flow in WSC mechanism, and each number in Figure 2 corresponds to the explanation below.

1. A thread requests a system call to External Kernel.
2. External Kernel buffers the request of system call to CA.
3. External Kernel checks whether the number of requests has reached the threshold. If the number of requests is less than the threshold, the thread scheduler is invoked to select the next thread from the ready threads in the process. If the number of request has reached the threshold, WSC mechanism sends the requests of system calls to the internal kernel to actually perform the system calls.
4. Internal Kernel accepts the requests of system calls and executes them one by one.
5. Internal Kernel stores the result of the system call to the address which Internal Kernel accepts as the third arguments of the system call. Also, Internal Kernel tells the thread, whose ID is accepted as the fourth argument, that it stores the result.

6. When Internal Kernel terminates executing all requests of system calls, External Kernel executes other threads. In other cases, WSC mechanism goes back to 3 and repeats this transaction.

WSC mechanism reduces overhead of system calls by decreasing the number of mode changes from user process to the kernel. Parameters and returned results of the buffered system calls under WSC mechanism are passed through CA of DRD to avoid frequent switches between the execution of user programs and that of the kernel.

**Table 2.** The values to calculate M (in clocks), and the estimated value of M

| processor (Hz) | $T_{gen}$ | $T_{sched}$ | $T_{sync}$ | $T_{req}$ | M |
|---|---|---|---|---|---|
| Celeron 500M | 63 | 31 | 21 | 31 | 3.4 |
| Pentium4 2.53G | 110 | 43 | 27 | 31 | 1.24 |
| Intel Core Duo 1.62G | 61 | 30 | 19 | 29 | 1.42 |

## 3 Evaluation: System Call Overhead

We evaluate the effectiveness of WSC mechanism on commodity processors. The test platform is built by extending Linux 2.6.14 on commodity PCs.

### 3.1 Estimation of the Effectiveness of WSC

First, we estimate the effectiveness of WSC mechanism by focusing on system call overhead. We compare the execution time of a program with normal system calls under the normal mechanism and that with split-phase system calls under WSC mechanism.

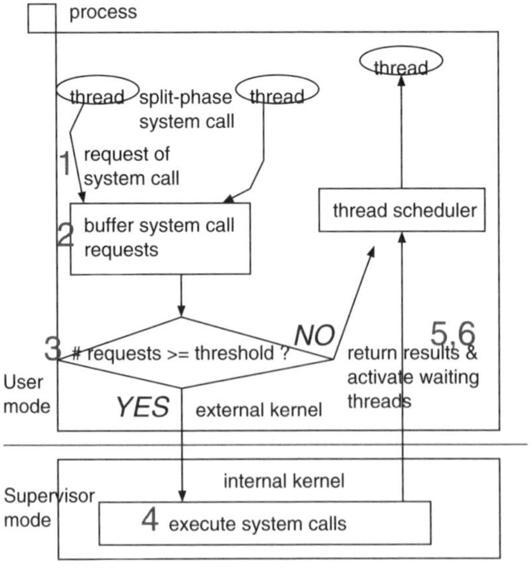


**Fig. 2.** Control flow in WSC mechanism

The total execution time of a program with N normal system calls under the normal mechanism, $T_{nor}$, is estimated as:

$$T_{nor} = T_{onor} + N \times (T_{sys} + T_{body}) + P_{nor} \tag{1}$$

where $T_{onor}$ is the execution time of the program portion excluding system calls under the execution of the normal system call mechanism, $T_{sys}$ is the setup and return cost of a single system call, and $T_{body}$ is the execution time of the actual body of the system call. In this estimation, we assume that we use the same system call and that there exist no penalties concerning memory hierarchies such as cache miss penalties and TLB miss penalties in $T_{onor}$ and $T_{body}$. $P_{nor}$ is the total penalties including cache miss penalties and TLB miss penalties during the execution of the normal system call mechanism.

Programs to which we can apply WSC mechanism are multithreaded and use split-phase style system calls. Additional thread management should be performed in this multithreaded program and we describe the overhead of this additional part as $T_{ek}$. $T_{ek}$ is estimated as:

$$T_{ek} = X \times T_{gen} + Y \times T_{sche} + Z \times T_{sync} \tag{2}$$

where X is the number of threads, $T_{gen}$ is the overhead to generate a single thread, Y is the number of times threads are scheduled, $T_{sche}$ is the overhead to schedule a thread, Z is the number of times synchronizations are tried and $T_{sync}$ is the overhead of a synchronization.

Although the execution of system call bodies will be aggregated, buffering system call request must be performed for each system call. We represent the overhead of buffering a single system call request as $T_{req}$. Thus, $T_{wsc}$, the total execution time of a program with N split-phase system calls under WSC mechanism, is estimated as:

$$\begin{aligned} T_{wsc} =& T_{owsc} + T_{ek} + N \times T_{req} + \\ & [N/M] \times T_{sys} + N \times T_{body} + P_{wsc} \end{aligned} \tag{3}$$

where $T_{owsc}$ is the execution time of the program portion excluding system calls, M is the number of system calls to be buffered for a single WSC (i.e. WSC threshold) and $P_{wsc}$ is the total penalties concerning memory hierarchies including cache miss penalties and TLB miss penalties during the execution under WSC. We assume none of such penalties exists in $T_{owsc}$ as in the estimation for $T_{onor}$ and $T_{body}$.

$\Delta T$, the difference between the execution time under the normal mechanism and that of under CEFOS with WSC is estimated as:

$$\begin{aligned} \Delta T =& T_{wsc} - T_{nor} \\ =& (T_{owsc} - T_{onor}) + \\ & \{T_{ek} + N \times T_{req} - (N - [N/M]) \times T_{sys}\} + \\ & (P_{wsc} - P_{nor}) \end{aligned} \tag{4}$$

We can say the performance is improved by WSC mechanism when $\Delta T < 0$. We estimate the value of M, the number of system calls to be buffered, to satisfy this condition. We assume the following conditions for the sake of simplicity: each program portion excluding system calls is the same, and each system call body is the same both

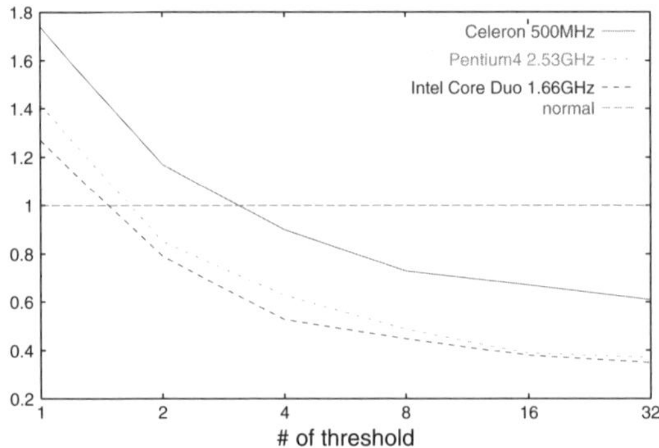


**Fig. 3.** Comparison of clock cycles (`getpid`)

in the normal version and in CEFOS version. Under these assumptions, we will only observe the difference of system call cost between the normal version and the CEFOS version. This assumption makes $T_{owsc} - T_{onor}$ and $P_{wsc} - P_{nor}$ amount to zero. We also assume X, Y and Z are equal to N. Thus, we can estimate the condition for M to satisfy $\Delta T < 0$ as:

$$M > \frac{T_{sys}}{T_{sys} - (T_{gen} + T_{sche} + T_{sync} + T_{req})} \tag{5}$$

We measured each value in (5) in order to calculate the value of M that satisfies the above condition as shown in Table 2 (we used the values of null call in Table 1 for $T_{sys}$) [1].

The performance on Pentium4 2.53GHz and Intel Core Duo 1.62GHz will be improved when M is larger 1. The performance on Celeron 500MHz will be improved when M is larger than 4. (Please note M is a natural number)

### 3.2 Performance Evaluation Using `getpid()`

The above estimation assumed each system call body is the same both in the normal version and in CEFOS version for the sake of simplicity. In this subsection, we examine our estimation by using `getpid()` as a system call to meet such an assumption. We measured the number of clocks for a number of `getpid()` system calls using the hardware counter. We executed 128 `getpid()` system calls in our experiments. We changed the threshold of WSC as 1, 2, 4, 8, 16 and 32 for the WSC version. We also measured the total time of successive `getpid()` system calls under the normal system call convention in unchanged Linux.

Figure 3 shows the comparison results of clock cycles for `getpid()` system calls. The x-axis indicates the threshold of WSC and y-axis the ratio of clock cycles of WSC versions compared with clock cycles under the normal system call convention in unchanged Linux. The lower y value indicates the better result of WSC.

[1] We omit the values of PowerPC G4 because of the problem of accuracy. However, the observed M for PowerPC G4 is 4 according to the experiment explained in the next subsection.

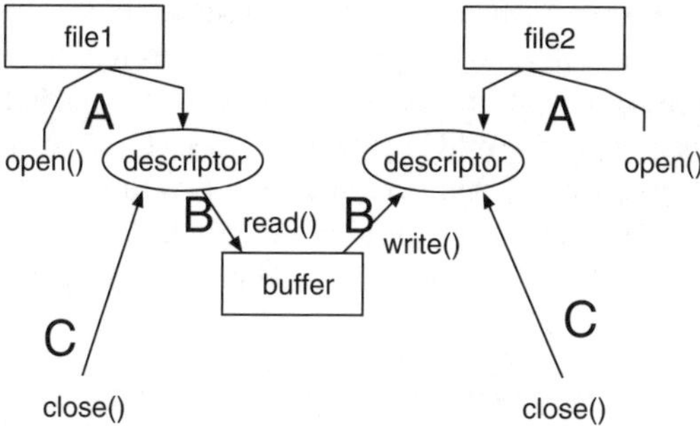


**Fig. 4.** Control flow in cp program

As seen from Figure 3, we have extra overhead when WSC threshold is 1, because of newly added load of $T_{gen}$, $T_{sche}$, $T_{sync}$ and $T_{req}$. However, we observe the effect of WSC when the threshold becomes 2 for Pentium4 2.53GHz and Intel Core Duo 1.66GHz and 4 for Celeron 500 MHz as we estimated in the previous estimation. The clock cycles in WSC versions are decreased as the threshold gets larger regardless of the processor type.

## 4 Evaluation: Locality of Reference

In the previous section, we evaluate the effectiveness of WSC mechanism in reducing overhead caused by system calls. In this section, we examine the effectiveness in exploiting locality of reference. We can expect high throughput when we can aggregate system calls which refer to the same code or data. The test platform is also built by extending Linux 2.6.14 on Pentium4 2.53 GHz.

### 4.1 `cp` Program

We use modified `cp` programs to evaluate the effectiveness of WSC mechanism in exploiting locality of reference. Figure 4 shows an overview of the control flow in `cp` program, and the symbols A, B and C in Figure 4 correspond to the ones in the explanation below.

A. one `open()` system call opens a file to read and the other `open()` system call opens another file to write, preparing a file descriptor for each file respectively.
B. `read()` system call reads up to designated bytes from the file descriptor into buffer, and then `write()` system call writes up to designated bytes to the file referenced by the file descriptor from buffer.
C. `close()` system calls close these files.

Thus, a `cp` program uses six system calls per transaction. We use a `cp` program called NORMAL version, which executes these six system calls in the order we show above, like `open()`, `open()`, `read()`, `write()`, `close()` and `close()`. We have to `open()` a file before executing `read()` or `write()`, and we have to specify the file descriptor, which is the result of `open()` system call, to execute `read()`, `write()`

and `close()`. Therefore, we cannot simply wrap these six system calls. We have to wrap two `open()` system calls and other four system calls respectively. Because of the additional overhead of using WSC mechanism that we mentioned in Figure 3, we cannot expect the effect when applying WSC mechanism to just one `cp` transaction. In fact, doing one `cp` in WSC version of one `cp` took about two times clock cycles compared to NORMAL version. Therefore, we consider doing multiple `cp`s in a program.

We use other four versions of `cp` program, and measure 11 portions of these 5 programs to observe:

I. whether WSC mechanism is effective or not in `cp` programs in total,
II. the difference between the effect of wrapping single type of system calls and that of wrapping various types of system calls, and
III. the effect of wrapping system calls which have the same code but refer to different data.

Figure 5 shows these 5 programs and 11 portions. "N" in Figure 5 is the number of `cp` transactions. Now, we explain each program and portion below. Then we explain why we choose these portions to examine the points of our interests above.

In Program 2 in Figure 5, we wrap every one of six kinds of system calls. We call this WSC+COLLECT version.

As a counterpart of this WSC+COLLECT, we also collect system calls of the same type in a block but execute the block with normal system call convention. We call this program as NORMAL+COLLECT version (Program 3 in Figure 5).

In addition, we implement WSC+RW and NORMAL+RW version (Program 4 and 5 in Figure 5), which change the order of `read()` and `write()` in WSC+COLLECT and NORMAL+COLLECT version.

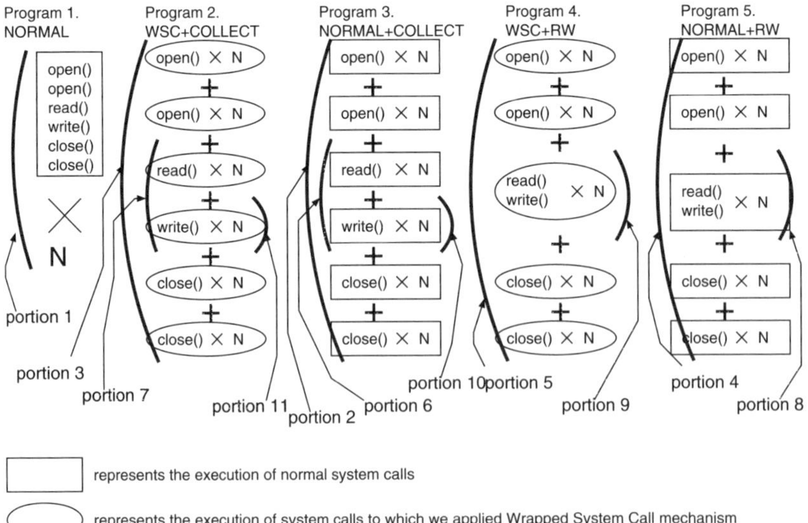


**Fig. 5.** Program and Portion we measure in `cp` program

Then, we show the explanation of 11 portions we measure.

1. from `open()` to `close()` of NORMAL.
2. from `open()` to `close()` of NORMAL+COLLECT.
3. from `open()` to `close()` of WSC+COLLECT.
4. from `open()` to `close()` of NORMAL+RW.
5. from `open()` to `close()` of WSC+RW.
6. `read()` and `write()` part of NORMAL+COLLECT.
7. `read()` and `write()` part of WSC+COLLECT.
8. `read()` and `write()` part of NORMAL+RW.
9. `read()` and `write()` part of WSC+RW.
10. only `write()` of NORMAL+COLLECT.
11. only `write()` of WSC+COLLECT

We measured only `write()` in portion 10 and 11 to observe the effect of wrapping system calls which refer to different data. While `read()` system call contains disk access time, `write()` system call buffers access to the disk and enables us to observe the effect of WSC mechanism excluding disk access time. Also, we implemented NORMAL+RW and WSC+RW and measured portion 6, 7, 8 and 9 to observe the effect of wrapping two types of system calls together. Then, we measured the whole `cp` in portion 1, 2 and 3 to examine if WSC mechanism is effective or not in total. Also, we measured portion 4 and 5 to examine the influence of wrapping `read()` and `write()` system calls on `cp` total.

We measure clock cycles and the number of events such as L1 cache misses in every portion. From these results, we investigate how WSC mechanism effects locality of reference from the view point of I, II and III above.

### 4.2 Performance Evaluation

Table 3 shows the result of `cp` programs. In this case, WSC threshold is 8 and we do `cp` transactions 100 times, which means N in Figure 5 is 100. The numbers in the row "portion" correspond to the numbers of the explanation we show in subsection 4.1. The row "#clocks" shows clock cycles, the row "L2$" shows L2 cache miss counts and rows "ITLB", and "DTLB" show the walk counts for ITLB and DTLB, respectively. We measured these events with a performance monitoring tool perfctr [8].

In `write()` sections (portion 10 and 11), the clock cycles for WSC+COLLECT `write()` are less than NORMAL+COLLECT `write()` in about 0.12 million cycles, which is reduction to 83 % in clock cycles. The reduction of this 0.12 million cycles by WSC mechanism is larger than the reduction estimated by using formula in section 3, which is about 0.072 million cycles for 100 `write()` system calls. We consider this improvement is achieved by enhanced locality of reference, therefore we measure the number of events concerning memory hierarchies. As we expected, we can see the reduction of L2 cache misses, ITLB walks and DTLB walks in WSC+COLLECT `write()` compared to NORMAL+COLLECT `write()`. Thus, we can say WSC mechanism is effective even when each system call refer to different data. We changed the threshold and measured portion 10 and 11 to compare the results with those of `getpid()`. Figure 6 shows the result, and we can see the same tendency as we see in Figure 3 that wrapping more than 2 system calls is effective in clock cycles in Pentium 4.

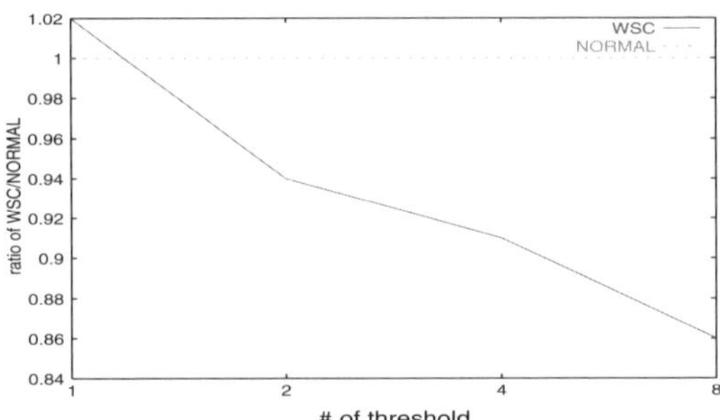


**Fig. 6.** Comparison of clock cycles (`write()`)

In read/write section (portion 6, 7, 8 and 9), we can see the effect of wrapping different system calls by comparing portion 6 with 7 and 8 with 9. Both clock cycles and number of events decrease in WSC version in both cases.

Finally, from portion 1, 2 and 3, we can say applying WSC mechanism to `cp` program is effective in total. When we compare portion 2 with 3 to ignore the difference of disk access pattern, the reduction in clock cycles is about 0.32 million cycles. As we see in `write()` system call, we can see the reduction of L2 cache misses, ITLB walks and DTLB walks. Therefore, we conclude that WSC mechanism for split-phase style system calls is effective in exploiting locality of reference. We can see the similar result from portion 4 with 5 and reach to the same conclusion.

**Table 3.** Results of `cp` program

| portion | #clocks | L2$ | ITLB | DTLB |
|---|---|---|---|---|
| 1. NOR | 2,884,325 | 7378 | 511 | 136 |
| 2. NOR+COLLECT | 2,588,200 | 8800 | 187 | 207 |
| 3. WSC+COLLECT | 2,262,523 | 7740 | 81 | 120 |
| 4. NOR+RW | 2,625,804 | 8264 | 227 | 200 |
| 5. WSC+RW | 2,431,758 | 8118 | 128 | 197 |
| 6. NOR+COLLECT read/write | 1,090,608 | 4385 | 112 | 69 |
| 7. WSC+COLLECT read/write | 876,703 | 3503 | 37 | 42 |
| 8. NOR+RW read/write | 1,096,045 | 3875 | 130 | 72 |
| 9. WSC+RW read/write | 985,227 | 3647 | 70 | 62 |
| 10. NOR+COLLECT write | 686,883 | 1779 | 93 | 28 |
| 11. WSC+COLLECT write | 569,206 | 1363 | 41 | 13 |

## 5 Conclusions

In this paper, we discussed our WSC mechanism in CEFOS. While CEFOS is based on a dataflow-like fine-grain multithreading model, WSC mechanism is effective in improving throughput even on commodity platforms which have no explicit support to dataflow-like fine-grain multithreading.

Today, many investigation have been made about utilizing multithreading processor, such as SMT. Many of them tackle with memory hierarchy problem because cache conflict often occurs under the condition where several threads run concurrently. One effective solution to this problem is improving the scheduling of thread, which is conventional Pthread, to utilize CPU resources more effectively[9]. On the other hand, our work split conventional thread and control the thread in user process. Thus, we have more chances to schedule fine-grained threads more flexibly with smaller overhead.

In `cp` program, the combination of our split-phase style system calls and WSC mechanism is effective in improving throughput by reducing mode changes and penalties concerning memory hierarchies such as L2 cache misses and TLB walks.

Recently, the overhead of system call and context switch is increasing on commodity processors. Besides, we think the tendency continues that latency of memory access becomes bottleneck, which is coming from the gap between processor speed and memory speed. Therefore, we think WSC will be more effective in the future, which can reduce the overhead of system call and context switch and enhance the locality of reference. We believe this will contribute to higher throughput of internet server and large-scale computation in the future. Our future work includes collecting more data from other processors and exploiting the effect of SYSENTER/SYSEXIT command in x86 architecture.

## References


1. Culler, D.E., Goldstein, S.C., Schauser, K.E., von Eicken, T.: Tam – a compiler controlled threaded abstract machine. Journal of Parallel and Distributed Computing 18, 347–370 (1993)
2. Kusakabe, S., et al.: Parallel and distributed operating system cefos. IPSJ ISG Tech. Notes, 99(251), 25–32 (1999)
3. McVoy, L., Staelin, C.: lmbench: Portable tools for performance analysis (1996), `http://www.bitmover.com/lm/lm-bench`
4. Behren, R., et al.: Revising old friends: Capriccio: scalable threads for internet services. In: Proc. of the 19th ACM symposium on Operating systems principles, pp. 268–281 (2003)
5. Thomas, E.A., et al.: Scheduler activation: Effective kernel support for the user-level management of parallelism. In: Proc. of the 13th ACM Symp. on OS Principles, pp. 95–109 (1991)
6. Purohit, A., et al.: Cosy: Develop in user-land, run in kernel-mode. In: Proc. of HotOS IX: The 9th Workshop on Hot Topics in Operating Systems, pp. 109–114 (2003)
7. Taniguchi, H.: Drd: New connection mechanism between internal kernel and external kernel. Tran. of IEICE, J85-D-1(2) (2002)
8. Petterson, M.: Perfctr (nd), `http://user.it.uu.se/~mikpe/linux/perfctr/`
9. Snavely, A., Tullsen, D.: Symbiotic jobscheduling for a simultaneous multithreading processor. In: 9th International Conference on Architectural Support for Programming Languages and Operating Systems, pp. 234–244 (2000)

# PART IV

# Information Systems and Data Management

# Adding More Support for Associations to the ODMG Object Model

Bryon K. Ehlmann

Department of Computer Science, Southern Illinois University Edwardsville
Edwardsville, IL 62026 U.S.A.
behlman@siue.edu

**Abstract.** The Object Model defined in the ODMG standard for object data management systems (ODMSs) provides referential integrity support for one-to-one, one-to-many, and many-to-many associations. It does not, however, provide support that enforces the multiplicities often specified for such associations in UML class diagrams, nor does it provide the same level of support for associations that is provided in relational systems via the SQL references clause. The Object Relationship Notation (ORN) is a declarative scheme that provides for the specification of enhanced association semantics. These semantics include multiplicities and are more powerful than those provided by the SQL references clause. This paper describes how ORN can be added to the ODMG Object Model and discusses algorithms that can be used to support ORN association semantics in an ODMG-compliant ODMS. The benefits of such support are improved productivity in developing object database systems and increased system reliability.

**Keywords:** ODMG Object Model, OODB systems, Constraint management, Object Relationship Notation (ORN).

## 1 Introduction

An *object data management system* (ODMS) allows objects created and manipulated in an object-oriented programming language to be made persistent and provides traditional database capabilities like concurrency control and recovery to manage access to these objects. An *object database management system* (ODBMS), one type of ODMS, stores the objects directly in an *object database*. An *object-to-database mapping* (ODM), another type of ODMS, stores the objects in another database system representation, usually relational [3].

The de facto standard for ODMSs is ODMG 3.0 [3], which was defined by the Object Data Management Group (ODMG) consisting of representatives from most of the major ODMS vendors. This standard defines an Object Model to be supported by ODMG-compliant ODMSs. The model defines the kinds of object semantics that can be specified to an ODMS. These semantics deal with how objects can be named and identified and the properties and behavior of objects. They also deal with how objects can relate to one another, which is the focus of this paper.

J. Filipe, B. Shishkov, and M. Helfert (Eds.): ICSOFT 2006, CCIS 10, pp. 257–269, 2008.

In addition to supporting generalization-specialization relationships, the Object Model supports one-to-one, one-to-many, and many-to-many binary relationships between object types. These are the non-inheritance, or structural, types of relationships, which are termed *associations* in the Unified Modeling Language (UML) [16]. For example, a one-to-many association between carpools and employees can be defined in the Object Model. A carpool object is defined so that it can reference many employee objects, and an employee object is defined so that it can reference at most one carpool.

The Object Model prescribes that the ODMS automatically enforce referential integrity for all defined associations. This means that if an object is deleted, all references to that object that maintain associations involving that object must also be deleted. This ensures that there are no such references in the database that lead to nonexistent objects.

What has just been described is the extent of support for associations in the Object Model. What is lacking is some additional, easily implementable support for associations that could significantly improve the productivity of developing object database systems and the reliability of those systems.

For example, the Object Model, like the relational model, does not support the specification of precise *multiplicities*. Such association constraints are almost always present in the diagrams used to model databases—the traditional Entity-Relationship Diagram (ERD) [4], where multiplicities are termed *cardinality constraints*, and the UML class diagram [16]. For example, the multiplicity for the Employee class in the carpool–employee association may be given as 2..* in a class diagram, meaning that a carpool must be related to two or more employees. Such association semantics, documented during conceptual database design, are sometimes lost during logical database design unless supported by the logical data model, e.g., the Object Model. If not supported, to survive, they must be resurrected by the programmer during implementation and for object databases translated into cardinality checks on collections and into exception handling code within relevant create and update methods.

The Object Model also does not support association semantics that are equivalent to those supported in standard relational systems via the references...on delete clause of the create table statement in SQL [5]. Such semantics would, for instance, allow one to declare an association between objects such that if an object is deleted, all related objects would be automatically deleted by the ODMS, i.e., an on delete cascade. For example, if an organization in a company were deleted, all subordinate organizations would be implicitly deleted. Such an association semantic is required for an ODMS to provide support for *composite objects*.

Object Relationship Notation (ORN) was developed to allow these kinds of semantics, and others often relevant to associations, to be better modeled and more easily implemented in a DBMS [6], [8], [9]. ORN is a declarative scheme for describing association semantics that is based on UML multiplicities.

In this paper we give a brief overview of ORN and show how the ODMG Object Model can be extended to include ORN. We also discuss and illustrate algorithms that are available and can be used by an ODMG-compliant ODMS to implement the association semantics as specified by ORN. The extension is very straightforward, and the algorithms are relatively simple. The end-result is an enhanced Object Model that supports more powerful association semantics—in fact, more powerful than those supported by relational systems without having to code complex constraints and triggers [6]. By extending models with ORN and providing the required mappings between them—UML class diagram to Object Model to ODMS implementation—we facilitate a model-driven development approach and gain its many advantages [13].

The specific benefits here are a significant improvement in the productivity of developing object database applications and an increase in their reliability. Productivity is improved when translations from class diagram models into object models are more direct and when programmers do not have to develop code to implement association semantics. Currently, many developers working on many database applications must implement, test, and maintain custom code for each type of association, often "reinventing the wheel." Reliability is increased when the ODMS is responsible for enforcing association semantics. Currently, developers sometimes fail to enforce these semantics or inevitably introduce errors into database applications when they do.

The remainder of this paper is organized as follows: section 2 gives a brief overview of ORN and related work, section 3 shows how the ODMG Object Model can be extended with the ORN syntax and describes ORN semantics in terms of this model, section 4 discusses and illustrates algorithms that can be used to implement ORN semantics in an ODMS that is based on the extended Object Model, and section 5 provides concluding remarks. A complete set of ORN-implementing algorithms is available on the author's website [12].

## 2 ORN and Related Work

ORN describes association semantics at both the conceptual, i.e., data modeling, and logical, i.e., data definition, levels of database development, and can be compared to other declarative schemes.

For data modeling, ORN has been integrated into ERDs and UML class diagrams [10]. ORN extends a class diagram by allowing binding symbols to be given with multiplicity notations. The bindings indicate what should happen when links between related objects are destroyed, either implicitly because of object deletions or explicitly. They indicate, for instance, what action the DBMS should take when destroying a link would violate the multiplicity at one end of an association. The binding symbols (or the lack of them) provide important semantics about the relative strength of linkage between related objects and define the scope of *complex objects*. For example, the association between a carpool, a complex object, and its riders can be specified in an ORN-extended class diagram to indicate that if the number of riders falls below two, either because an employee leaves the company (an employee object is deleted) or just leaves the carpool (a link between an employee and a carpool is destroyed), the carpool should be deleted.

For database definition, ORN has been implemented within the Object Database Definition Language (ODDL). ODDL is a language used to define classes, attributes, and relationships to a prototype ODMS named Object Relater *Plus* (OR+) [7]. OR+ closely parallels ODMG and is built on top of Object Store [15]. The integration of ORN into ODDL allows a direct translation of association semantics from an ORN-extended class diagram into the database definition language and enables these semantics to be automatically maintained by the DBMS. Using ORN, the semantics for an association between employees and carpools as previously described can be both modeled and implemented by the notation |~X~<2..*-to-0..1>. No programming is needed.

In [6], the power of ORN in describing association semantics is compared to that of other declarative notations proposed for various object models and to that of the references clause of SQL. The comparison reveals that the most unique aspect of ORN, and what accounts for its ability to specify a larger variety of association types, is that it provides for the enforcement of upper and lower bound multiplicities and allows delete propagation to be based on these multiplicities. It also provides a declarative scheme at a conceptual level of abstraction that is independent of database type, object or relational. ORN can also be compared to extensions to the ER model that others have suggested to specify or enforce association semantics, or *structural integrity constraints* [1], [2], [14]. These extensions, however, are more procedural in nature.

# 3 Adding ORN to ODL

## 3.1 Associations in ODL

In the ODMG Object Definition Language (ODL), which defines the ODMG Object Model, an association is defined by declaring a relationship *traversal path* for each end of the association. A traversal path provides a means for an object of one class to reference and access the related objects of a *target class* (which is the same class in a recursive relationship). Access to many target class objects requires the traversal path declaration to include an appropriate collection type, usually a set or list, that can contain references of target class type. Access to at most one target class object requires the declaration to include a reference of target class type.

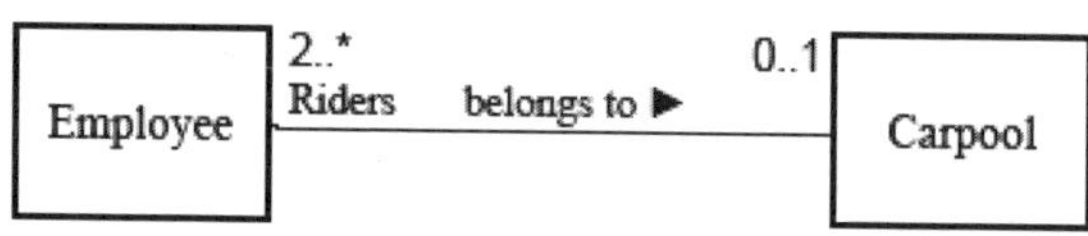


**Fig. 1.** Class diagram for employee–carpool association

```
class Employee {
   ...
   relationship Carpool           carpool
                                  inverse Carpool::riders;
   ...
};
class Carpool {
   ...
   relationship set<Employee> riders
                                  inverse Employee::carpool;
   ...
};
```

**Fig. 2.** ODL for employee–carpool association

A traversal path declaration must also include the name of its inverse traversal path. For example, the one-to-many relationship between carpools and employees, discussed earlier and modeled by the class diagram in Fig. 1, would be declared in ODL as shown in Fig. 2. The 2.* multiplicity given in the class diagram must be implemented by application code.

### 3.2 Adding ORN Syntax

Adding ORN to the Object Model is relatively straightforward. Essentially, ODL is extended to allow an *<association>* to be given for each declared relationship. The syntax for an *<association>*, which is the syntax for ORN, is given in Fig. 3, and the ORN-extended ODL syntax is given in Fig. 4.

To illustrate the syntax and semantics of ORN in the context of the Object Model, a database containing the employee–carpool association as well as two other associations is modeled by the ORN-extended class diagram given in Fig. 5. In such a diagram, the ORN bindings for a class (or role) in an association are given as stereotype icons at the association end corresponding to that class (or role). When no binding symbols are given for an association end (or role), default bindings are assumed, the semantics of which will be defined later.

The database modeled in Fig. 5 is implemented by the ORN-extended ODL given in Fig. 6.

If an *<association>* is not given for a relationship in ODL (see Fig. 4), the default *<association>* is <0..1-to-0..1> for a one-to-one, <0..1-to-*> for a one-to-many, and <*-to-*> for a many-to-many relationship. These defaults give relationships the same semantics as they have in the existing Object Model.

An *<association>* given for a relationship need only be given for one of the traversal paths. If given for both, the *<association>*s must be inverses of each other. For example, an *<association>*, if given for riders in Fig. 6, must be given as <0..1-to-2..*> |~X~.

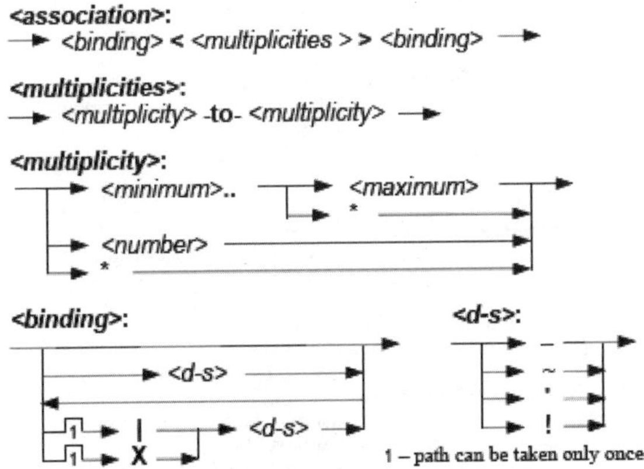


**Fig. 3.** ORN syntax diagrams

```
<rel_dcl>              ::= relationship
                           <target_of_path> <identifier>
                           inverse <inverse_traversal_path>
                           [ <association> ]
<target_of_path>       ::= <identifier>
                           | <coll_spec> < <identifier> >
<inverse_traversal_path> ::= <identifier> :: <identifier>
```

**Fig. 4.** Updated BNF for a relationship in ODL

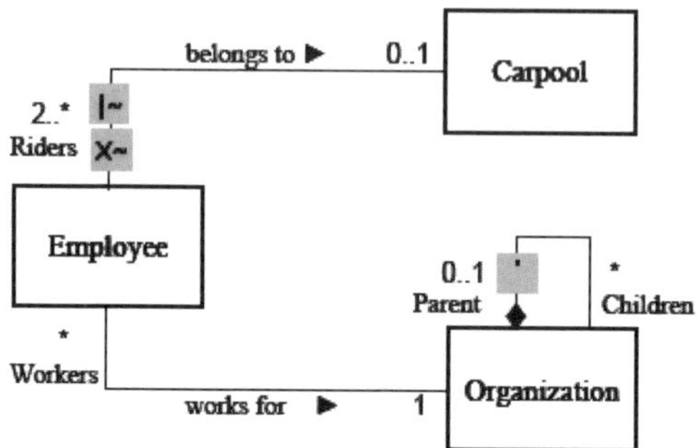


**Fig. 5.** ORN-extended UML class diagram

When an *<association>* is given for a traversal path *tp* in class *C*, the multiplicity and binding given after the -to- apply to *tp* and to the target class, the multiplicity and binding given before the -to- apply to the inverse *tp* and to class *C*. For example, in Fig. 6, the multiplicity 0..1 and default bindings apply to the traversal path carpool and the target class Carpool, and the multiplicity 2..* and binding |~X~ apply to the traversal path riders and class Employee. If the multiplicity given for a traversal path in an *<association>* implies "many," then the type of that traversal path must be a collection.

```
class Employee {
   relationship Carpool          carpool
                                 inverse Carpool::riders
                                 |~X~<2..* -to-0..1>;
   relationship Organization     organization
                                 inverse Organization::workers
                                 <*-to-1>;
   ...
};
class Carpool {
   relationship set<Employee>  riders
                               inverse Employee::carpool;
   ...
};
class Organization {
   relationship set<Employee>    workers
                               inverse Employee::organization;
   relationship Organization     parent
                               inverse Organization::children;
   relationship set<Organization> children
                               inverse Organization::parent
                               '<0..1-to-*>;
   ...
};
```

**Fig. 6.** ODL for class diagram shown in Fig. 5

The last issue to address in extending ODL is association inheritance. In the Object Model, a relationship can be inherited by a class via the extends relationship. For example, the declaration **class** SalesPerson **extends** Employee **{** ... **}** would mean that the SalesPerson class inherits the attributes, relationships, and behavior of the Employee class. Thus, the carpool traversal path as declared in the Employee class in Fig. 6 would be inherited by the SalesPerson class, allowing sales people to join carpools. When a relationship is inherited by a class, all of the semantics defined by its *<association>*, given or defaulted, are also inherited.

And, of course, the semantics of all *<association>*s defined in the ODL—defaulted, given, or inherited—must be maintained by the ODMS.

### 3.3 ORN Semantics in ODL Context

The semantics of the *<multiplicity>*s in an *<association>* are identical to those of the multiplicities defined in UML [16]. The semantics of the *<binding>*s are given in Table 1.

Previous papers have described ORN semantics conceptually in terms of ER and class diagrams, e.g. [10]. The reader may review these papers for a more detailed discussion of ORN. Here, we focus more on describing ORN semantics in terms of the Object Model, or ODL. Thus, instead of "association links" being conceptually "created" and "destroyed," "relationship references" (or, alternatively, "traversal path references") are "formed" and "dropped." Dropping a relationship or traversal path reference also means dropping the corresponding inverse reference (in the inverse traversal path). Also, bindings and multiplicities are now associated with traversal paths as well as with the related classes. This is convenient for identifying bindings and multiplicities in recursive relationships since the subject class and related class, now called the "target class," are the same. Traverse path names can be equated to role names given in UML class diagrams. In Table 1, traversal path names $tp_A$ and $tp_B$ are also role names in the class diagram for relationship *R*.

As indicated in Table 1, association semantics are derived from multiplicity semantics and the semantics of the given bindings. For example, in the I~X~<2..*-to-0..1> association between employees and carpools, the I~ symbol in the *<binding>* for the Employee class means (from Table 1): on delete of an Employee object, a carpool reference (see Fig. 6) can be implicitly dropped; however, when this violates the multiplicity 2..*, the target Carpool object must be implicitly deleted. The X~ symbol means: a carpool reference can be explicitly dropped; however, when this violates multiplicity 2..*, the target Carpool object must be implicitly deleted. The multiplicity 2..* is violated when a reference to one of just two employees in a carpool, i.e., one of just two references in the set riders, is dropped. The default *<binding>* for the Carpool class means (again, from Table 1): on delete of a Carpool object, a reference in riders (see Fig. 6) can be implicitly dropped provided this does not violate multiplicity 0..1, and a reference in riders can be explicitly dropped provided this does not violate multiplicity 0..1. A 0..1 multiplicity is never violated by dropping a reference in riders (or a carpool reference for that matter).

**Table 1.** ORN binding semantics for the Object Model

***<binding>* semantics:**
Given in terms of a subject class *A* in a relationship *R* having multiplicity *m* and *<binding> b*. *R* relates objects in class *A* to objects in a related, or target, class *B*.

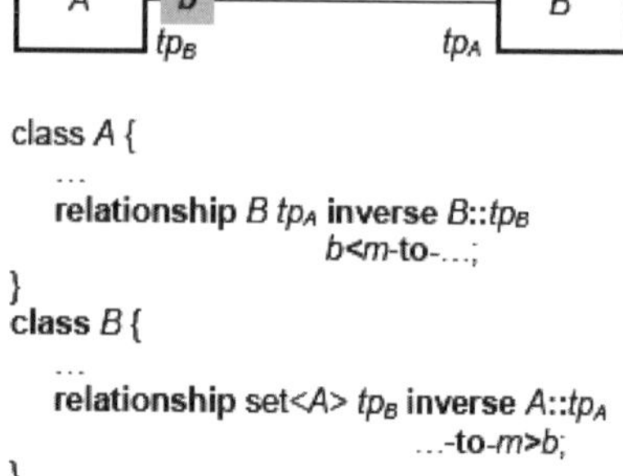

A
m
b
tpB
R ▶
tpA
B
class A {
...
relationship B tpA inverse B::tpB
b<m-to-...;
}
class B {
...
relationship set<A> tpB inverse A::tpA
...-to-m>b;
}


*R* is implemented in the Object Model by a traversal path $A::tp_A$ and inverse traversal path $B::tp_B$. The type of $tp_A$ is *B* or some suitable collection of *B* (shown here as *B*), and the type of $tp_B$ is *A* or some suitable collection of *A* (shown here as set<*A*>). The binding for traversal path $tp_B$ is *b* and the multiplicity is *m*. Forming or dropping a $tp_A$ reference from an *A* object to a *B* object includes forming or dropping, respectively, its corresponding inverse $tp_B$ reference from object *B* to the *A* object.

A **|** in *b* symbolizes a "cut" and denotes the implicit destruction, i.e., dropping, of an existing $tp_A$ reference that must occur on deletion of an *A* object. An **X** in *b* symbolizes an eXplicit "X out" and denotes the explicit destruction, i.e., dropping, of a $tp_A$ reference.[†] Deletion of a *A* object succeeds only if all existing relationship references from/to that object are implicitly destructible. Also, the deletion of an *A* object or the dropping of a $tp_A$ reference succeeds only if all required implicit deletions succeed.

***<d-s>* semantics:**
A *<d-s>*, i.e., destructibility symbol, in *b* has the meaning given below. If a *<d-s>* is given after a **|**, this meaning applies to implicit destruction; if given after an **X**, it applies to explicit destruction; and if given alone, it applies to both. If no *<d-s>* is given for implicit destruction, explicit destruction, or both, it implies default destructibility for whichever.

(No *<d-s>* given) *Default destructibility.* A $tp_A$ reference can be dropped provided this does not violate *m*.*

- \- *Minus destructibility.* A $tp_A$ reference cannot be dropped.
- ~ *Propagate destructibility.* A $tp_A$ reference can be dropped; however, when this violates *m*, the target object must be implicitly deleted.
- ' *Prime destructibility.* A $tp_A$ reference can be dropped; however, after being dropped, an implicit delete must be attempted on the target, i.e., subordinate, object. This implicit deletion is required if and only if its failure and resultant undoing would violate *m*.*
- ! *Emphatic prime destructibility.* A $tp_A$ reference can be dropped, but the target, i.e., subordinate, object must be implicitly deleted.

† - A $tp_B$ reference change done as a single operation that replaces an *A* object with another is not treated as an explicit reference destruction relative to class *A*.
* - The check for a violation is deferred until the end of the current complex object operation.

Below are more of the association semantics that are modeled in Fig. 5 and implemented in Fig. 6. They are described both conceptually and, within brackets, in terms of Object Model.

- If an employee [Employee object] is deleted, the link to the employee's organization is implicitly destroyed [the object's organization reference to its target Organization object is implicitly dropped] (default binding and * multiplicity).
- If an organization [Organization object] is deleted, all descendant organizations [Organization objects recursively referenced via children] are implicitly deleted (' binding); however, an organization is not deleted if it has any employees [if workers references any Employee objects] (default binding and 1 multiplicity).

- If a link between organizations is destroyed [if a children reference (or its inverse parent reference) is dropped], the child organization and all descendant organizations [Organization objects recursively referenced via children] are implicitly deleted (' binding); however, again, an organization is not deleted if it has any employees (default binding and 1 multiplicity).

## 4 Implementing ORN

The implementation of ORN semantics in an ODMG-compliant ODMS is described by giving the algorithms required to create and delete objects and form and drop relationship references. These operations become complex object operations in the context of ORN. This means they may no longer involve just one object or relationship reference but may involve many objects, relationships, and relationship references in the scope of a complex object.

In [12], we give the algorithms for these operations by providing all related pseudocode, with commentary, for the ObjectFactory::new() and Object::delete() methods, which are associated with an object, and the *C*::form_*tp*() and *C*::drop_*tp*() methods, which are associated with a declared traversal path *tp* in a user-declared class *C*. These methods are defined as part of the Object Model (see Chapter 2 of [3]).

In this section, due to space constraints, we discuss these algorithms in general and illustrate them by giving the algorithm for just the Object::delete() method. The pseudocode shown in this section is about one quarter of that given in [12].

The algorithms have been developed by reverse engineering the code for implementing ORN within OR+. This is the same code executed when one uses the ORN Simulation, a web-based, prototype modeling tool [11]. Thus, the algorithms are well-tested but have a slightly different wrapping.

Their implementation of ORN semantics is noncircular and unambiguous in the presence of association cycles. By noncircular, we mean the processing of traversal paths always terminates. By unambiguous, we mean the results of a complex object operation are independent of the order in which traversal paths and the references in these paths are processed. This is true as long as an *<association>* does not have a |– binding on just one end and is discussed in detail and proved in [9].

As stated in the introduction, the algorithms are relatively simple; however, they depend on the ODMS implementation supporting a nested transaction capability. Nested transactions are needed to implement the semantics of the ' (prime) binding and are desirable so that the system can check multiplicity violations at the end of a complex object operation, undoing the operation upon any exception and thus making the complex object operation atomic. The Object Model defines a Transaction Model, which does not provide nested transactions. So, before giving the algorithms for the complex object operations in [12], we extend the Transaction Model to support nested transactions, at least for the purpose of implementing the ODMS. We assume such support for nested transactions and give algorithms for transaction methods, focusing on the actions required to support ORN semantics.

All methods are assumed to execute in the context of a opened database *d*, and methods new(), delete(), form_$tp_A$(), and drop_$tp_A$() are assumed to execute within the scope of a user-defined transaction.

The pseudocode that expresses the algorithms is some mixture of ODL, C++, Java, and English. We have tried to stick as close as possible to the conventions of ODL.

Indention indicates control structure, with appropriate end's often used to terminate compound statements. The try...handle...end handle control structure for exception handling is similar to Java's try {...} catch {...}. Methods for a class are introduced with a header of the form Method *<variable>*.*<method name>*( ... ), where the *<variable>* is used in the body of the method to refer to the object on which the method is invoked, i.e., the implicit parameter and this object in C++ and Java. A *<method name>* begins with an underscore if it is to be invoked only by the ODMS implementation.

The algorithms are expressed using the variables defined in Table 1.

Fig. 7 gives the delete() method and two methods that it uses, _try_delete() and _enforce_binding(). The given delete() replaces the primitive delete() method as currently defined in the Object Model.

The remainder of this section briefly explains the pseudocode in Fig. 7. For a more detailed explanation and for the pseudocode of all methods invoked by the delete() algorithm, see [12].

The algorithm for delete() uses these functions:

Type(*o*) – the type, or class, of object *o*, which is the most specific type of *o* in any type hierarchy.

Inverse(*tp*) – the inverse traversal path of *tp*.

LbM(*tp*) – the lower bound multiplicity for *tp* in the *<association>* for the relationship represented by traversal path *tp*.

ImpB(*tp*) – the implicit destructibility binding for *tp* (minus any | symbol) in the *<association>* for the relationship represented by traversal path *tp*.

Refs(*o.tp*) – the number of references in *o.tp*, which, if *tp* is a collection, is the cardinality of the collection, i.e., *o.tp*.cardinality() and, if *tp* is a reference, is 0 if nil and 1 if not.

The delete() method provides a nested transaction that embeds the complex object operation, permitting its effects on the database to be undone if an exception occurs.

```
Method a.delete() raises(IntegrityError)
// Delete complex object a.
   t = Transaction::current();
   Transaction nt(t);
   nt.begin()
   try
      a._try_delete();
      nt.commit();
   handle IntegrityError
      nt.abort();
      raise IntegrityError;
   end handle

Method a._try_delete() raises(IntegrityError)
// Try to delete complex object a.
   t = Transaction::current();
   if t._deleted(a) then exit;
   t._mark_for_deletion(a);
   A = Type(a);
   for each traveral path tpA in A do
      for each target object b referenced by a.tpA do
         a._primitive_drop_tpA (b);
         tpB = Inverse(tpA);
         b._enforce_binding(ImpB(tpB), tpB);
      end for
   end for
   a._primitive_delete();

Method b._enforce_binding(binding, TraversalPath tpB)
                                        raises(IntegrityError)
// Enforce given binding for traversal path tpB in target
//     object b.
   t = Transaction::current();
   case binding
      nil:  if Refs(b.tpB) < LbM(tpB) then
                t._check_path_at_commit(b, tpB);
      – :   raise IntegrityError;
      ~ :   if Refs(b.tpB) < LbM(tpB) then b._try_delete();
      ' :   if Refs(b.tpB) < LbM(tpB) then
                t._check_path_at_commit(b, tpB);
            Transaction nt(t);
            nt.begin()
            try
               b._try_delete();
               nt.commit();
            handle IntegrityError
               nt.abort();
            end handle
      ! : b._try_delete();
   end case
```

**Fig. 7.** Method delete() in interface Object

The _try_delete() method is an indirectly recursive method that may result in the implicit deletion of many objects that are related directly or indirectly to the object upon which it is invoked, designated here as *a*. Its invocation on an object must be dynamically bound to the method on the class representing the object's most specific type. This ensures that _try_delete() processes all traversal path instances involving the object.

The method first checks that object *a* has not already been marked for deletion by invoking the _deleted() method on the current transaction. If it has, _try_delete() simply exits. If not, it marks object *a* for deletion by invoking _mark_for_deletion().

The outer **for each** loop traverses every traversal path $tp_A$ defined in (or inherited by) class *A*. For each such path in object *a*, the inner **for each** traverses all references in the traversal path. The purpose here is to attempt to implicitly drop each reference to a target object *b* (including the inverse reference to *a*) so that object *a* can be deleted. The code first drops each such reference by invoking the _primitive_drop_$tp_A$ method on *a*, which drops $a.tp_A$'s reference to *b* and $b.tp_B$'s reference to *a*. It then invokes the method _enforce_binding() on the target object *b* to enforce the implicit destructibility binding ImpB($tp_B$) for the inverse traversal path $tp_B$.

The last step of _try_delete() actually deletes the object but only if none of the _enforce_binding() invocations raise an exception.

The _enforce_binding() method is assumed for simplicity to be defined in the interface Object. The method for one class in a relationship must be accessible to the other class. The method enforces the destructibility binding semantics specified in Table 1. Here, *b* denotes the implicit parameter and $tp_B$ denotes the explicit parameter since _enforce_binding() is invoked on a target object to enforce the binding for the inverse traversal path in that target object. It is invoked after a reference to target object *b* and its inverse reference in the traversal path $tp_B$ have been dropped by the caller. The **case** statement executes the appropriate code for the given *binding*. The method _check_path_at_commit() is invoked to ensure that a lower bound constraint is rechecked at the end of the complex object operation, i.e., within commit() of the current, nested transaction.

## 5 Conclusions

In this paper, we have proposed adding ORN to the ODMG Object Model and have referenced, illustrated, and discussed algorithms for implementing ORN semantics in an ODMS. The shortcomings of our proposal are that the Object Model is made slightly more complex and ODMS implementations must include a nested transaction capability. Despite these shortcomings and regardless of whether or not ORN is added to the ODMG standard, we believe that vendors should strongly consider including ORN as an extended feature to their ODMSs. We conclude by summarizing the reasons:

- ORN is a simple notation that allows the database developer to specify a variety of association semantics, which define the scopes of complex and composite objects.
- The extended ODL would facilitate a straightforward mapping of association semantics from a conceptual database model, expressed as an ORN-extended UML class diagram, to the logical database model, expressed in the ODL.

- The ODMS would provide the same support for associations that is provided by relational DBMSs via the SQL references clause plus support even more powerful association semantics—for instance, "referential actions" that are based on multiplicities, and also three flavors of "on delete cascade," all of which are highly desirable in describing association semantics.
- If no *<association>* is given for a traversal path, the default *<association>* corresponds to current system capabilities. Thus, adding ORN is a pure extension requiring no changes to the underlying Object Model capabilities.
- The implementation of this extension is relatively simple as shown by the algorithms we have made available and their implementation in OR+.
- The benefits are increased database development productivity and improved database integrity as much less code needs to be developed and maintained by database application developers.

## Acknowledgements

This work was partially supported by the NSF co-operative agreement HRD-9707076.

## References

1. Balaban, M., Shoval, P.: MEER – A EER model enhanced with structure methods. Information Systems 27(4), 245–275 (2002)
2. Bouzeghoub, M., Metais, E.: Semantic modeling and object oriented databases. In: Proc. 17th Int'l VLDB Conference, Barcelona, Spain, pp. 3–14 (1991)
3. Cattel, R.G.G., Barry, D.K., Berler, M., Eastman, J., Jordan, D., Russell, C., Schadow, O., Sta-nienda, T., Velez, F.: The Object Database Standard: ODMG 3.0, San Mateo, CA. Morgan Kaufmann, San Francisco (2000)
4. Chen, P.P.: The entity-relationship model: towards a unified view of data. ACM Transactions on Database Systems 1(1), 1–36 (1976)
5. ANSI. Information technology - Database languages - SQL, Parts 1-4, American National Standards Institute (ANSI) (2003), New York, http://www.ansi.org
6. Ehlmann, B.K., Riccardi, G.A.: A comparison of ORN to other declarative schemes for specifying relationship semantics. Information and Software Technology 38(7), 455–465 (1996)
7. Ehlmann, B.K., Riccardi, G.A.: Object Relater Plus: A Practical Tool for Developing Enhanced Object Databases. In: Proc. 13th Int'l Conference on Data Engineering, Birmingham, England, pp. 412–421 (1997)
8. Ehlmann, B.K., Rishe, N., Shi, J.: The formal specification of ORN semantics. Information and Software Technology 42(3), 159–170 (2000)
9. Ehlmann, B.K., Riccardi, G.A., Rishe, N., Shi, J.: Specifying and enforcing association semantics via ORN in the presence of association cycles. IEEE Transactions on Knowledge and Data Engineering 14(6), 1249–1257 (2002)
10. Ehlmann, B.K., Yu, X.: Extending UML class diagrams to capture additional association semantics. In: Proc. 20th IASTED Int'l Conf. on Applied Informatics, Innsbruck, Austria, pp. 395–401 (2002)

11. Ehlmann, B.K.: A data modeling tool where associations come alive. In: Proc. 21st IASTED Int'l Conf. on Modelling, Identification, and Control, Innsbruck, Austria, pp. 66–72 (2002), http://www.siue.edu/~behlman
12. Ehlmann, B.K.: Algorithms for the implementation of ORN in an ODMG-compliant ODMS (2006), http://www.siue.edu/~behlman
13. Mellor, S.J., Clark, A.N., Futagami, T.: Guest editor's introduction: Model-Driven Development. IEEE Software 20(5), 19–25 (2003)
14. Lazarevic, B., Misic, V.: Extending the entity-relationship model to capture dynamic behavior. European Journal of Information Systems 1(2), 95–106 (1991)
15. Progress Software. ObjectStore Interprise. Bedford, MA: Progress Software (2006), http://www.objectstore.com/datasheet/index.ssp
16. OMG. Unified Modeling Language (UML) Specification. Version 2.0. Object Management Group (OMG) (2005), http://www.uml.org

# Measuring Effectiveness of Computing Facilities in Academic Institutes: A New Solution for a Difficult Problem

Smriti Sharma[1] and Veena Bansal[2]

Department of Industrial & Management Engineering
Indian Institute of Technology, Kanpur, U.P., India
smriti.sh@gmail.com, veena@iitk.ac.in

**Abstract.** There has been a constant effort to evaluate the success of Information Technology in organizations. This kind of investment is extremely hard to evaluate because of difficulty in identifying tangible benefits, as well as high uncertainty about achieving the expected value. Though a lot of research has taken place in this direction, but not much is written about evaluating IT in non-profit organizations like educational institutions. Measures for evaluating success of IT in such kind of institutes are markedly different from that of business organizations. The purpose of this paper is to build further upon the existing body of research by proposing a new model for measuring effectiveness of computing facilities in academic institutes. As a baseline, Delone & McLean's model for measuring the success of Information System [2], [3] is used, as it is the most pioneering model in this regard.

**Keywords:** Computing Facilities, Usability, Functional Utility, User Satisfaction, Individual and Organizational Impact.

## 1 Introduction

Given the crucial role of education in development and the expansion of Information and Communication technology in the global economy, the role of IT in education cannot be ignored. Of late there has been a major surge in the use of IT in the territory of education. This, at the same time, has raised the questions- How effective is IT in academic institutions? How to measure the effectiveness/ success of IT in educational institutions? Effectiveness is concerned about the impact of the information provided in helping users do their job. It is important to evaluate the impact of the IT on the organization as a whole rather than looking at the quality of the system, user satisfaction or by looking at a narrow financial perspective of the evaluation.

The difficulties in effectively evaluating the impact of information systems are widely acknowledged in the IS literature [2], [8], [9].

Evidence suggests that poor performance of the IS function is a serious inhibitor to good business performance [1]. Better use of information, both internal and external, relates positively to profitability [7].

J. Filipe, B. Shishkov, and M. Helfert (Eds.): ICSOFT 2006, CCIS 10, pp. 270–278, 2008.

A lot of research has been undertaken in this regard to develop frameworks for measurement of Information Systems' success. Economic and quantitative measures for the success of IS, however, are difficult to obtain. Researchers and practitioners alike often rely on subjective assessment and surrogate measures, such as end-user computing satisfaction (EUCS) instrument.

Saunders and Jones [4] developed the "IS Function Performance Evaluation Model" which was used to describe how measures should be selected from the multiple dimensions of the IS function relative to specific organizational factors and based on the perspective of the evaluator.

The model proposed by Delone et al [2], [3] to measure the effectiveness of Information System is the most pioneering work in this regard. DeLone and McLean Information Systems (IS) Success Model is a framework and model for measuring the complex-dependent variable in IS research. It concludes with a model of "temporal and causal" interdependencies between their six categories of IS success- *Information Quality*, *System Quality*, *Use*, *User Satisfaction*, *Individual Impact*, and *Organizational Impact*.

Their model depicts the relationships of the 6 IS success dimensions. They contend that System Quality and Information Quality singularly and jointly affect both Use and User Satisfaction. Additionally, the amount of Use can affect the degree of User satisfaction. Use and User Satisfaction are direct antecedents of Individual Impact; and lastly, this impact on individual performance should eventually have some Organizational Impact. This model was later on validated by many researchers including Seddon and Kiew [6], who tested the causal structure of the model.

Inspite of being the most complete and a better known model some shortcomings have been sighted in this model by researchers. It does not take into consideration the effect of extraneous variables both internal and external to the organization. They themselves accept that it is necessary to include the organization type and its environment into context before applying this model.

In the light of the above argument, we have made an attempt to modify Delone and McLean's model to make it relevant for measuring the effectiveness of computing facilities in academic institutes. Information Quality and System Quality have been replaced by Usability and Functional Utility. Use construct is omitted from the proposed model. Measures for evaluating success of IT in such kind of institutes are markedly different from that of business organizations. Therefore, for capturing Individual Impact and Organizational Impact measures suitable in the context of academic institutes have been introduced.

## 2 Proposed Model

Following modifications have been proposed in the Delone & McLean's model.

**Replacing System Quality and Information Quality.** We are concerned with measuring effectiveness of all the computing facilities of an academic institute unlike [2] where focus is on an individual Information System. Therefore, System Quality and Information Quality have been replaced by Usability and Functional Utility.

**Omission of Use Construct.** A main criticism of Delone and McLean has centered on the *Use* construct. It is considered to be an inappropriate measure of IS success. Its implication is that if a system is used, it must be useful, and therefore successful. Take the example of an expensive design software, which is used only by handful of students. If this software helps these students to produce some excellent research work, it will be considered as an asset for the institute, irrespective of the number of students using it. Hence, Use construct was considered as inappropriate in this context.

Taking the points mentioned above into consideration, the proposed model includes the following five constructs- Usability, Functional Utility, User Satisfaction, Individual Impact and Organizational Impact. The relationship between the constructs is as shown in Fig. 1.

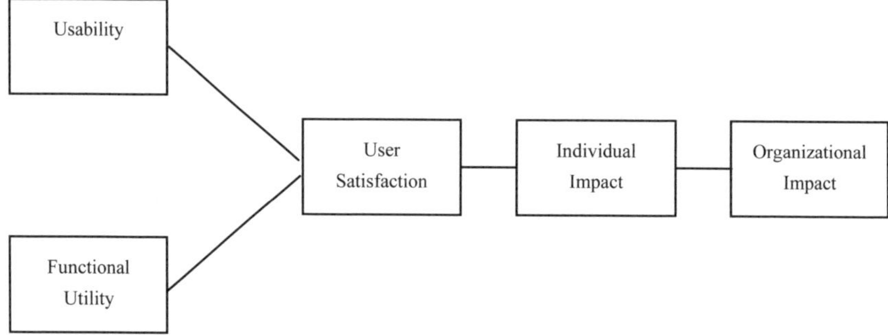


**Fig. 1.** Proposed model

This model shows the interdependent nature of success categories used.

Usability measures the extent to which the computing facilities match user characteristics and the skills for the tasks concerned. Functional Utility focuses on how well the computing facilities meet the requirements of the users. It also measures the availability, accuracy and up-to–datedness of the information obtained from the use of computing facilities. User satisfaction is the most extensively used single measure for IS evaluation [2]. End-user's feelings of satisfaction arise when he or she combines his or her perception of and valuation of discrepancy regarding desires and expectations from the use of computing facilities. Individual Impact and Organizational Impact indicate the impact of computing facilities on individual performance and organizational performance, respectively. Measures used for Individual Impact are concerned with evaluating the impact of computing facilities on an individual in learning, course work, research work, planning and decision making, communication and overall productivity. Likewise, Measures of Organizational Impact evaluate the impact of comporting facilities on the organizational as a whole in the following respects- innovation, research quality, pass rate/grades, decision making, image of the institute, capacity in terms of students, and overall productivity of the institute.

## 3 Model Validation

Aim of testing this model was to provide an empirical evidence for the relationships between the five constructs used in the proposed model. We conducted a self-administered survey to collect the primary data from the target population, which consisted of students and faculty of five academic institutes.

For the survey, a questionnaire was designed based on discussions with students and faculty and literature. Respondents were asked to fill the questionnaire in the context of computing facilities used in their institutes.

Questionnaire contained five sets of questions to measure the five constructs of the model.

Questions were framed by discussions with students and faculty of various academic institutes and available literature. To evaluate the first construct Usability, a set of four questions was used. For measuring Functional Utility six questions were framed. Four questions on Overall Satisfaction were from Seddon and Yip [5]. To measure Individual Impact and Organizational Usability measures the extent to which the computing facilities match user characteristics and Impact group of five and six questions were used, respectively.

Likert scale was used for measurement in which respondents indicate a degree of agreement or disagreement with each of a series of statements about the stimulus objects. Each statement has been assigned seven response categories, ranging from 1 to 7. One signifies strong agreement, and seven means strong disagreement.

### 3.1 Data Collection

Questionnaires were administered personally to the students and faculty of the aforementioned institutes. Total of 500 Questionnaires were distributed, out of which, 411 completed questionnaires were returned by the respondents. After screening of questionnaires to identify illegible, incomplete, or ambiguous responses, 31 questionnaires were rejected. Total, 380 questionnaires were found suitable for data analysis. Treatment of missing values was done by substituting a neutral value.

### 3.2 Data Analysis and Results

To establish the model, three regression models have been used:

- Multiple regression model with Usability and Functional Utility as independent variables and User Satisfaction as dependent variable.
- Simple regression model with User Satisfaction as independent variable and Individual Impact as dependent variable.
- Simple regression model with Individual Impact as independent variable and Organizational Impact as dependent variable.

Using the abbreviations

$X_1$ = Usability
$X_2$ = Functional Utility
$X_3$ = User Satisfaction
$X_4$ = Individual Impact
$X_5$ = Organizational Impact

the following linear regressions are considered

$$X_3 = b_{3.12} + b_{31.2} X_1 + b_{32.1} X_2 \quad (1)$$
$$X_4 = b_{4.3} + b_{43} X_3 \quad (2)$$
$$X_5 = b_{5.4} + b_{54} X_4 \quad (3)$$

The eq. (1) represents a multiple linear regression and (2) and (3) are simple linear regressions, hereafter called Simple Regression 1 and Simple Regression 2 respectively.

Here $b_{3.12}$ , $b_{4.3}$ and $b_{5.4}$ are constants; $b_{43}$ and $b_{54}$ are regression coefficients ,$b_{31..2}$ and $b_{32.1}$ are partial regression coefficients. The suffix after the dot refers to the variable held constant.

### 3.3 Hypotheses

The hypotheses to be tested are as follows:

H1: The partial regression coefficient $b_{31..2} > 0$

It is assumed that if the user finds the computing facilities easy to use, perceived usefulness of the system will increase for him. This subsequently, will result into increased User Satisfaction.

H2: The partial regression coefficient $b_{32..1} > 0$

Increase in Functional Utility will result into increased usefulness for the user and hence increased satisfaction. The more the facilities meet the requirements of the user the more will be the User Satisfaction

H3: The regression coefficient $b_{43} > 0$

This hypothesis states that if a student is more satisfied with the computing facilities then it will have a more positive Individual Impact e.g. better learning or communication with students/faculty.

H4: The regression coefficient $b_{54} > 0$

Higher Individual Impact will result into higher Organizational Impact e.g. a positive effect of computing facilities on learning of individual students will result into overall improvement in pass rate/ grades of the institute.

Data analysis was done using SPSS.

**Table 1.** Cronbach's alpha

| | No. of Items | Cronbach alpha |
|---|---|---|
| Usability($X_1$) | 4 | .6790 |
| Functional Utility($X_2$) | 6 | .8479 |
| User Satisfaction($X_3$) | 4 | .8497 |
| Individual Impact($X_4$) | 6 | .8772 |
| Organizational Impact($X_5$) | 7 | .8796 |

High Cronbach's alpha for all the variables in Table 1, except for Usability, which is marginally less, is an indication of high internal consistency. Low value for Usability can be attributed to lower number of items used to measure it.

**Table 2.** Pearson Correlation matrix

| | $X_1$ | $X_2$ | $X_3$ | $X_4$ | $X_5$ |
|---|---|---|---|---|---|
| Usability($X_1$) | 1 | | | | |
| Functional Utility($X_2$) | .562 | 1 | | | |
| User Satisfaction($X_3$) | .602 | .815 | 1 | | |
| Individual Impact($X_4$) | .551 | .774 | .817 | 1 | |
| Organizational Impact($X_5$) | .537 | .722 | .769 | .812 | 1 |

Table 2 shows the Pearson Coefficient of Correlation between all the variables. Pearson's correlation coefficient (r) is a measure of the strength of the association between the two variables.

The coefficient of correlation between the constructs Usability and Functional Utility is low, which indicates their independence. The coefficients of correlation are high for the constructs Functional Utility and User Satisfaction; User Satisfaction and Individual Impact; Individual Impact and Organizational Impact as suggested by the model. However, it is on the lower side for the constructs Usability and User Satisfaction, which suggests that dependence of User Satisfaction is higher on Functional Utility as compared to Usability.

**Table 3.**

| | $R^2$ | Adjusted $R^2$ | F (p-value) |
|---|---|---|---|
| Multiple Regression | .695 | .693 | 428.747 (0) |
| Simple Regression 1 | .667 | .666 | 757.880 (0) |
| Simple Regression 2 | .659 | .658 | 729.581 (0) |

The high values of t and F- statistic in all the cases strongly support the rejection of the Null hypotheses, that the regression coefficients are zero. The regression coefficients except for $b_{31.2}$ have high positive values. Also the 95% confidence intervals are small. The coefficients of determination show reasonably good fit. All the above results tend to validate the model and support all the four hypotheses.

## 4 Conclusions

Results obtained from path analysis of the survey data provide considerable empirical evidence for the model. Results show strong dependence of User Satisfaction on Usability and Functional Utility; Individual Impact on User Satisfaction and Organizational Impact on Individual Impact. All the four Hypotheses assumed in the beginning of the research are found be true.

An implication of the model is that because of the causal nature of these dimensions, Usability, Functional Utility and User Satisfaction are sufficient to measure the effectiveness of computing facilities.

On the basis of the small piece of work done in this thesis, it is strongly recommended that every academic institution should undergo through this self screening or self assessment process. This model can be used by academic institutes to get regular feedbacks about their computing facilities, which will help them in continuous improvements.

An attempt has been made to include all the suitable measures of each construct. However, there is a scope of including new measures for each of the constructs. More questions can be added to the questionnaire to measure each of these constructs, including both positive and negative statements to check the consistency of the respondents. Finally, inclusion of other constructs in the model can be investigated.

**Table 4.**

| | Path | | Unstandardized Coeff. | | Std. Coeff | t (p-value) | 95% Conf. Bounds | |
|---|---|---|---|---|---|---|---|---|
| | from | to | Coeff | Std. Er. | | | Lower | Upper |
| H1 | Usability | User Satisfaction | .228 | .037 | .211 | 6.126 (0) | .155 | .302 |
| H2 | Functional Utility | User Satisfaction | .737 | .036 | .696 | 20.234 (0) | .666 | .809 |
| H3 | User Satisfaction | Individual Impact | .779 | .028 | .817 | 27.530 (0) | .723 | .835 |
| H4 | Individual Impact | Organizational Impact | .798 | .030 | .812 | 27.011 (0) | .740 | .856 |

## References

1. Carlson, W.M., McNurlin, B.C.: Do you measure UP? Computerworld 26(49), 95–98 (1992b)
2. Delone, W.H., McLean, E.R.: Information systems success: the quest for dependent variable. Information Systems Research 3, 60–95 (1992)

3. DeLone, H.W., McLean, R.E.: The DeLone and McLean model of information systems success: A ten-year update. Journal of Management Information Systems 19(4), 9–30 (2003)
4. Saunders, C.S., Jones, J.W.: Measuring performance of the information systems function. Journal of Management Information Systems 8(4), 63–82 (1992)
5. Seddon, P.B., Yip, S.K.: An empirical evaluation of user information satisfaction UIS, measures for use with general ledger accounting software. Journal of Information Systems, 75–92 (1992)
6. Seddon, P.B., Kiew, M.-Y.: A partial test and development of Delone and Mclean's model of IS success. In: Fifteenth Annual International Conference of Information Systems (ICIS) (1994)
7. Strassman, P.A.: The Business value of Computers: An executive Guide. Information Economic Press, New Canaan, CI (1990)
8. Willcocks, L., Lester, S.: Beyond the IT Productivity Paradox. European Management Journal 14(3), 279–290 (1996)
9. Willcocks, L.: Investing in Information Systems: Evaluation and Management, 1st edn. Chapman & Hall, Boca Raton (1996)

## Appendix: Survey on Computing Facilities in Academic Institutes

This questionnaire uses a seven-point scale. The scale represents a spectrum. 1 signifies that you strongly agree with the given statement, and 7 means you strongly disagree. For each question tick the number that reflects what you think about each statement. Computing facilities refer to computer hardware, software and network of your institute.

PART A: Usability

| | |
|---|---|
| 1 Computing facilities are easy to use. | 1 2 3 4 5 6 7 |
| 2 Computing facilities are user friendly. | 1 2 3 4 5 6 7 |
| 3 It is easy to acquire skills for using the Computing facilities. | 1 2 3 4 5 6 7 |
| 4 It requires lot of effort to use the Computing facilities | 1 2 3 4 5 6 7 |

PART B: Functional Utility

| | |
|---|---|
| 1 Computing facilities meet most of your requirements. | 1 2 3 4 5 6 7 |
| 2 The content of information obtained with the help of computing facilities meets your requirements. | 1 2 3 4 5 6 7 |
| 3 Computing facilities are available whenever required. | 1 2 3 4 5 6 7 |
| 4 You can get in touch with sufficient sources of information by using computing facilities. | 1 2 3 4 5 6 7 |
| 5 Computing facilities enable you to obtain accurate information. | 1 2 3 4 5 6 7 |
| 6 Computing facilities enable you to obtain up-to-date information. | 1 2 3 4 5 6 7 |

## PART C: User Satisfaction

| | | | | | | | |
|---|---|---|---|---|---|---|---|
| 1 Computing facilities meet your information processing and computational needs. | 1 | 2 | 3 | 4 | 5 | 6 | 7 |
| 2 Computing facilities are fast enough. | 1 | 2 | 3 | 4 | 5 | 6 | 7 |
| 3 Computational facilities are effective. | 1 | 2 | 3 | 4 | 5 | 6 | 7 |
| 4 Overall,you are satisfied with the computing facilities. | 1 | 2 | 3 | 4 | 5 | 6 | 7 |

## PART D: Individual Impact

| | | | | | | | |
|---|---|---|---|---|---|---|---|
| 1 Computing facilities help you in learning. | 1 | 2 | 3 | 4 | 5 | 6 | 7 |
| 2 Computing facilities help you in course work. | 1 | 2 | 3 | 4 | 5 | 6 | 7 |
| 3 Computing facilities help you in research work. | 1 | 2 | 3 | 4 | 5 | 6 | 7 |
| 4 Computing facilities help you in planning and decision making. | 1 | 2 | 3 | 4 | 5 | 6 | 7 |
| 5 Computing facilities help you in communication with teachers and students. | 1 | 2 | 3 | 4 | 5 | 6 | 7 |
| 6 Computing facilities help you in improving your overall productivity. | 1 | 2 | 3 | 4 | 5 | 6 | 7 |

## PART E: Organizational Impact

| | | | | | | | |
|---|---|---|---|---|---|---|---|
| 1 Computing facilities help in encouraging innovation. | 1 | 2 | 3 | 4 | 5 | 6 | 7 |
| 2 Computing facilities help in improving research quality. | 1 | 2 | 3 | 4 | 5 | 6 | 7 |
| 3 Computing facilities help in improving overall pass rate/grades. | 1 | 2 | 3 | 4 | 5 | 6 | 7 |
| 4 Computing facilities help in better decision making. | 1 | 2 | 3 | 4 | 5 | 6 | 7 |
| 5 Computing facilities help in improving the image of the institute. | 1 | 2 | 3 | 4 | 5 | 6 | 7 |
| 6 Computing facilities help you in increasing capacity in terms of students. | 1 | 2 | 3 | 4 | 5 | 6 | 7 |
| 7 Computing facilities help in improving overall productivity of the institute. | 1 | 2 | 3 | 4 | 5 | 6 | 7 |

# Combining Information Extraction and Data Integration in the ESTEST System

Dean Williams and Alexandra Poulovassilis

School of Computer Science and Information Systems, Birkbeck College, University of London
Malet Street, London WC1E 7HX, U.K.
{dean,ap}@dcs.bbk.ac.uk

**Abstract.** We describe an approach which builds on techniques from Data Integration and Information Extraction in order to make better use of the unstructured data found in application domains such as the Semantic Web which require the integration of information from structured data sources, ontologies and text. We describe the design and implementation of the ESTEST system which integrates available structured and semi-structured data sources into a virtual global schema which is used to partially configure an information extraction process. The information extracted from the text is merged with this virtual global database and is available for query processing over the entire integrated resource. As a result of this semantic integration, new queries can now be answered which would not be possible from the structured and semi-structured data alone. We give some experimental results from the ESTEST system in use.

**Keywords:** Information Extraction, Data Integration, ESTEST.

## 1 Introduction

The Semantic Web requires us to be able to integrate information from a variety of sources, including unstructured text from web pages, semi-structured XML data, structured databases, and metadata sources such as ontologies. Other applications exist which also need to make used of heterogeneous data that is structured to varying degrees as well as related free text. For example, in UK Road Traffic Accident reports data in a standard structured format is combined with free text accounts in a formalised subset of English; in crime investigation operational intelligence gathering, textual observations are associated with structured data relating to people and places; and in Bioinfomatics structured databases such as SWISS-PROT [1] include comment fields containing related unstructured information.

Data integration systems provide a single virtual global schema over a collection of heterogeneous data sources that facilitates global queries across the sources [2,3]. Such systems are able to integrate data occurring in a variety of structured and semi-structured formats but, to our knowledge, they have not so far attempted to include unstructured text. In Information Extraction (IE) systems, pre-defined entities are extracted from text and this data fills slots in a template using shallow NLP techniques [4]. Data integration and IE are therefore complementary technologies and we argue

J. Filipe, B. Shishkov, and M. Helfert (Eds.): ICSOFT 2006, CCIS 10, pp. 279–292, 2008.

that a system that combines them can provide a basis for applications that need to integrate information from text as well as structured and semi-structured data sources. Our ESTEST system integrates the schemas of structured data sources with ontologies and other available semi-structured data sources, creating a virtual global schema which is then used as the template and a source of named entities for a subsequent IE phase against the text sources. Metadata from the data sources can be used to assist the IE process by semi-automatically creating the required input to the IE modules. The templates filled by the IE process will result in a new data source which can be integrated with the virtual global schema. The resulting extended virtual global database can subsequently used for answering global queries which could not have been answered from the structured and semi-structured data alone.

The rest of this paper is structured as follows: Section 2 describes the architecture of ESTEST and use of existing data integration and IE software. Section 3 give the design and implementation of our ESTEST system, alongside an example that illustrates its usage. Section 4 presents results from initial experiments in the Road Traffic Accident application domain. Section 5 compares and contrasts our approach with related work. Finally, in Section 6 we give our conclusions and plans for future work.

## 2 Background

The ESTEST *Experimental Software to Extract Structure from Text* system is implemented as a layer over the AutoMed [5] data integration toolkit and the GATE [6] IE framework, making use of their facilities. We now briefly describe GATE and AutoMed.

**AutoMed.** In data integration systems, several data sources, each with an associated schema, are integrated to form a single virtual database with an associated global schema. Data sources may conform to different data models and therefore need to be transformed into a common data model as part of the integration process. AutoMed is able to support a variety of common data models by providing graph-based metamodel, the Hypergraph Data Model (HDM). AutoMed provides facilities for specifying higher-level modeling languages in terms of this HDM e.g. relational, entity-relational, XML. These specifications are stored in AutoMed's Model Definitions Repository (MDR). A generic wrapper for each data model is provided, with specialisations for interacting with specific databases or repositories e.g. a set of relational wrappers for interacting with the common relational DBMS. The schemas of such data sources can be extracted by the appropriate wrapper and are stored in AutoMed's Schemas & Transformations Repository (STR). AutoMed provides a set of primitive transformations that can be applied to schemas e.g. to add, delete and rename schema constructs. AutoMed schemas are therefore incrementally transformed and integrated by sequences of such transformations (termed *pathways*).

Add and delete transformations are accompanied by a query (expressed in a functional query language, IQL [7]) which specifies the extent of the added or deleted construct in terms of the rest of the constructs in the schema. These queries can be used to translate queries or data along a transformation pathway [8]. In particular, queries expressed in IQL can be posed on a virtual integrated schema, are reformulated by AutoMed's Query Processor into relevant sub-queries for each data source, and are sent to the data source wrappers for evaluation. The wrappers interact with the data sources for the evaluation of these sub-queries, and with the Query Processor for post-processing of sub-query results.

The queries supplied with primitive transformations also provide the necessary information for these transformations to be automatically *reversible*, e.g. an add transformation is reversed by a delete transformation with the same arguments. AutoMed is therefore defined as a *both-as-view (BAV)* data integration system in [8] which gives an in-depth comparison of BAV with the other major data integration approaches, Global-As-View (GAV) and Local-As-View (LAV) [2].

One of the main advantages of using AutoMed for ESTEST rather than a GAV or LAV-based data integration system is that, unlike GAV and LAV systems, AutoMed readily supports the evolution of both source and integrated schemas. This is because it allows transformation pathways to be extended, so that if a new data source is added, or if a data source or integrated schema evolves, then the entire integration process does not have to be repeated and instead the schemas and transformation pathways can be 'repaired'.

As a pre-requisite for the development of ESTEST we have made two extensions to the AutoMed toolkit which were required for ESTEST but which are also more generally applicable. Firstly, we have extended AutoMed to include support for data models used to represent ontologies. We have modeled RDF [9] graphs and associated RDFS [10] schemas in the HDM. A corresponding AutoMed wrapper for such data sources has been implemented using the JENA API [11]. Although only RDF/S data sources are currently supported, the use of JENA means that additional ontology models can easily be added as specialisations of the current wrapper, similarly to the specialised AutoMed relational wrappers for specific RDMBS. Secondly, we have a developed the AutoMed HDM Store, a native HDM repository that is used for storing the data that is extracted by ESTEST from text sources.

**GATE.** The GATE system [6] provides a framework for building IE applications. It includes a wide range of standard components and also allows the integration of bespoke components. Applications are assembled as pipelines of components which are used to process collections of documents. Applications can be built and run either as standalone Java programs or through the GATE GUI. A pattern matching language called *JAPE* [12] is provided for constructing application specific grammars. Some standard JAPE grammars are provided with GATE e.g. for finding names of people in text. The result of running an application is a collection of annotations over a text. For example in the string "RAN IN FRONT OF BUS" an annotation might state that from the seventeenth to nineteenth character there is a reference to a public service vehicle.

## 3 The ESTEST System

The ESTEST system supports an incremental approach to integrating text with structured and semi-structured data sources, whereby the user iterates through a series of steps as new information sources need to be integrated and new query requirements arise. The ESTEST approach is described in [13,14]. This paper describes an implementation of that approach, as well as giving some experimental results.

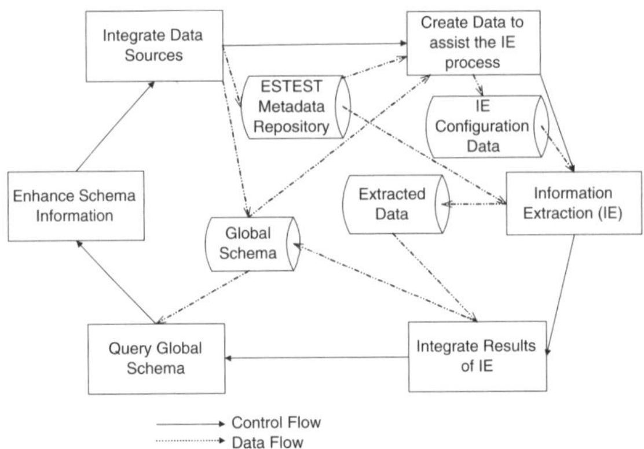


**Fig. 1.** Overview of the ESTEST system

We use as our running example, a simple example application based on Road Traffic Accident reports. In the UK, accidents are reported using a format known as STATS-20 [15]. In STATS-20, a record exists for each accident, following which there are one or more records for the people and the vehicles involved in the accident. The majority of the schema consists of coded entries, and detailed guidance as to what circumstances each of the codes should be used accompanies these. A textual description of the accident is also reported, expressed in a stylised form of English. An example of the textual description collected for a specific accident might be "FOX RAN INTO ROAD CAUSING V1 TO SWERVE VIOLENTLY AND LEAVE ROAD", where "V1" is short for "vehicle 1" and is understood to be the vehicle which caused the accident. The schema of the structured part of the STATS-20 data is well designed and there have been a number of revisions during its several decades of use. However, there are still queries that cannot be answered via this schema alone.

In our running example we suppose that an analysis of the road traffic accidents caused by animals is required, including the kind of animal causing the obstruction.

Figure 1 illustrates the main components of the ESTEST system and each of these is described in turn in sections 3.1 to 3.5 below.

### 3.1 Integrate Data Sources

The *ESTEST Integrator* component is configured with the collection of data sources available to the user. Each of these has an associated *ESTEST Wrapper* instance (there is an ESTEST Wrapper for each data model supported by AutoMed). The ESTEST Wrapper makes use of the corresponding AutoMed wrapper in order to construct a data source schema within the AutoMed STR. However, the ESTEST Wrapper is also able to extract additional metadata from the data source, which is stored in the *ESTEST Metadata Repository* — see Figure 1. The ESTEST wrapper also transforms the AutoMed representation of a data source schema into the ESTEST data model described below.

The first step for the Integrator is to iterate through the data sources and use the associated wrapper to construct an initial schema within the AutoMed STR.

In our example, two data sources are assumed to be available: AccDB is a relational database holding the relevant STATS-20 data and AccOnt is a user-developed RDF/S ontology concerning the type of obstructions which cause accidents. Figure 3 shows the AccOnt RDFS schema and some associated RDF triples. The AccDB database consists of three tables:

```
accident(acc_ref,road,road_type,hazard_id, acc_desc)
vehicle(acc_ref,veh_no,veh_type)
carriageway_hazards(hazard_id,hazard_desc)
```

In the accident table, each accident is uniquely identified by an acc_ref, the road attribute identifies the road the accident occurred on, and road_type indicates the type of road. The hazard_id contains the carriageway hazards code and this is a foreign key to the carriageway_hazards table. We assume that the multiple lines of the text description of the accident have been concatenated into the acc_desc column. There may be zero, one or more vehicles associated with an accident and information about each them is held in a row of the vehicle table. Here veh_reg uniquely identifies each vehicle involved in an accident and thus acc_ref,veh_no is the key of this table. The Integrator calls on the wrappers to create a relational schema for AccDB and RDF/S schema for AccOnt.

The data sources are each now converted from their model specific representation into the *ESTEST data model*. The table below shows the constructs of the ESTEST data model and their representation in the HDM. We see that the ESTEST model provides *concepts* which are used to represent anything that has an extent i.e. instance data. Concepts are represented by HDM nodes e.g. ⟨⟨fox⟩⟩, ⟨⟨animal⟩⟩, and are structured into an isA hierarchy e.g. ⟨⟨isA, fox, animal⟩⟩. Concepts can have *attributes* which are represented by a node and an unnamed edge in the HDM e.g. the attribute ⟨⟨animal, number_of_legs⟩⟩. We note that in AutoMed's IQL query language instances of modeling constructs within a schema are uniquely identified by their *scheme*, enclosed within double chevrons, ⟨⟨ ... ⟩⟩.

| **ESTEST Data Model** | |
|---|---|
| **Construct** | **HDM Representation** |
| Concept: ⟨⟨c⟩⟩ | Node: ⟨⟨c⟩⟩ |
| Attribute: ⟨⟨c,a⟩⟩ | Node: ⟨⟨a⟩⟩<br>Edge: ⟨⟨_,c,a⟩⟩ |
| isA: ⟨⟨isA,c1,c2⟩⟩ | Constraint ⟨⟨c1⟩⟩ $\subseteq$ ⟨⟨c2⟩⟩ |

The two data source schemas in our example are automatically converted into their equivalent ESTEST representation when their ESTEST wrapper is called by the Integrator. The representation of AccDB is shown in Figure 2 and that of AccOnt in Figure 3.

Similarly, the Integrator now calls each ESTEST wrapper to collect metadata from its data source. This metadata is used to assist in finding correspondences and for suggesting to the user which schema constucts could be of use in the later IE step (see section 3.3 below). As well as type information, Word Forms and Abbreviations are collected as described below. This metadata is stored in the ESTEST Metadata Repository.

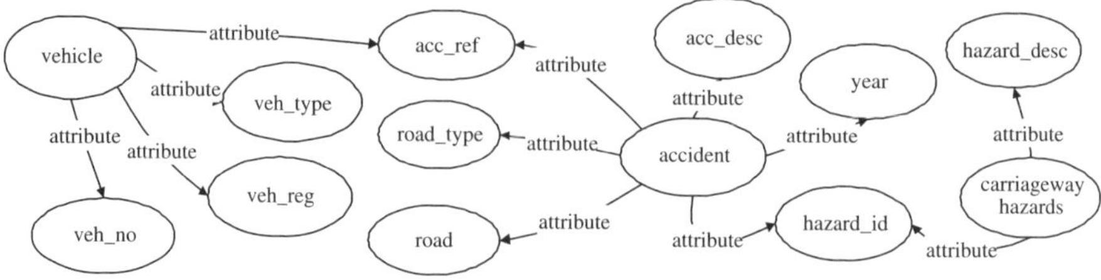


**Fig. 2.** The AccDB schema represented in the ESTEST model

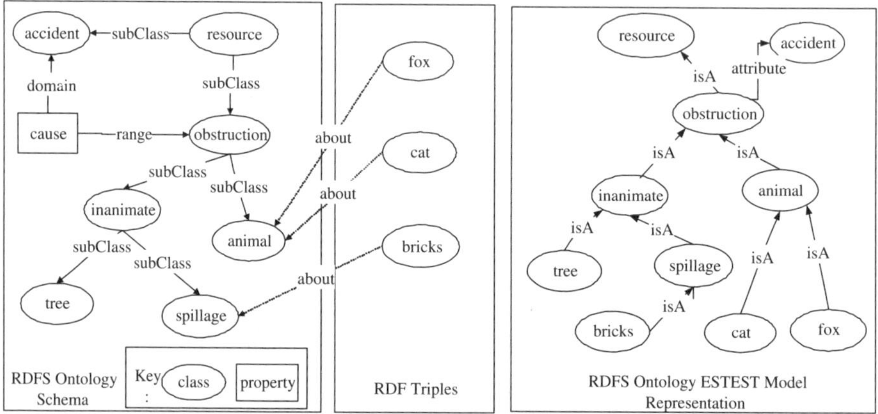


**Fig. 3.** AccOnt Ontology and its representation in the ESTEST model

*Word Forms* are words or phrases which represent a concept. Word forms associated with concepts are important in ESTEST because of their use in the IE process. The ambiguity inherent in natural language means that word forms can be associated with many concepts. The, often limited, textual clues available in the data source schemas are used by ESTEST to find word forms. These clues include schema object identifiers and comment features such as the 'remarks' supported by the JDBC database API. In database schemas, a number of informal naming conventions are often seen, for example naming database columns "acc_identifier" or "accIdentifier". The Integrator recognises a number of these conventions and breaks identifiers returned by the wrappers into their component parts.

*Abbreviations* are often used in database schema object identification, and the Integrator has an extendable set of heuristics for parsing common naming conventions e.g. a rule exists for identifying when an abbreviation of the table name is prefixed onto the column names in a relational table e.g. the abbreviation "acc" in the column "acc_ref" of the "accident" table used in the example. The user can also enter abbreviations commonly used in the application domain. Abbreviations are combined with full word forms to generate possible combinations.

There are a number of alternative sources for expanding the word forms for a concept when required: manually entered, from database description metadata, from lower down the concept isA hierarchy, or from the WordNet [16] natural language ontology. The Integrator is able to use these to expand the number of word forms for a concept as required. A confidence measure is assigned to the word form depending on its source.

Type metadata is also extracted from the data sources and is used to suggest which schema constructs might be text to be processed by the IE component or used as sources of named entities. Named entity recognition is one of the core tasks in information recognition and involves looking up flat-file lists containing the known instances of some type. ESTEST extends this by identifying concepts in the database schema that are likely to be sources of named entity types and uses either their extent or the word forms associated with the concept for the ESTEST named entity recognition component.

In our running example, abbreviations are suggested e.g. "acc" for "accident", "veh" for "vehicle". The Integrator generates alternative word forms for the schema names and presents them to the user e.g. for the schema construct ⟨⟨accident, acc_ref⟩⟩ the associated word forms are "accident reference", "accident" and "reference". The user is now asked if they wish to expand the current word forms of each concept. Suppose that for now only the concept ⟨⟨animal⟩⟩ is selected to be expanded and solely from the schema. In this case the word forms "cat" and "fox" are added to ⟨⟨animal⟩⟩ from the isA hierarchy.

In order to complete the integration and develop the global schema, the Integrator next attempts to find correspondences between concepts in different source schemas using the gathered metadata. The user can accept or amend the list of suggested equivalent schema objects. The corresponding schema constucts are now renamed to have the same name. Using the facilities provided by AutoMed, each of the ESTEST model schemas is incrementally transformed until they each contain all the schema constucts of all the source schemas — these are the union schemas shown in Figure 5. An arbitrary one of these schemas is designated finally as the *global schema*. In our simple example just "accident" from the domain ontology and "accident" from the schema is suggested as a correspondence by the Integrator and is accepted by the user. No additional manual correspondences are supplied and the initial global schema shown in Figure 4 is created. Finally, as the later IE process will require a repository to store the extracted information, an additional data source is created for this purpose by the Integrator. The schema of this new data source is the HDM representation of the ESTEST global schema just derived. This new data source is integrated into the global schema in the same way as the other data sources.

For our example, the resulting network of transformed and integrated schemas is shown in Figure 5 — for each data source there is now an AutoMed schema and correspondences between schema constructs in the data sources have been identified. There is a pathway to an equivalent ESTEST model schema and then on to a union schema. Any one of these union schema can be used as the global schema for the entire network.

## 3.2 Create Metadata to Assist the IE Process

The *ESTEST Configuration* component now makes use of the global schema and collected metadata to suggest basic information extraction rules, such as macros for named entity recognition, and to create templates to be filled from concepts in the schema which have attributes of unknown value. In IE systems, templates are filled by

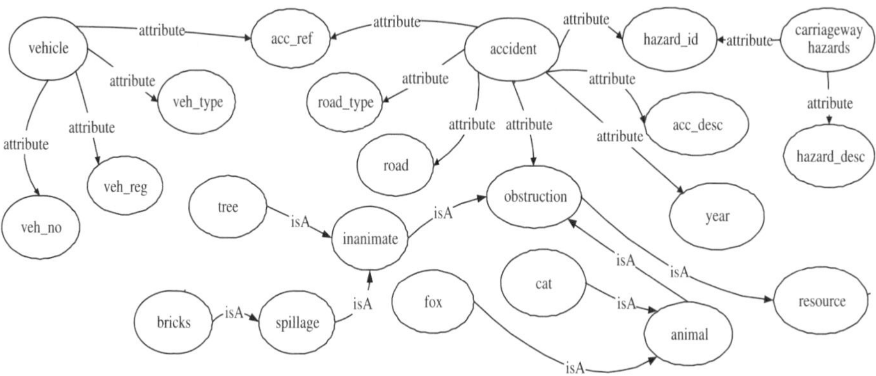


**Fig. 4.** Initial ESTEST global schema

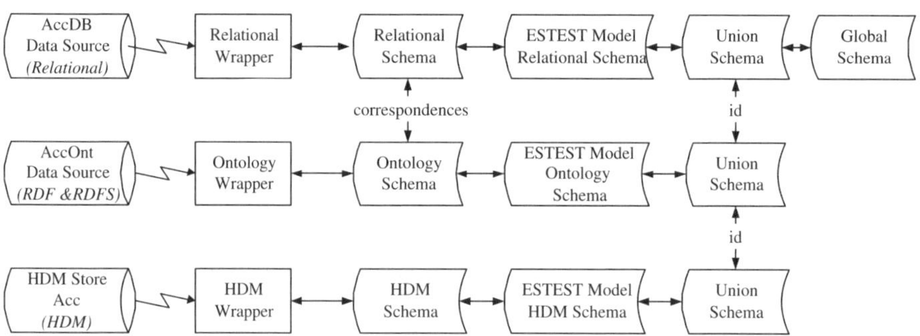


**Fig. 5.** AutoMed Schema Network for the Road Traffic Accident Example

annotations over text (the template slots). ESTEST extends this notion of templates in a number of ways. Firstly, some slots in a template are pre-filled from the structured part of the source data and these values are used to attempt to disambiguate when multiple annotations are suggested over the same text. Secondly, ESTEST makes use of type information contained in the metadata to validate suggested annotations. Finally, ESTEST extends the idea of the relationship that exists between annotations. Normally this only goes as far as linking slots to templates. There is Annotation Schema idea in GATE but this is restricted to defining the annotation features which are allowed and is used to drive the manual entry of annotations. In our system, we link annotation types to concepts in the global schema. In this way the structural relationships between annotation types can be used and extracted annotations can be linked to related instances of other concepts in the source data.

The user confirms or amends the automatically produced configuration. In our example, the user selects just ⟨⟨animal⟩⟩ from the list of suggested named entity sources. the Configuration component identifies the ⟨⟨accident⟩⟩ concept as a template. As the ⟨⟨obstruction⟩⟩ concept has no extent, the IE step will attempt to find values for this. A macro for ⟨⟨animal⟩⟩ is created:

```
Macro: ANIMAL
  ({Lookup.minorType == animal })
```

and a stub JAPE rule for ⟨⟨obstruction⟩⟩. The user enhances this stub as follows:

```
Rule: OBSTRUCTION1 (({Token.string == "RUNS"} |
  {Token.string == "WALKS"}|
  {Token.string  == "JUMPS"})
  (SPACE)?
  ({Token.string == "INTO"} |
  {Token.string == "ONTO"} |
  {Token.string  == "IN FRONT OF"} )
  (ANIMAL) ) :obstruction -->
        :obstruction.obstruction =
        {kind = "Obstruction",
                  rule = "OBSTRUCTION1"}
```

### 3.3 Information Extraction Component

Now the GATE-based ESTEST IE component is run. The following standard components GATE are configured and assembled into a pipeline: *Document Reset* which ensures the document is reset to its original state for reruns; the *English Tokeniser* splits text into tokens such as strings and punctuation; *Sentence Splitter* divides text into sentences; the *JAPE Processor* takes the grammar rules developed above and applies them to the text. Also used is our bespoke *Schema Gazetteer* component which extends the gazetteers used in the Named Entity recognition core IE task. The annotation type is linked to a concept in the global schema and instances are found either by presenting a query to the global schema for the extent concept across the data sources, or the word forms previously associated with the concept are obtained.

To illustrate, suppose in our example the AccDB database contains just three accidents with the following descriptions:

| **AccDB Accident Descriptions** | |
|---|---|
| **Acc Ref** | **Description** |
| A001234 | FOX RUNS INTO ROAD CAUSING V1<br>TO SWERVE VIOLENTLY<br>AND LEAVE ROAD OFFSIDE |
| A005678 | UPPERTON ROAD LEICESTER<br>JUNCTION SYKEFIELD AVENUE<br>V1 TRAV SYKEFIELD AVE FAILS<br>TO STOP AT XRDS AND<br>HITS V2 TRAV UPPERTON RD V2<br>THEN HITS V3 PKD ON OS<br>OF UPPERTON RD |
| A009012 | ESCAPED KANGAROO JUMPS IN<br>FRONT OF V1 |

When the IE component has run the following annotations are found:

| Annotations | | | |
|---|---|---|---|
| **Annotation** | **Start** | **End** | **Literal** |
| ANIMAL | 1 | 3 | FOX |
| OBSTRUCTION | 1 | 13 | FOX RUNS INTO ROAD |

### 3.4 Integrate Results of IE

The data extracted from the text is now stored in the HDM repository. In our example an instance id is generated "#1" and new instances of HDM nodes ⟨⟨fox⟩⟩, ⟨⟨animal⟩⟩ and ⟨⟨obstruction⟩⟩, are added with value [#1]. Edges ⟨⟨isA,fox,animal⟩⟩ and ⟨⟨isA,animal, obstruction⟩⟩ have the value of the pair{#1,#1}.Edge⟨⟨attribute,accident,obstruction⟩⟩ has value {A001234,#1}. Queries on the global schema now include the new fact that a fox caused accident A001234.

### 3.5 Remaining ESTEST Phases

The user can now pose queries to the global schema the results of which will include the new data extracted from the text. The global schema may subsequently be extended by new data sources being added, or new schema constucts identified and added to it. The user may also choose to expand the number of word forms associated with schema concepts. Following any such changes, the process is then repeated and new data extracted from the text.

In our example, suppose the user suspects that the results may not be complete and expands the word forms from WordNet for ⟨⟨animal⟩⟩. The Integrator suggests a mapping between this schema object and a WordNet *synset* (a set of word forms with the same meaning) which the user confirms. Word forms are then obtained by descending the WordNet hypernym relations. The list obtained includes "kangaroo" and as a result re-running the Configuration and IE components now produces similar additional annotations for the third accident report, and an additional fact in the HDM store of KANGAROO as an obstruction for accident A009012.

## 4 Experiments with Road Traffic Accident Data

There are a number of variables which affect the performance of the ESTEST system including: the availability of structured data sources relating to the text, the degree of similarity between the text instances, the amount of effort spent by the user configuring the system to the specific application domain, and the domain expertise of the user. In order to provide some initial confirmation of the potential of our approach we have experimented with a data set of Road Traffic Accident reports consisting of 1658 accident reports. These were six-months worth of reports from one of Britain's 50 police forces. We identified five queries of varying complexity which cannot be answered by the STATS-20 structured data alone. These queries are shown in the table below.

| RTA Queries | |
|---|---|
| Q1 | Which accidents involved red traffic lights? |
| Q2 | How many accidents took place 30-50m of a junction? |
| Q3 | How many accidents involve drunk pedestrians? |
| Q4 | Which accidents were caused by animals? |
| Q5 | How many resulted in a collision with a lamppost? |

The ESTEST system was configured using a randomly chosen set of 300 reports.

We then ran the system over the remaining 1358 unseen reports and compared the results to a subsequent manual examination of each report. The table below shows these results in terms of the actual number of relevant reports, the recall (the number of correctly identified reports as a percentage of all the correct reports) and the precision (the number of correctly identified reports as a percentage of all the identified reports).

**Results of RTA Query Experiments**

| **Query** | **Relevant Reports** | **Recall** | **Precision** |
|---|---|---|---|
| Q1 | 25 | 84% | 95% |
| Q2 | 89 | 99% | 99% |
| Q3 | 9 | 78% | 100% |
| Q4 | 14 | 86% | 86% |
| Q5 | 15 | 88% | 93% |

Where appropriate the queries combined structured and text results, for example STATS-20 structured data includes a flag used to indicate whether the accident took place within 20 meters of a junction but does not reveal the distance otherwise. In the query "How many accidents took place 30-50m of a junction?" accidents with the 20 meters flag set were discarded and only the remaining reports with relevant IE results considered. Configuring the system took 5 hours to develop a domain ontology and enhancing the generated stub rules for the IE process. These results are promising even with the short time spent configuring the system by a non-domain expert user.

## 5 Related Work

An alternative to rule-based IE is text mining, in which some NLP process creates a structured data set from the text and then this is mined in order to discover patterns in the structured data [17]. Our ESTEST system, in contrast, is driven by specific new querying requirements and the potential of making use of new data sources. None the less, a text mining extension might be a useful addition to ESTEST for some application domains e.g. the SWISS-PROT database, while [18] has shown the potential for combining IE and Text Mining in general.

Recent developments in IE have included moves to support the challenges of knowledge management [19] and applications for the semantic web e.g. support for ontologies has recently been added to GATE [20]. A common starting point for recent research is the limitations of the traditional core Named Entity Recognition task in IE where annotations are assigned a type from a list of types names, and recent work is moving towards the idea of *Semantic Annotation* [21] where the annotation is linked to a concept in an ontology. This is similar to the ESTEST approach of linking annotations to a concept in a virtual global schema.

The developers of the Knowledge and Information Management (KIM) platform [22] believe that a lightweight ontology that provides structure but few axioms is sufficient for the IE task. Our ESTEST data model can similarly be thought of as a lightweight ontology providing a taxonomy and the facility for concepts to have attributes. KIM is based on an ontology of 'everything' (KIMO) pre-defined to include concepts and entities from the common IE tasks; annotations that are found by the IE system are treated separately from the knowledge in the ontology. In contrast, ESTEST develops the global schema from available structured data sources specific to the application, and seeks to expand the instance data and schema incrementally by adding to the previously known data and schema. Also, ESTEST makes use of already known instances of structured data which relate to the specific text being processed in order to assist in finding slot values, and it uses other schema constucts and word forms obtained from metadata as sources for named entities.

Finally, previous work on making use of the text in UK Road Traffic Accident reports used expert knowledge of the sub-language in the reports to code description logic-based grammar rules [23]. ESTEST makes use of the structured data as well as the text to answer queries and the approach is more generally applicable to other application domains as it does not depend on a specific restricted sub-language.

## 6 Conclusions and Future Work

We have described the ESTEST system, which combines techniques from Data Integration and Information Extraction in order to make integrated use of heterogeneous data that is structured to varying degrees as well as related free text. ESTEST does this by extracting metadata from data sources to support their integration into a virtual global schema expressed in the ESTEST data model. This schema and its associated metadata are used to semi-automatically configure an IE process. The newly extracted information from the text is merged into the virtual global database which can then be used to answer new queries that could not have been answered before. The user can extend the global schema, add new data sources, enhance the IE configuration, and rerun as required.

This approach is novel in a number of ways. First, to our knowledge this is the first time that a Data Integration system has been extended to include support for free text. Second, ESTEST uses schema integration techniques to integrate available structured data into a global schema which is used as a light-weight ontology for semantic annotation — this is a realistic application-specific alternative to building ontologies from scratch. Third, ESTEST uses the global schema to semi-automatically configure the IE process, thereby reducing the configuration overhead of the IE process.

We have given a simple example illustrating the operation of ESTEST on Road Traffic Accident reports. Initial experimental results in the same domain indicate the approach is able to increase the utility of text stored alongside structured data and that the system's performance is at least comparable with standalone IE systems.

Based on this preliminary evaluation we have identified three areas for further enhancement of ESTEST: developing its schema matching component by using IE on the textual schema metadata; exploring identifier disambiguation techniques to assist with the co-referencing task in IE; and improving the support for enhancing the global schema with new structure found in the text.

To investigate how generally applicable our approach is, we will then evaluate the enhanced ESTEST system in the crime investigation and semantic web application domains. In this evaluation, we will consider the results obtained by ESTEST users who are experts in the given application domain. These results will be compared to both any existing approaches (such as manual inspection or keyword search) employed by these expert users and to results obtained by them using a stand-alone IE system.

Finally, in order to support the end user, we will analyse the requirements of a workbench for accessing the ESTEST components and functionality via a graphical user interface.

## References


1. Bairoch, A., Boeckmann, B., Ferro, S., Gasteiger, E.: Swiss-Prot: Juggling between evolution and stability. Brief. Bioinform. 5, 39–55 (2000)
2. Lenzerini, M.: Data Integration: A Theorectical Perspective. In: Proc. PODS 2002, pp. 247–258 (2002)
3. Halevy, A.Y.: Data Integration: A Status Report. In: Weikum, G., Schöning, H., Rahm, E. (eds.) BTW, GI. LNI, vol. 26, pp. 24–29 (2003)
4. Appelt, D.: An introduction to Information Extraction. Artificial Intelligence Communications 12, 161–172 (1999)
5. AutoMed Project (2006), `http://www.doc.ic.ac.uk/automed/`
6. Cunningham, H., Maynard, D., Bontcheva, K., Tablan, V.: GATE: A framework and graphical development environment for robust NLP tools and applications. In: Proc. of the 40th Anniversary Meeting of the Association for Computational Linguistics (2002)
7. Poulovassilis, A.: A tutorial on the IQL query language. Technical report, AutoMed Project (2004)
8. McBrien, P., Poulovassilis, A.: Data integration by bi-directional schema transformation rules. In: Proc. ICDE 2003, pp. 227–238 (2003)
9. Lassila, O., Swick, R.: Resource description framework (RDF) model and syntax specification. W3C Recommendation (1999), `http://www.w3.org/TR/REC-rdf-syntax/`
10. Brickley, D., Guha, R.: RDF vocabulary description language 1.0: RDF schema. W3C Recommendation (2004), `http://www.w3.org/TR/rdf-schema/`
11. McBride, B.: Jena: A semantic web toolkit. IEEE Internet Computing 6, 55–59 (2002)
12. Cunningham, H., Maynard, D., Tablan, V.: JAPE: a Java Annotation Patterns Engine. In: Research memorandum, 2nd edn. University of Sheffield (2000)
13. Williams, D., Poulovassilis, A.: An example of the ESTEST approach to combining unstructured text and structured data. In: Proc. of the Database and Expert Systems Applications (DEXA 2004), pp. 191–195, IEEE Computer Society, Los Alamitos (2004)
14. Williams, D.: Combining data integration and information extraction techniques. In: Proc. Workshop on Data Mining and Knowledge Discovery, at BNCOD 2005, pp. 96–101 (2005)

15. UK Department for Transport: Stats20: Instructions for the completion of road accident report form (1999), http://www.dft.gov.uk
16. Fellbaum, C. (ed.): WordNet An Electronic Lexical Database. MIT Press, Cambridge (1998)
17. Tan, A.H.: Text mining: The state of the art and the challanges. In: Proc. of the PAKDD 1999 Workshop on Knowledge Discovery from Advanced Databases, pp. 65–70 (1999)
18. Nahm, U.Y., R.M.: Using Information Extraction to aid the discovery of prediction rules from text. In: Proc. of the KDD-2000 Workshop on text Mining, pp. 51–58 (2000)
19. Cunningham, H., Bontcheva, K., Li, Y.: Knowledge Management and Human Language: Crossing the Chasm. Journal of Knowledge Management 9, 108–131 (2005)
20. Bontcheva, K., Tablan, V., Maynard, D., Cunningham, H.: Evolving GATE to Meet New Challenges in Language Engineering. Natural Language Engineering 10, 349–373 (2004)
21. Kiryakov, A., Popov, B., Ognyanoff, D., Manov, D., Kirilov, A., Goranov, M.: Semantic Annotation, Indexing, and Retrieval. In: Fensel, D., Sycara, K.P., Mylopoulos, J. (eds.) ISWC 2003. LNCS, vol. 2870, pp. 484–499. Springer, Heidelberg (2003)
22. Popov, B., Kiryakov, A., Ognyanoff, D., Manov, D., Kirilov, A.: KIM - a semantic platform for information extraction and retrieval. Nat. Lang. Eng. 10, 375–392 (2004)
23. Wu, J., Heydecker, B.: Natural language understanding in road accident data analysis. Advances in Engineering Software 29, 599–610 (1998)

# Introducing a Change-Resistant Framework for the Development and Deployment of Evolving Applications

Georgios Voulalas and Georgios Evangelidis

Department of Applied Informatics, University of Macedonia
156 Egnatia St., Thessaloniki, Greece
voulalas@uom.gr, gevan@uom.gr

**Abstract.** Software development is an R&D intensive activity, dominated by human creativity and diseconomies of scale. Current efforts focus on design patterns, reusable components and forward-engineering mechanisms as the right next stage in cutting the Gordian knot of software. Model-driven development improves productivity by introducing formal models that can be understood by computers. Through these models the problems of portability, interoperability, maintenance, and documentation are also successfully addressed. However, the problem of evolving requirements, which is more prevalent within the context of business applications, additionally calls for efficient mechanisms that ensure consistency between models and code, and enable seamless and rapid accommodation of changes, without interrupting severely the operation of the deployed application. This paper introduces a framework that supports rapid development and deployment of evolving web-based applications, based on an integrated database schema. The proposed framework can be seen as an extension of the Model Driven Architecture targeting a specific family of applications.

**Keywords:** Model-driven Development, Meta-Models, Evolving Business Applications, Application Generators, Application Deployment Platforms, Reflectional Programming.

## 1 Introduction

Information systems are one of the most effective ways for the enterprises to deal with challenges of today's dynamic, competitive environment. The enterprise may be a commercial business, a government agency or an academic institution. A vast majority of these information systems are long-lived, multi-step applications that support mission-critical business processes spanning multiple enterprise applications, corporate departments, and business partners.

Why have these process-driven applications become so prevalent? There are certainly many reasons but the most apparent ones are [12]:

- Today's economic challenges have forced enterprises to look for new efficiencies by automating processes untouched by their existing enterprise systems. Packaged enterprise applications such as ERP systems manage only typical processes such as material resource planning and financial reporting.

J. Filipe, B. Shishkov, and M. Helfert (Eds.): ICSOFT 2006, CCIS 10, pp. 293–306, 2008.

- The rigidity of packaged applications (ERP, CRM, etc.) nullifies what many firms regard as their competitive advantage, i.e., their unique business processes.
- Processes are embedded in ERP and other monolithic systems. Embedding processes in software is a bad idea, since they cannot be easily changed, combined with others, or integrated for collaboration.
- Business processes extending behind the firewall and over the Internet have created new opportunities for companies to achieve channel efficiencies by creating new business processes and extending existing ones to customers, trading partners and suppliers.
- The emergence of the Application Service Provision model in the late '90s has created new prospects in setting enterprise collaboration infrastructures (e.g., e-marketplaces). New business models have arisen, like the e-Business Service Provision model, which introduces an intermediate player that delivers business development services through dynamically adaptive software solutions for inter-organizational process automation & improvement.

The response to these challenges is similar: companies are looking for technology solutions to improve enterprise processes, leverage existing infrastructure and create new ways to compete. The fact that they can obtain powerful computational resources and reliable, high-performance network infrastructures at low cost enables them to focus solely on the development of efficient and sophisticated software solutions.

Still, software development is an area in which we are struggling with a number of major problems. The most important problems are [7]:

**The Productivity, Documentation, and Maintenance Problem.** The software development process includes a number of phases: (a) Conceptualization and requirements elicitation and gathering, (b) Analysis and functional description, (c) Architectural specification and design, (d) Implementation, (e) Testing, and, (f) Deployment. Whether we use an incremental and iterative process, or the traditional waterfall process, documents and diagrams are produced during the first three phases. The connection between those artefacts and the code fades away as implementation progresses. Changes widen the gap, since they are usually done at the code level only, due to time restrictions. The idea of Extreme Programming (XP) has rapidly become popular, since it is built upon the fact that the code is the driving force of software development and thus the phases that should accumulate the major effort are coding and testing. However, having just code and tests makes maintenance of a software system very difficult. Practically speaking, analysis and design artefacts are required, but to be really productive they should not be just static, paper representations. They have to stay in high cohesion with the code throughout the software lifecycle, they should elevate technologists above the lower level complexities that are imposed by the available (with continuously increased complexity) technologies, and they need to be eligible as input in forward-engineering operations.

**The Portability Problem.** The software industry has a special characteristic that makes it stand apart from most other industries. Each year, and sometimes even faster, new technologies are being invented and becoming popular (e.g., Java, CORBA, UML, XML, J2EE, .NET, and Web Services). The new technologies offer concrete benefits for companies and many of them cannot afford to lag behind. As a consequence, the investments in previous technologies lose value, and existing systems have to be ported to the new technology in order for interoperability (with systems built with the new technology) restrictions to be completely wiped out.

**The Interoperability Problem.** Software systems rarely live isolated. Most systems need to communicate with other, often legacy, systems.

**The Evolution Problem.** The management of evolution in information systems is a dominant requirement. This is even stronger in business applications, due to the dynamic nature of business domains. In [11] the following factors that drive information system evolution are listed:

"*A change in the universe of discourse*": The application world is continually evolving. A viable application system should accommodate these changes.

"*A change to the interpretation of facts about the universe of discourse and the manner in which the task is realized in a system*": People are not able to precisely express the desired functionality of a large-scale application system. Only experience from using the system will enable them to properly formulate the needs and requirements.

"*Changes in the form of updates to effect upgrades to the functionality or scope of a system*": People do not know in advance all the desired functionality of a large-scale application system. Only experience from using the system will enable them to realize and express all needs and requirements.

"*Changes in the form of updates to effect efficiency improvements*". For example, the restructuring of database elements in order for faster information retrieval to be achieved.

In order for evolution to be handled efficiently the following objectives should be met:

- changes should be seamlessly incorporated without the need of restructuring the existing application,
- analysis and design artefacts should be updated in order for changes to be reflected,
- the operation of the deployed application should not be interrupted, or at least interruption should minimized, and
- access to old business objects within their right context should be supported, i.e., at any time an old business object should be able to be easily retrieved and examined through the specific version of the application that produced and manipulated it, in order for user to be able to trace back to former business data.

This paper introduces a new framework for the development and deployment of web-based business applications. In Section 2 we introduce a composition framework that singles out four essential constituents for every business application. In Section 3, we present the Model Driven Architecture (MDA) and the modern practices brought out by Microsoft. In Section 4, we discuss the areas of the MDA that will take advantage of the proposed framework. In section 5, the framework is introduced. The last section provides a conclusive summary of the paper and identifies our future research plan.

## 2 Defining the Puzzle

The four coordinates that drive software production and evolution within an enterprise, a business network or even a marketplace are the following:

- **Flow of Events (Workflow):** Every business application incorporates a workflow model that indicates the flow of activities & information, how involved roles interact and the conditions mastering the flow. When applications are developed with generic development platforms (e.g. J2EE, .NET etc), there are several software engineering techniques to capture & design such flows (e.g. activity diagrams), but during implementation the workflow model gets embedded in the code.
- **Object Processing:** Every process incorporates business objects that are created, routed, processed and archived within its activities. These objects transfer, among the involved actors, the information that is necessary for the execution of the process. With generic development platforms, there is enough flexibility to implement components managing any structured information.
- **Enterprise Modelling:** Besides workflows and data processing logic, every business application incorporates mechanisms for Enterprise Modelling, organizational relationships establishment and role assignment services. It should be noted that Enterprise Modelling often indicates the optimum manner that applications should be utilized within an organization. In typical applications developed with generic development platforms, Enterprise Modelling is limited to user administration, authentication & authorization services, but the development environment itself provides the opportunity to develop models as complex as one wishes.
- **Integration:** Integration with third-party information systems, either workflow or ERP systems, custom applications or embedded systems (e.g. applications embedded in manufacturing equipment), is also essential for process automation. Here, a combination of XML standards, WEB Services and object-oriented techniques for mastering the complexity of integration requirements is very essential. Unfortunately, the majority of application development environments consider system integration as simple data import & export, and usually such implementations allow for limited interoperability.

Those four coordinates will help us to define the core model of our proposed framework in Section 5.

## 3 MDA and Microsoft Software Factories

MDA [7], [8], [9] is a framework for software development defined by the OMG. The MDA development lifecycle is not very different from the traditional lifecycle; they both involve the same phases. One of the major differences has to do with the nature of the artefacts that are produced during the development process. The artefacts are models that can be understood and processed by computers. The following three models are at the heart of the MDA.

**Platform Independent Model (PIM).** This model is the first to be defined and is a model with a high level of abstraction that is independent of any implementation technology. Within a PIM, the system is modelled from the aspect of how it best supports the business requirements.

**Platform Specific Model (PSM).** In the next step, the PIM is transformed into one or more PSMs. A PSM specifies the system (or part of the system) in terms of the implementation details defined by one specific implementation technology.

**Code.** The final step in the development is the transformation of each PSM to code. Because a PSM fits its technology rather closely, this transformation is relatively straightforward.

For many specifications, PIM and PSMs are defined in UML, making OMG's standard modelling language a foundation of the MDA.

In contrast to traditional development, MDA transformations are always executed by tools. Many tools are able to transform a PSM into code; there is nothing new to that. What's innovative in MDA is that the transformation from PIM to PSM is automated as well (Fig. 1).

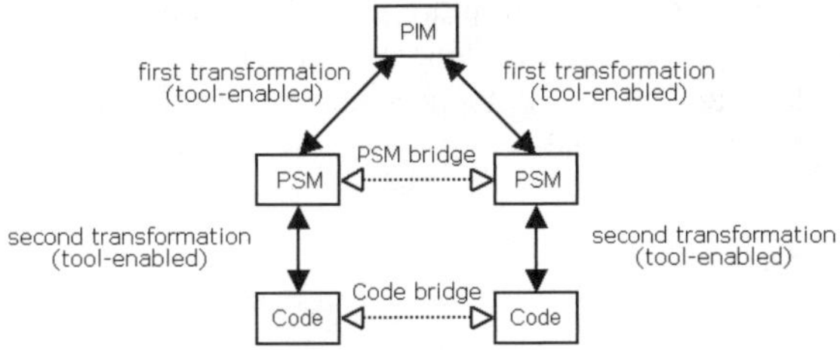


**Fig. 1.** Models, transformations & bridges in the MDA development process

Let us now clarify how MDA responds to the challenges presented in the previous section.

**Productivity, Documentation and Maintenance.** In MDA the focus for a developer shifts to the development of a PIM. The PSMs that are needed are produced automatically, and code is in turn generated automatically from the PSMs. Developers can shift focus from code to PIM, thus paying more attention to eliciting requirements and resolving the business problems. This results in systems that fit much better with the needs of the end users, and are developed in less time. The PIM fulfils the function of high-level documentation that is needed for any software system. The PIM is not frozen after writing, since changes made to the system will eventually be made by changing the PIM and regenerating the PSMs and the code. In the MDA approach the documentation at a high level of abstraction will naturally be available; this makes maintenance easier.

**Portability.** Portability is achieved by focusing on the development of PIMs that are by definition platform independent.

**Interoperability.** When PSMs are targeted at different platforms, they cannot directly talk to each other. Concepts from one platform should be transformed into concepts used in another platform. MDA addresses this problem by generating not only the PSMs, but the necessary bridges between them as well.

**Evolution Management.** The PIM is a live artefact that depicts precisely the system throughout its lifecycle, since all changes made to the system are eventually made by changing the PIM and regenerating the PSMs and the code.

On the other side, Microsoft has recently introduced Domain Specific Languages (DSLs) with its own modelling environment, Visual Studio 2005 Team System (VSTS). DSLs [4] are programming languages dedicated to specific problems and consisting of their own built-in abstractions and notations. DSLs underpin Microsoft's concept of software factories, that are planned modules of tools, content and processes used to build applications in specific domains like healthcare, human resources or enterprise resource planning. Microsoft has chosen the term "software factory" in order to emphasize upon reusable assets and tooling for supporting them. The software industry welcomed the new approach, however many are still cautious, mainly due to the displacement of the UML and the fact that since software is an R&D and not a production activity, it is difficult to apply manufacturing principles. Undoubtedly, narrowing the domain enables to more precisely define the features of the target family and facilitates the definition of languages, patterns, frameworks and tools that automate the development of its members. One early backer for the DSL and Software Factories approach is Borland.

## 4 Rethinking MDA

MDA is a complete framework that enables organizations to respond efficiently to the augmentative requirements of modern software projects.

The current status of the framework is mainly shaped by the availability of support tools and therefore presents the following deficiencies [7]:

- Though OMG has defined the mapping standards between the three models (the PIM, the PSM and the code), it has yet to define how to implement the models. This task has been left to the software development tool vendors currently supporting the MDA initiative. Although many of these vendors have implemented parts of the MDA, few have done so in its entirety. In order for users to fully benefit from MDA, vendors need to implement all of MDA, i.e., implement all three coordinates, and ensure that their tools are standards-based and business model-driven.
- Tools should automatically transform higher-level platform-independent models into lower-level platform-specific models and generate code automatically. Current tools are not sophisticated enough to fully provide the transformations from PIM to PSM and from PSM to code. The developers need to manually improve the transformed PSM and / or code models.
- The extent to which portability can be achieved depends on the automated transformation tools that are available. For popular platforms, a large number of tools will undoubtedly be available. For less popular platforms, the user may have to

use add-on tools that support transformation definitions, or write proprietary transformation definitions.
- Cross-platform interoperability can be realized by tools that generate both the PSMs and the bridges between them. Existing tools are not so advanced to cope with this requisite.

Undoubtedly, it is a matter of time before software vendors overcome the above-mentioned limitations. However, there exist a number of areas that can be improved. More specifically, MDA fails to:

- **Ensure Consistency between the Produced Code and the Preceding Models.** Even if vendors succeed in building transformation tools that fully generate the required code based on the specifications modelled in the PSMs, one cannot guarantee that developers will not interfere manually with the generated code. Consequently, the consistency between the three cornerstone models is unstable.
- **Cope Efficiently with the Problem of Evolving Requirements.** In MDA, every new change requires code to be regenerated and recompiled, and the final application to be redeployed. What's more, the arbitrary realization of changes may create gaps between the three models. Last but not least, MDA can provide access to data that have been manipulated by previous versions of the application, only by maintaining different installations of the applications, approach that is a neither practical, nor elegant.

Those limitations are inherent to the MDA's comprehensiveness, since it is very difficult to elaborate on a more sophisticated solution while in parallel coping with all types of applications.

## 5 The Proposed Framework

Motivated by the above-mentioned findings related to the MDA paradigm, its core principles, and the latest practices adopted by Microsoft and Borland, we introduce an innovative extension for the realization of a development and deployment framework targeted to web-based business applications. The proposed framework (depicted in Fig. 2) will be structured on the basis of a universal database schema (meta-model).

Development will be supported by components (modelling tools) that will elicit functional specifications from users and transform them in formal definitions, and by data structures (part of the meta-model) that will be utilized for the storage of the definitions.

Deployment will be supported by generic components (meta-components) that will be dynamically configured at run-time according to the functional specifications provided during development, and by application-independent data structures (part of the meta-model) that will hold all application-specific data.

The following two statements outline the philosophy of the proposed solution:

- No code (SQL, Java, C++, JSP, ASP, etc.) will be generated for the produced applications; just run-time instances of generic components will be created.

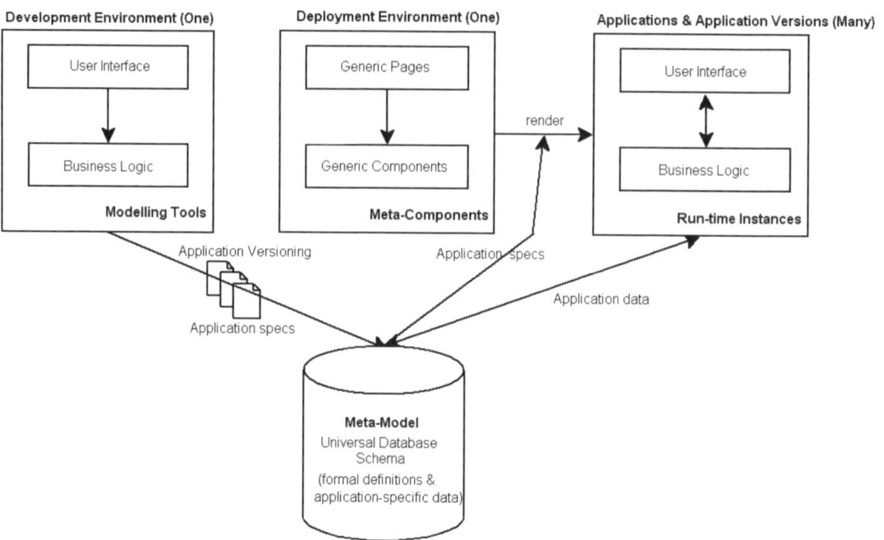


**Fig. 2.** Structure of the Proposed Development and Deployment Framework

- There will always exist one deployed application, independently of the actual number of running applications. Application-specific behaviour will be rendered by this universal application according to the functional definitions that are maintained in the database. In other words, functional and presentation specifications are shifted from the middle and front tier respectively to the database tier (taking as basis a 3-tier approach that is the most outstanding architectural paradigm). Response to business changes is instant, simply through the manipulation of data tuples.

More specifically, the proposed framework includes the models that are described below.

## 5.1 Domain Model

The Domain Model is a business-oriented model that maps to the MDA Platform Independent Model and covers the coordinates presented in Section 2. It defines the structure of the data that the application is working on (objects, attributes, and associations), along with their behavioural aspect (methods) and business rules. It is mainly structured on the basis of the Object-Oriented paradigm, augmented with the extensions introduced by the Object Constraint Language [10], [3] for the description of constraints that govern the modelled objects, plus elements from an acceptable business rules classification scheme [1], [2], [6], with the Ross method [1] being the prevalent. Therefore, its main entities are:

- Business objects. Business objects are created, routed, processed and archived within the different business activities. They carry the information that is necessary for the execution of a process. *Example: Travel Application, Accommodation Proposal, Air Ticket, and Traveller.*

- Status: Each business object passes through different statuses during its lifecycle. *Example: Un-submitted, Submitted, and Rejected (for the travel application).*
- Attributes: Define the static aspect (information) of a business object. *Example: Cost (numeric), Notes (alphanumeric), and Check-out date (for the Accommodation Proposal).*
- Methods: Define the dynamic aspect (behaviour) of a business object. *Example: Submit, Approve, and Reject (for the Travel Application).*
- Association: Represents structural relationship between business objects that exist for some duration (in contrast with transient links that, for example, exist only for the duration of an operation). *Example: A Travel Application is associated with one or more Accommodation Proposals.*
- Argument: A parameter required for the execution of method. *Example: Submission notes and priority are arguments of the 'submit' method.*
- Term: A noun or noun phrase with an agreed upon definition. A term is essentially an object or attribute that is included in a business rule. *Example: Air Ticket, fare.*
- Fact: A complete statement connecting terms (via verbs or prepositions) into sensible, business-relevant observations. A fact is essentially a business-significant association. *Example: A Travel Application is associated with at least one Traveller.*
- Computation Rule: Provides an algorithm for arriving at the value of a term. A computation rule is essentially a business-significant method. *Example: The total cost of a Travel Application is computed as the air tickets fare plus the accommodation cost.*
- Pre-condition: A condition that must hold before executing an operation. It typically evaluates one or more attributes. *Example: The 'submit' method can only be executed upon those travel applications that are un-submitted.*
- Post-condition: Defines either the return value of a method or modifications on the value of component attributes that must be performed. *Example: The status of a Travel Application changes to 'submitted' after the execution of the 'submit' method.*
- Guard: Force the execution of operations anytime triggers (i.e. all attributes involved in the guard condition) get a specific state. *Example: Each time an Accommodation Proposal gets approved by the travellers (i.e., its status changes to 'approved') the status of the associated Travel Application is updated.*
- Invariant Constraint: A condition that must always hold as long as the system operates. It typically constraints the value of an attribute. *Example: The value of the attribute 'numberOfPassengers' should always be greater than zero.*

Besides business rules and data processing logic, every business application incorporates mechanisms for enterprise modelling, business relationships establishment, role assignment, and personnel administration. Thus, the Domain Model embraces an additional component, named Enterprise Model, which covers inter-organizational and intra-organizational aspects. The main entities of this sub-model are:

- Business Role: In each process, one or more business roles are identified. Example: Corporation, Travel Agency.

- Enterprise: The organization that participates in the process by undertaking a specific business role. In the case of business applications limited to the enterprise scope, only one organization exists. In the case of business networks or e-marketplaces multiple organizations exist. Example: Corporation X, Travel Agency Y.
- Business Units: Departments, branches or affiliated companies of an enterprise. Example: The accounting department of corporation X
- Partnership: Cooperation relationships established between enterprises (applies only to business networks and e-marketplaces). Example: The Partnership that has been established between corporation X and travel agency Y within the CTP (supposing that an e-marketplace that enables the cooperation of travel agencies with corporate customers exists).
- Partner: An enterprise that participates in a partnership by playing an undertaking business role. Example: The travel agency Y in the previous partnership.
- Employee: A person employed by an enterprise. Employees usually belong to business units. Example: Mr. X.
- Role: Represents the responsible actor for the fulfilment of a set of activities (methods implemented by business objects). An activity can be optionally associated with more than one role. Example: Traveller, Travel Arranger, Travel Agent, and Travel Administrator.
- User: An employee that has access to the business application. A user is associated with one or more roles. Example: Mr. X that access the business application as traveller.

Although the entities included in the Enterprise model can be implemented as instances of the meta-entities of the core Domain Model, we have selected to handle them separately for reasons of performance. Thus, instead of dynamically configuring the meta-entities to render the desired functionality, we utilize standard entities. This differentiation stems from the fact that the mechanisms implemented by the Enterprise Model can be specified in advance, as they are common among all business applications.

Specifications included in the Domain Model will be stored in a database. The database schema should embrace the proposed structure and include all identified entities (Business Object, Method, Rule, etc.).

As for modelling language, UML including OCL will be extensively utilized within the Domain Model. However there is need for a specialization of UML for modelling inter- and intra-organizational aspects, which means that a new UML profile focused on the Enterprise Model should be defined.

### 5.2 Application Model

The Application Model maps to the MDA Platform Specific Model and focuses on the targeted platform. The Application Model contains the following three sub-models:

- **Presentation Model:** It pictures the overall structure of the presentation elements. Display pages are defined for every business object based on the identified attributes.

Input pages that elicit the information required for the execution of the methods are defined based on the specified methods and arguments. Pages are interrelated according to the identified object associations. In order for the model to include every presentation detail, the domain model should include exhaustive information, such as the conditions under which attributes are hidden / displayed, the controls that should be used for the selection of values (radio-button or selection list), formatting properties for currencies and dates, etc.

- **Business Logic Model:** Suppose that we select the Java 2 Standard Edition (J2SE) as target platform. All objects and terms will be mapped to the 'java.lang.Object' class. Alphanumeric attributes will be mapped to 'java.lang.String' class. A method (or piece of a method) that returns part of an alphanumeric will be mapped to the 'substring' method that is implemented by the 'java.lang.String' class. Similarly, a computation rule will be mapped to a set of primitive methods supported by the target platform that will be invoked in specific order in order for the rule to be propagated. In general, all elements included in the Domain Model will be mapped to fundamental elements of the target programming language. Note that the mapping of the elements of the Enterprise Model to the elements of the target language will be much more direct, since the Enterprise Model is not a meta-model (i.e. included entities are predefined).
- **Data Model:** Based on the identified objects, their attributes and the way they associated, a data model is structured. Only persistent objects (i.e. objects that need to "survive") are mapped to database structures. The discrimination between persistent and transient objects is captured in the domain model. Note that since the part of the data model that covers the data needs of the Enterprise Model has predefined structure, only the mapping to the selected database system specs (data types, etc.) has to be conducted for it.

### 5.3 Operation Model

The Operation model consists of the following building blocks.

- **Presentation Model Instance:** Run-time instances of generic presentation elements (e.g., Java Server Pages or Active Server Pages that obey to specific Cascading Style Sheets).
- **Business Logic Model Instance:** Run-time instances of the generic functional components (meta-objects) that render the behaviour of an application-specific object. The exact process is the following: application specifications are retrieved from the database at run-time and the generic components are configured dynamically in order to expose the specified functionality by utilizing reflectional adaptation techniques (reflection is the process by which a program can modify its own behaviour and is supported by many object-oriented programming languages). For each different technology utilized at Application Level (J2SE, .NET, J2EE), different components should exist. Practically speaking, every programming language that supports reflectional behaviour can be utilized.
- **Data Model Instance:** The part of the unified database schema that will hold the realizations of the business object instances (e.g., realizations of the travel applications, orders, products, etc.). The database schema will be independent of the applications, i.e., its structure will be fixed. In [15] a framework for dynamically

evolving database environments is introduced. Similar to our approach it is based upon a database structure that is independent of applications. Changes to the data structure of the application result to record modifications, instead of changing the schema itself. In comparison to our approach the specific research effort focuses only to the data side of applications.

Note that the three sub-models included in the Application Model are not transformed to code at operation level, except for the part of the Business Logic Model that originates from the Enterprise Model. Instead, the definitions that they include are coupled with the generic components (presentation elements, functional components, and database) in order for the required functionality to be rendered.

### 5.4 Discussion

Note that the three sub-models included in the Application Model are not transformed to code at operation level, except for the part of the Business Logic Model that originates from the Enterprise Model. Instead, the definitions that they include are coupled with the generic components (presentation elements, functional components, and database) in order for the required functionality to be rendered.

The proposed framework responds to the challenges identified in Section 4 as follows:

- **Consistency between the Produced Code and the Preceding Models.** Since no code is generated and the middle model is generated automatically in its entirety, all changes are realized through the Domain Model.
- **Efficient Handling of Evolving Requirements.** Having shifted the functional and presentation specifications from the middle and front tier respectively to the database tier we can easily achieve evolution management by applying standard data versioning techniques. In case the static (attributes) or dynamic (methods) definition of a business object is modified this results in modifications to the underlying data instances, i.e., we can deal with changes at deployment time without recompiling and redeploying the application. What's more we can, at anytime, refer to a previous version of an application and examine old data in their real context by retrieving the corresponding data instances from the database, without the need of maintaining multiple installations.

What's more, in full compliance with the MDA principles, the framework enhances productivity by incorporating application generation features through the elicitation of high-level, formal definitions that are automatically transformed to low-level technical specifications, and supports portability through the Application Model that can be theoretically supported by any programming language that supports reflection and by any database system.

## 6 Conclusions and Further Research

In this paper we examine the development and deployment of web-based business applications through a different perspective: our main aim is to elaborate on and limit the side-effects that are induced by the continuously changing requirements, while conforming to the principles introduced by the MDA paradigm and retaining its undisputable advantages, i.e., improved productivity, efficient documentation, effective maintenance, production, portability, and interoperability. For this reason, we suggest transferring the functional specifications of the application from the components (code) to the database and utilizing them at run-time in order to configure generic components. The development and deployment platform will be based upon a unified database schema. The generic components will be built with the use of a programming language that supports reflection. These meta-components will be configured at run-time in order to render the application-specific functionality. Dynamic functional specifications will let end-users deal with changes at deployment time without recompiling and redeploying the application. What's more, with simple data versioning techniques that enable the retrieval of previous specifications, the operation of previous versions of an application will be feasible through the same, unique installation. Last but not least, since all changes pass through the Domain Model, the consistency between the three cornerstone models will not be compromised.

It should be clear that our goal is to present an interesting perspective that could somehow extend the MDA framework and not replace it. Besides, one can easily identify a set of drawbacks in comparison with the MDA framework:

- The proposed framework has narrower scope, since it focuses on web-based business applications.
- MDA handles efficiently integration with other systems, while the current formulation of the proposed framework supplants the specific coordinate.
- Indisputably, a solution that is build upon a meta-model and extensively utilizes reflection requires increased computational resources compared to a traditional one.

The first constraint is enforced by the fact that is practically infeasible to create a generator that can produce any application [5], [14] and is in compliance with the latest developments as pictured by the initiatives undertaken by major software players. This is the main reason for considering and evaluating this framework as an extension of the MDA that targets on a specific group of applications. The third drawback is minor, since the availability of powerful computational resources encourages the elaboration of sophisticated solutions. Working towards a 'lighter' solution, we will consider adopting partial behavioural reflection [13]. We also plan to address the issue of interoperability.

Future research will focus on:

- **Extending the Framework** with a coordinate that will cover the need for cross-platform interoperability. This coordinate will be structured on the basis of the Web Services paradigm.
- **Elaborating on a New UML Profile** for the modelling of business entities.
- **Implementing the Required Infrastructure.** After finalizing the structure of the framework and identifying all main entities, we have to elaborate on the database schema. Performance issues should be seriously taken into account in the selection of the adopted data-modelling paradigm (relational, object-relational, object). The next step will be the specification and implementation of the meta-components along with the components that will support the development process. The derived prototype will verify the viability and efficiency of the proposed solution.

## References


1. Business Rules Forum 2004 Practitioners Panel. The DOs and DON'Ts of Business Rules, http://www.brcommunity.com/b230.php?zoom_highlight=panelists
2. Butleris, R., Kapocius, K.: The Business Rules Repository for Information Systems Design. In: ADBIS Research Communications, pp. 64–77 (2002)
3. Coronato, A., Cinquegrani, M., Giuseppe, D.P.: Adding Business Rules and Constraints in Component Based Applications. In: Meersman, R., Tari, Z., et al. (eds.) CoopIS 2002, DOA 2002, and ODBASE 2002. LNCS, vol. 2519, pp. 948–964. Springer, Heidelberg (2002)
4. Greenfield, J.: Software Factories: Assembling Applications with Patterns, Models, Frameworks, and Tools (2004), http://msdn.microsoft.com/library/default.asp?url=/library/en-us/dnbda/html/softfact3.asp
5. Guerrieri, E.: Case Study: Digital's Application Generator. IEEE Software 11(5), 95–96 (1994)
6. Herbst, H.: Business Rules in Systems Analysis: a Meta-Model and Repository System. Inf. Syst. 21(2), 147–166 (1996)
7. Kleppe, A., Warmer, S., Bast, W.: MDA Explained. The Model Driven Architecture: Practice and Promise, ch. 1. Addison-Wesley, Reading (2003)
8. Miller, J., Mukerji, J.: Model Driven Architecture – A Technical Perspective (2001), http://www.omg.org/cgi-bin/doc?ormsc/2001-07-01
9. Miller, J., Mukerji, J.: Technical Guide to Model Driven Architecture: The MDA Guide v1.0.1 (2003), http://www.omg.org/cgi-bin/doc?omg/03-06-01
10. OMG. Object Constraint Language Specification (2003), http://www.omg.org/cgi-bin/doc?ptc/2003-10-14
11. Roddick, J.F., Al-Jadir, L., Bertossi, L.E., Dumas, M., Estrella, F., Gregersen, H., Hornsby, K., Lufter, J., Mandreoli, F., Mannisto, T., Mayol, E., Wedemeijer, L.: Evolution and Change in Data Management - Issues and Directions. SIGMOD Record 29(1), 21–25 (2000)
12. Smith, H., Fingar, P.: Business Process Management: The Third Wave – Business Process Management Systems. Meghan-Kiffer Press (2002)
13. Tanter, E., Noye, J., Caromel, D., Cointe, P.: Partial behavioral reflection: spatial and temporal selection of reification. In: OOPSLA, pp. 27–46 (2003)
14. Wu, J.-H., Hsia, T.-C., Chang, I.-C., Tsai, S.-J.: Application Generator: A Framework and Methodology for IS Construction. In: 36th Annual Hawaii International Conference on System Sciences (IEEE - HICSS), pp. 263–272 (2003)
15. Yannakoudakis, E.J., Tsionos, C.X., Kapetis, C.A.: A new framework for dynamically evolving database environments. Journal of Documentation 55(2), 144–158 (1999)

# Smart Business Objects for Web Applications: A New Approach to Model Business Objects

Xufeng (Danny) Liang and Athula Ginige

School of Computing and Mathematics, University of Western Sydney, Sydney, Australia
danny@scm.uws.edu.au, a.ginige@uws.edu.au

**Abstract.** At present, there is a growing need to accelerate the development of web applications and to support continuous evolution of web applications due to evolving business needs. The object persistence capability and web interface generation capability in contemporary MVC (Model View Controller) web application development frameworks and model-to-code generation capability in Model-Driven Development tools has simplified the modelling of business objects for developing web applications. However, there is still a mismatch between the current technologies and the essential support for high-level, semantic-rich modelling of web-ready business objects for rapid development of modern web applications. Therefore, we propose a novel concept called Smart Business Object (SBO) to solve the above-mentioned problem. In essence, SBOs are web-ready business objects. SBOs have high-level, web-oriented attributes such as email, URL, video, image, document, etc. This allows SBO to be modelled at a higher-level of abstraction than traditional modelling approaches. A lightweight, near-English modelling language called SBOML (Smart Business Object Modelling Language) is proposed to model SBOs. We have created a toolkit to streamline the creation (modelling) and consumption (execution) of SBOs. With these tools, we are able to build fully functional web applications in a very short time without any coding.

**Keywords:** Business Object, Modelling Language, Web Engineering, Rapid Development.

## 1 Introduction

Web programming languages (such as PHP, Python, Perl, ASP, Java, etc) and database technologies have been around for a long time with major web applications developed using them. However, with the increased time-to-market pressure, we can no longer afford the time to work with rows and columns in sophisticated databases, and create business web-based applications from scratch. Thus, there is a growing need to rapidly develop web applications that can evolve meeting the ever-changing business needs. One of the challenges in developing web applications is to minimise the gap between the development domain and the actual problem domain. This has led to investigate ways of creating better modelling techniques that empowers users to express their mental model at a higher-level of abstraction. Further to find smarter tools that can capture and convert, for implementation, those models into software

J. Filipe, B. Shishkov, and M. Helfert (Eds.): ICSOFT 2006, CCIS 10, pp. 307–322, 2008.

objects in order to create powerful web applications (i.e. by executing those models). Our work builds on early work done by Reenskaug in MVC (Model-View-Controller): a modelling approach to bridge the gap between users' mind and computer data [18], [17]. Moreover, empowering users and allowing trained end users to maintain or even enhance existing applications is a cost-effective way to support web application evolution [23].

The OO (Object-Oriented) paradigm provides us with techniques to build software applications by mapping real world objects directly into software objects. In the past, object mapping techniques have proven to be successful in software engineering projects [6]. These techniques provided a natural correlation between real world objects and objects in the software and database domain. Additionally, OO design techniques are applicable of handling the domain evolution [4]. The encapsulation concept in OO provides us with a systematic way to handle software evolution. System behaviours are encapsulated inside the objects as methods. This provides a means for software evolution to be handled gracefully by delegating responsibilities to objects. The ability to systematically handle software evolution makes object orientation a suitable technique for implementing web applications.

However, the traditional object concept has a low level of abstraction and has been designed for use by software developers. On the contrary, business objects are "business-focused" software objects modelled to represent real world business entities [11]. They operate at a higher level of abstraction than software objects. Business objects offer representations of organisational concepts, such as resources and actors, which collaborate with one another in order to achieve business goals [5]. Maamar and Sutherland [13] state that business objects provide "an insight into what aspects of a business should be delegated, how these aspects may evolve, and what will be the effect of specific changes", and through business objects, "managers and users can understand each other by using familiar concepts and creating a common model for interactions". Thus, an important attribute that distinguishes business objects from traditional software objects is the fact that they can be understood by both business (business managers and users) and software (software developers and the software itself). They are considered as the bridge between software developers and domain experts.

While object-orientation has been long proven suitable for building business applications, existing web development tools and frameworks do not accommodate the need for high-level modelling and rapid development of web-based business applications. What is required instead are business objects that make provision for web interfaces and behaviours, we call those web-ready business objects. Web-ready business objects should have associated conventions such as:

- Providing a file upload facilities for documents or other binary media contents
- Displaying URL as hypertext links, emails addresses as mailto hypertext links
- Rendering calendars to assist user to enter date information
- Showing interactive maps for location related attributes (e.g. address)
- Offering the suitable media players for video content.

In order to speed up the development of web-based business applications, we need business objects that incorporate those conventions. These conventions are imperative directives that contribute to "web-readiness" of business objects.

As a consequence, web-ready business objects should embrace semantic-rich, web-oriented attributes. These web-oriented attributes have high-level, semantic-rich abstract data types (ADT) such as: email, URL, image, video, document, and date. These abstract data types require special validation logic, content handling methods, and presentation mechanisms. An image attribute for example, we need to validate its filename appropriately, provide an upload facility to record the image's filename to the database and store the actual image file to a preconfigured location on the server (assuming that we are not storing binary data inside the database), and render the file content as image to the web browser (via the `<img>` tag if HTML is used). Web-oriented attributes can affect different layers of a web application. The benefit of being able to program using abstract data types is well understood in programming (see [12]). Over decades, programmers have taken advantage of language-provided, built-in data types, such as "integer", to perform normal operations, such as arithmetic calculations, without worrying about the underlying low-level instructions that are required to be carried out by the machine. Similarly, in the context of web applications, we need the direct support for using richer and higher-level abstract data types in order to represent web-oriented attributes of business objects.

At present the responsibility of handling these web-oriented richer data types is passed down to the applications logic, based on primitive data types, such as "string" or "text". For example, an email address attribute is not considered as type "email", but type "string" or "text". As a consequence, web developers need to craft the same regular expression for validating the email address from users' input and customise the necessary web templates to render the email attribute as an email hypertext link (mailto) in every web application they build. This is mainly due to the fact that "email" is not a built-in, language-provided data type. The missing notion of web-ready business objects does not only decelerate web application developments, but also poses impediments to business objects being modelled at a higher level of abstraction. We cannot simply model: "Employee has photo", and expect a file upload facility is provided for updating the photo attribute and a correctly displayed image of the uploaded photo is rendered for viewing.

Thus, in this paper, we propose a Smart Business Object (SBO) concept. SBO is designed to empower users by addressing the issues in modelling and building web applications. SBO uses representations of business objects and their attributes to achieve a higher-level of design abstraction in web applications leading to faster development. We will demonstrate the concept of SBO through the use of the lightweight SBO Modelling Language (SBOML) to model a SBO and create different views of SBO as web applications.

## 2 Related Work

Recent MVC web development frameworks such as [2], [7], [20] and Model-Driven Development tools such as [1], [14], [21] provided the capability of auto generating basic web user interfaces for CRUD (Create Retrieve Update Delete) operations for user-defined persistent objects. The built-in capabilities of object persistence and web presentation UI generation in contemporary tools or frameworks have simplified the process of developing business objects for the web. However, the real-world semantics and the high-level abstraction required for rapid modelling and developing of business objects for enterprise web applications is still missing.

Most current tools rely on low-level database column types to determine the web presentation UI for the corresponding business object attributes. For example, an attribute is rendered as a textbox if its column type in the database is ‘text’. However, the semantics offered by database column type is insufficient for defining business objects in web applications. It is tedious and unproductive for developers having to craft the same regular expression to validate the email attribute of a business object. For example, in Ruby on Rails, each time we need to validate an email address attribute of a business object, we need to code the same regular expression:

```
class Employee < ActiveRecord::Base

  validates_presence_of :first_name, :last_name, :email
  validates_format_of :email,:with => /^([^@\s]+)@((?:[-a-z0-
9]+\.)+[a-z]{2,})$/

end
```

Modelling business objects in most model-driven tools via UML class diagram variants are also low-level and lack high-level semantics suitable for modelling web-ready business objects. For example, we have to model “Employee has email:string” and then customise in different layers of an web application (such as presentation layer and domain layer (refers to the layers of enterprise application defined in [9])) in order to make the email attribute to be rendered as mailto hypertext link and to be validated properly from user inputs.

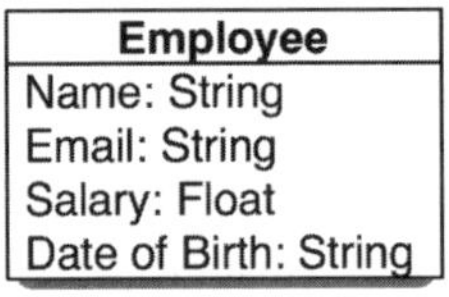

| **Employee** |
|---|
| Name: String<br>Email: String<br>Salary: Float<br>Date of Birth: String |

**Fig. 1.** An UML class diagram for an employee class

The Naked Object [15] address the problem of “behaviour completeness” in business objects. Approaches such as ARANEUS [3], WebML [8], and OOHDM [19] have focused on issues surrounding the modelling of content, navigation, and structure in web applications. Most of them have abstracted content into traditional business objects. However, none has looked at “web-readiness” of business objects and theirs potential in raising the level of abstraction in modelling business object in order to accelerate the development of web applications.

## 3 Smart Business Object

Smart Business Object (SBO) is a web-ready business object that supports semantic-rich, web-oriented attributes suitable for implementing web-based business applications. As previously mentioned, these attributes will support convention settings such as file upload facility for documents, displaying URLs as hypertext links, rendering a calendar to assist user to enter date information, etc. This enables SBOs to auto generate appropriate web interfaces that will accelerate the development of web-based business applications. Figure 2 is an example of rendering a class of SBO called "employee" as a web table with search capability. In this example, the contents of the email address attribute are displayed as mailto hypertext links and the content of the photo attribute is displayed as images.

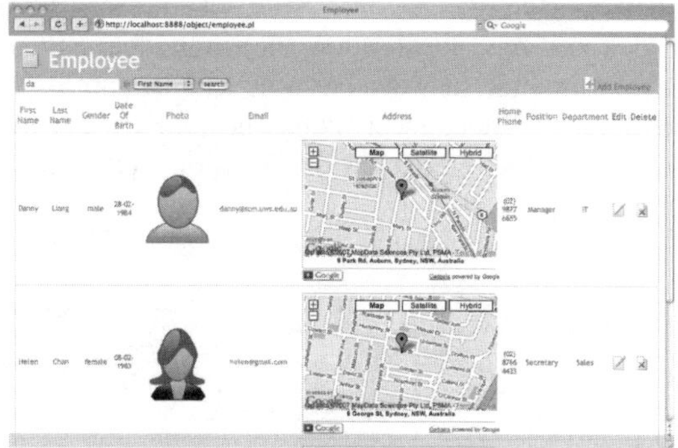


**Fig. 2.** Rendering an SBO as a table with search capability

To assist the generation of useful web user interfaces (such as in Figure 2) for rapid web application development, each SBO have a rich set of built-in methods (operations) for rendering commonly used web user interfaces, such as tables, forms, navigation menus, etc. These user interfaces allow end users to interact with SBOs via a web browser to perform CRUD operations or execute various custom methods of SBOs.

The one line of code in Perl used to generate the user interface in Figure 2 is as follow:

```
organisation::employee->render_as_table(create => 1, edit => 1,
delete => 1, search_form => 1);
```

In other words, render the "employee" SBO in the "organisation" namespace as a table, and allow user to have create, view, edit, delete, and search capability to the "employee" SBO.

Users can create fully functional web applications by modelling SBO and executing them to generate various web user interfaces. The concept of web-oriented attributes allows SBOs to be modelled at a high-level of abstraction than conventional modelling approaches.

### 3.1 High-Level Architecture of Smart Business Object

SBO is a lightweight component that can be easily integrated into existing web frameworks for building both data intensive and process intensive web applications. The SBO is layered on top of a persistent object layer (Figure 3). A persistent object layer is usually realised using ORM (Object Relational Mapping) technologies unless an OODBMS is used. The reference implementation of SBO uses ORM technologies and relational database to achieve object persistence. The Builder component is mainly responsible for modelling SBO. The interpreter for the SBOML lives inside the Builder component.

SBOs are organised by their namespaces. The relationships among SBOs are handled at the object level (as opposed to being at the database level). The advantage of this is that SBOs can establish relationships with other SBOs coming from physically diverse databases. Thus, the role of the Metaobject component is to maintain the relationship definition between SBOs. Moreover, custom SBO schemas can be used to control the behaviours of (e.g. look and feel, localisation, etc) individual SBOs. Thus we need to preserve the mapping information between customs schema and SBOs. This information is also maintained by the Metaobject component. Furthermore, the Metaobject component also maintains the credential information required to connect to the underlying data sources.

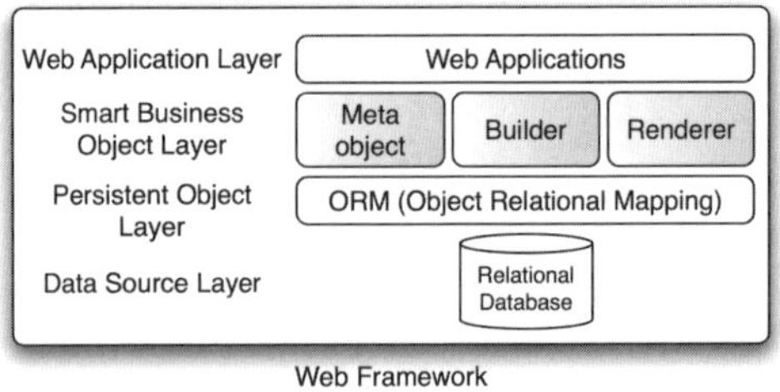


**Fig. 3.** Smart Business Object high-level architecture

As previously mentioned, SBO have a rich set of built-in methods (operations) for rendering commonly used web user interfaces. The Renderer component is responsible for rendering SBOs. It has a host of APIs (Application Programming Interfaces) to support the generation of various web user interfaces for SBOs. Each API provides a rich set of options to achieve fine grain control over the behaviours of the generated user interfaces. For example, in Figure 2, we have enabled create, edit, view, and delete access for the "employee" SBO. Each API utilises one or more templates. Thus, by specifying customised templates to the rendering APIs, we are able to achieve different look and feel for the generated user interfaces. If the default set of user interfaces are insufficient for certain application, we could extend the existing APIs (by subclassing them) or add new APIs.

### 3.2 The Smart Business Object Schema

The SBO has a default schema to control the global behaviours of all SBOs. The schema defines a set of default templates used by each rendering API. Thus, by specifying different templates in the default schema or specifying custom schemas, we can change the look and feel of various user interfaces generated by the Renderer.

Additionally, the schema defines the behaviours of each attribute type. By customising the attribute type definitions in the default schema or by specifying custom schemas, we can easily change the behaviours of existing SBO attributes or add new ones.

In turn, each attribute type definition defines a number of behaviours of an attribute, such as: validations, localisation, option values, and formatting and conversion of values. SBO generates the appropriate web user interfaces for its attributes based on their nominated attribute type given when the SBO was modelled. For SBO attributes whose attribute type is not explicitly defined during the time when the SBO is modelled, SBO will aggregate the meta-information of the underlying data source, such as the table definition of a database, and match them against the known attribute types defined in the specified schema to logically derive the most suitable (conventional) web user interfaces for those SBO attributes at run-time.

In this way, the modelling of SBO can be greatly simplified. For example, we can simply model "employee has email", then the generated "employee" SBO automatically and smartly considers its email attribute as being of the high-level type "email" without extra declaration. This feature adds smartness to SBOs. Thus, we are able to achieve a much higher level of abstraction than traditional modelling approaches.

A partial extract from the default SBO schema implemented in XML is given below:

```
<?xml version="1.0" encoding="UTF-8" ?>
<sbo version = '0.0.28'>
  ...
  <smartness>1</smartness>
  <table_template>
    table.tt
  </table_template>
  ...
  <attribute_definition>
    ...
    <attribute>
      <name>salary<name>
      <validate>MONEY</validate>
      <format>
        <to_ui>
          Renderer::_to_ui_money
        </to_ui>
      </format>
      <sort>NUM</sort>
      <default>0.00</default>
      <maxlength>14</maxlength>
    </attribute>
    <attribute>
      <name>photo<name>
      <type>file<type>
      <validate>FILENAME</validate>
      <sort>NAME</sort>
      <convert>
        <to_ui>
          Renderer::_to_ui_file
        </to_ui>
```

```
        <to_db>
          Renderer::_to_db_file
        </to_db>
      </convert>
      <format>
        <to_ui>
          Renderer::_to_ui_image
        </to_ui>
      </format>
    </attribute>
    <attribute>
      <name>gender<name>
      <options>
        <option>male</option>
        <option>female</option>
      <options>
    </attribute>
    ...
  </attribute_definition>
  ...
</sbo>
```

In the example schema, "<smartness>" defines whether SBO should automatically derive the high-level attribute types for its attributes. The "<table_template>" element defines that the "table.tt" template file will be used for the rendering API(s) responsible for rendering SBOs as a web table.

Different attribute types are defined within the "<attribute_definition>" element. For example, in the "gender" attribute, we have specified two option values: "male" and "female". For the "photo" attribute, we have specified various trigger functions to control the conversion and formatting behaviours. Firstly, we define that "photo" is a "file" type, such that a file upload input field is provided by default. Before saving the value from users' input (usually via web forms generate by the SBO), the "_to_db_file" function is triggered, such that the filename of the uploaded image file is saved in the underlying database and the actual binary image file is saved on a preconfigured location on the server. Similarly, during retrieval, each value of the "photo" attribute is sent to the "_to_ui_file" trigger function in order to construct the necessary URL path needed access the image file on the server. Then it is sent to the "_to_ui_image" function for formatting, such that the values are displayed as images on users' web browser (such as via the HTML <img> tag). The "employee" SBO in Figure 2 is rendered utilising various high-level attribute types defined in the default SBO schema, including "photo" attribute type that we have just discussed. We can always define new trigger function in order to handle special attribute types.

### 3.3 Smart Business Object Modelling Language

According to Pilone and Pitman [16], modelling is "a means to capture ideas, relationships, decisions, and requirements in a well-defined notation that can be applied to many different domains". Domain modelling is the building of an object model of the domain that incorporates both behaviour and data [9]. To streamline the modelling and creation of SBOs, we need a higher-level modelling language. SBOML (Smart Business Object Modelling Language) is a lightweight modelling language designed for modelling SBO.

SBOML is not proposed to be another object-oriented programming language or to extent existing OO concepts. Its main intention is to be a lightweight modelling language that leverage on existing, most commonly used (conventional) OO concepts that are suitable for building web based business applications. It brings OO concepts closer to users' mental model. It is designed to allow users to express their domain specific business objects in near natural language syntax.

In this section, we will use the following conventions to represent the formal construction of the SBOML:

- Keyword elements are emphasised in both bold and italic
- Normal style texts represent user-defined elements
- When an element consists of a number of alternatives, the alternatives are separated by a vertical bar ("|")
- Optional elements are indicated by square brackets ("[" and "]")
- An ellipsis ("...") indicates the omission of a section of a statement, typically refers to recursive statements.

The statement for defining SBO attributes, methods, and 'has' relationships between SBOs is as follow:

***in*** namespace, business object ***has*** attribute A [([***mandatory***] [***type***] [***which could be*** option a ***or*** | ***and*** option b])], [***might have***] [***many***] another business object [(***has*** attribute B, attribute C, yet another business object (***has*** ...))]... ,[ ***use*** method A (***method name type from*** location [***option is*** value,... ] [***with*** attribute A, attribute B,... | ***with*** attribute A ***as*** parameter name abc , attribute B ***as*** parameter name ...]), service B...]

The "***in***" clause defines the namespace where the subsequent business object(s) are created within. If the namespace does not exist, a new namespace is created. The "***has***" clause defines the attributes of the intended SBO, or 'has' relationships with another intended SBO. The optional "***use***" clause defines the methods (operations) of the SBO. We first explain the statement by referencing to a simple example:

```
in organisation, employee has first name, last name, gender,
date of birth, photo, email, address, home phone, position (has
title, description)
```

Literally, we have just defined an "employee" SBO and a "position" SBO where "employee" has a "position". When the above statement is executed, and we can directly render the "employee" SBO to the web, such as to generate a web form for adding new employees (creating new "employee" SBO instances).

As previously mentioned, by default, a SBO predicts its attribute types by matching the attribute name against the defined attributes types in the default SBO schema. Thus, when rendered as a web form (Figure 4), the "employee" SBO automatically:

**Fig. 4.** Rendering the "employee" SBO as a web form

- Enforce first name, last name, date of birth, and email address attributes as mandatory fields and enforce the appropriate validation rules to all corresponding fields
- Provide the "male" and "female" option values to the gender attribute according to SBO schema
- Provide a calendar to assist users for date entry for the "Date of Birth" attribute and present the date according to users' locale setting
- Provide a file upload facility for the photo attribute to upload binary image file
- List the available positions as options items (assuming that we have previously created some "Position" SBO instances), due the relationship established between the "employee" SBO and the "position" SBO

We can always overwrite the default settings, and explicitly declare the attribute type. For example:

```
in organisation, employee has first name, last name,…,
department (mandatory name which could be IT or Sales)
```

By default, all defined attributes are optional, except for whose attribute types are defined as mandatory in the specified SBO schema or due to the requirement of the underlying data source (such as a NOT NULL column of a database table). In the example, the "*mandatory*" keyword enforces that the value of the department attribute cannot be empty (i.e. a mandatory field on a web form). The "name" specifies the type of attribute. Thus, could be any attribute type defined in the default SBO schema or in any custom SBO schema. The "*which could be … or | and*" clause allows users to specify the possible value set of an attribute. In case of the department attribute in the example, option values are "IT" and "Sales". The "*and*" keyword implies that multiple selection values are allowed (checkboxes are used instead of radio buttons).

The "*many*" keyword indicates a "has many" relationship, in UML terms, the cardinality is [1..*]. In combination with the "*might have*" keyword, i.e. "*might have many*", then the cardinality becomes [0..*]. For example:

```
in organisation, employee has first name, last name,…, might
have many office (has room number, building id)
```

SBO can easily aggregate local functions or remote service as its methods (operations). This enables SBO to be seamlessly integrated with workflow engines and SOA (Service-Oriented Architecture) to develop more complex process oriented business web applications. This can be achieved using the "`use`" clause. For example:

```
in organisation, employee has first name, last name,…, use
Notify HR (notify_HR from http://10.10.10.2/notify.wsdl with
first name as param_first_name, last name as param_last_name)
```

In the example, "`Notify HR`" is the name of the method for the employee SBO, and "`notify_HR`" is the actual name of the remote method "`from`" the WSDL file located at "`http://10.10.10.2/notify.wsdl`". In the reference implementation, the SBO support Web Services and XML-RPC for remote invocation. Thus, the "`type`" keyword could be: Local (for executing local application APIs), Web Service, or XML-RPC. The "`with`" keyword is used to indicate the mapping of the attributes of the SBO to the required parameters of the remote method. In the example, when the "`notify_HR`" method is executed, the value of the first name and last name of an "employee" SBO instance is passed to the "`param_first_name`" parameter and the "`param_last_name`" parameter respectively. Depending on the nature of the remote method, more arguments, such as URI, may be required to identify and execute the remote method, thus the clause within the squarely blanket allows user to specify key-value pairs for any optional argument that is needed.

After incorporating the changes to example shown in Figure 4, when we retrieve an instance of the "employee" SBO and render it as a web form again, it would generate the screen shown in Figure 5. Now, users can assign a department, multiple offices to the employee and click the "notify HR" button to execute a web service.

We can use the following statements to explicitly define relationships among existing SBOs. For "has" relationships, we could use either:

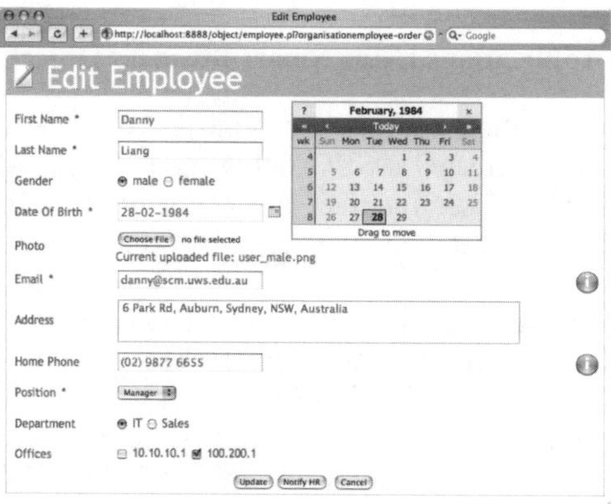


**Fig. 5.** Rendering the "employee" SBO as a web form with a department, multiple offices and a new method

```
in namespace, business object has| might have another business
object [as attribute X] [via yet another business object]
```

or

```
business object in namespace has| might have another business
object in another namespace [as attribute X] [via yet another
business object in yet another namespace]
```

The second construct allows SBOs to establish relationships across namespaces. The "*as*" clause is to nominate a specific attribute as a reference to a foreign SBO instance. For example:

```
in organisation, employee might have employee as supervisor
```

In the above example, the "employee" SBO has a self-referential "has" relationship, such that an employee may have a supervisor, which is also an employee.

The "*via*" clause is a shorthand for specifying "many-to-many" relationships. For example:

```
in organisation, employee has car via company car rental
```

Similarly, to define "is a" relationships (inheritance) between SBOs, we can use either of the following statements:

```
in namespace, child business object is parent business object
```

or

```
child business object in namespace is parent business object in
another namespace
```

For example:

```
in organisation, employee is person
```

### 3.4 Creating Web Applications Using Smart Business Object

The reference implementation of SBO is deployed on a web framework called CBEADS© [10]. We have created a SBO lightweight toolkit, which consists of the SBO Builder and the SBO User Interface Generator on the CBEADS© framework. They are designed to streamline the creation (modelling) and consumption (execution) of SBOs.

The SBO Builder (Figure 6) allows users to model and create SBOs and relationships among them using the SBOML. The SBO Builder can also auto-generate a graphical representation of any modelled SBOs (Figure 7).

The SBO User Interface Generator (Figure 8) allows users to easily create applications on the CBEADS© framework by rendering SBOs using the SBO rendering APIs. It also allows users to customise various options supported by the SBO rendering APIs. The SBO toolkit allows fully functional web applications to be created without any coding.

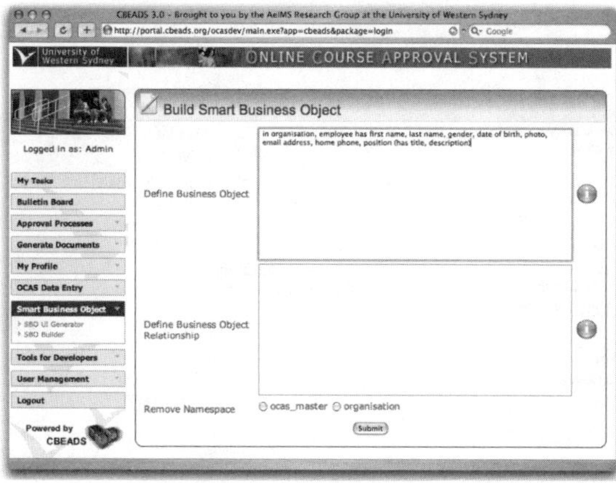


**Fig. 6.** The SBO Builder

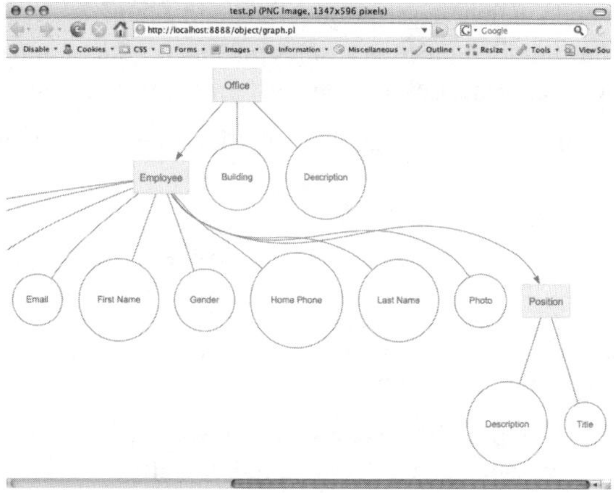


**Fig. 7.** An auto-generated graph of the modelled SBOs

### 3.5 Creating a Customer Relationship Management (CRM) Application

In this section, we will demonstrate how we can use SBOs to generate a lightweight CRM application on the CBEADS© framework. According to the actual business requirements, we need to first identify the actors and their actions, for example:

- Potential customers can make enquiries about products, request sales people to visit them to discuss about products, and make purchases
- Sales persons need to keep track of customers, visits, and sales orders.

Next, we need to identify the necessary business objects:

- A customer has first name last name, email address, phone number, and address
- A sales person has first name, last name, and phone number
- An enquiry has title, question, answer, and date

**Fig. 8.** The SBO User Interface Generator

- A product has code, name, description, price, and enquiries
- A sales order has number, customer, sales person, products, and total amount
- A visit has title, date, time, description, customer, sales person, and sales order.

Using the SBO Builder tool, we can generate those business objects by expressing them in SBOML:

```
in crm, visit has title, date, time, description, customer (has
first name, last name, email address, phone number, address),
sales person (has first name, last name, phone number), sales
order (has number(mandatory alphanumeric), many product (has
code, name, description, price, many enquiry (has title,
question, answer, date)), total amount)
```

The above SBOML expression models all the identified business objects at the same time. However, we can also model them individually.

**Fig. 9.** Sales management function assigned to sales people

Lastly, we generate various views (such as Figure 9) of the SBOs purely using the SBO User Interface Generator and assigned them to the system user groups defined in CBEADS©. Similarly, we can easily extend the CRM application by modelling additional business objects, such as suppliers or competitors, to meet the evolving business needs. Once the business objects and the views of the business objects based on the actions actors need to perform are identified, we can quickly generate fully functional web-based applications using the SBO toolkit and the CBEADS© framework. Thus, the overall development time can be greatly reduced.

## 4 Conclusions

In this paper, we have introduced the Smart Business Object concept. The SBOs support semantic-rich, web-oriented attributes. We have presented a modelling language that allows users to express their mental model at a higher-level of abstraction. We have created a tool that generates web-ready Smart Business Objects from the high-level models. We have demonstrated the significant benefits of utilising Smart Business Object in web application development such as ability to model the application based on high level business domain objects and very rapid development of the application using the tools that we have created.

We have implemented several industry projects using SBOs. A significant project is an enterprise level application; the Online Course Approval System (OCAS) [22] developed for University of Western Sydney (UWS). The use of SBOs greatly reduced the low level modelling activities such as creating ER diagrams and database schemas and enabled us to rapidly develop OCAS.

## References


1. AndroMDA, Cutting Edge MDSD/MDA Toolkit (2005)
2. Apple WebObjects 5 Reviewer's Guide (2001)
3. Atzeni, P., Gupta, A., Sarawagi, S.: Design and Maintenance of Data-Intensive Web Sites. In: Schek, H.-J., Saltor, F., Ramos, I., Alonso, G. (eds.) EDBT 1998. LNCS, vol. 1377. Springer, Heidelberg (1998)
4. Barstow, D., Arango, G.: Designing software for customization and evolution. In: Proceedings of the 6th international workshop on Software specification and design (1991)
5. Caetano, A., Silva, A.R., Tribolet, J.: Using roles and business objects to model and understand business processes. In: Symposium on Applied Computing, Santa Fe, New Mexico. ACM Press, New York (2005)
6. Casey, R.M.: Object Mappings in a Software Engineering Project. Software Engineering Notes - ACM SIGSOFT 24 (1999)
7. Catalyst Welcome to Catalyst Development (2005)
8. Ceri, S., Fraternali, P., Bongio, A.: Web Modeling Language (WebML): a Modeling Language for Designing Web Sites. In: WWW9 Conference (2000)
9. Fowler, M.: Patterns of Enterprise Application Architecture. Addison-Wesley Professional, Reading (2002)

10. Ginige, J.A., Silva, B.D., Ginige, A.: Towards End User Development of Web Applications for SMEs: A Component Based Approach. In: Lowe, D.G., Gaedke, M. (eds.) ICWE 2005. LNCS, vol. 3579, pp. 489–499. Springer, Heidelberg (2005)
11. Lhotka, R.: Expert One on One Visual Basic.NET Business Objects. Wrox Press Ltd, Birmingham (2003)
12. Liskov, B., Zilles, S.: Programming with Abstract Data Types. In: Symposium on Very High Level Programming Languages (1974)
13. Maamar, Z., Sutherland, J.: Toward intelligent business objects. Communications of the ACM 43 (2002)
14. OpenMDX openMDX - the leading open source MDA platform (2005)
15. Pawson, R., Matthews, R.: Naked Objects. John Wiley and Sons Ltd, Chichester (2002)
16. Pilone, D., Pitman, N.: UML 2.0 in a Nutshell. O'Reilly Media, Inc., Sebastopol (2005)
17. Reenskaug, T.: MODELS - VIEWS - CONTROLLERS (1979a)
18. Reenskaug, T.: THING-MODEL-VIEW-EDITOR:an Example from a planning system (1979b)
19. Rossi, G., Garrido, A., Schwabe, D.: Navigating between objects. Lessons from an object-oriented framework. ACM Computing Surveys (CSUR) 32 (2000)
20. Ruby on Rails, Web development that doesn't hurt. Ruby on Rails (2005)
21. Tangible Engineering, Tangible Architecture (2005)
22. University of Western Sydney, Online Course Approval System (OCAS). University of Western Sydney (UWS) (2006)
23. Wulf, V., Jarke, M.: The Economics of End-User Development. Communications of ACM 47 (2004)

# A Data Mining Approach to Learning Probabilistic User Behavior Models from Database Access Log

Mikhail Petrovskiy

Faculty of Computational Mathematics and Cybernetics, Moscow State University
Vorobjevy Gory, Moscow, Russia
michael@cs.msu.su

**Abstract.** The problem of user behavior modeling arises in many fields of computer science and software engineering. In this paper we investigate a data mining approach for learning probabilistic user behavior models from the database usage logs. We propose a procedure for translating database traces into representation suitable for applying data mining methods. However, most existing data mining methods rely on the order of actions and ignore time intervals between actions. To avoid this problem we propose novel method based on combination of decision tree classification algorithm and empirical time-dependent feature map, motivated by potential functions theory. The performance of the proposed method was experimentally evaluated on real-world data. The comparison with existing state-of-the-art data mining methods has confirmed outstanding performance of our method in predictive user behavior modeling and has demonstrated competitive results in anomaly detection.

**Keywords:** User behavior modeling, Data mining, Database access logs, Probabilistic models.

## 1 Introduction

User behavior modeling is one of the most important and interesting problems needed to be solved when developing and exploiting modern software systems. By *user behavior modeling* we mean discovering patterns of user activity and constructing predictive models based on precedent behavior information. These models allow forecasting next user action on the basis of the current activity. Primarily such technique was oriented to the commercial applications in recommendation systems [12], [8]. At present time the area of its application is significantly wider. These methods play a great role in computer security systems [4], [6], where they are used for detecting malicious or unqualified user actions. Besides, recently user behavior modeling is applied for analysis, understanding and optimization of the architecture and business logic of various software systems. Models of user behavior can help to improve the UI usability, to optimize the database structure and data cashing strategy, to detect hidden use-cases, etc. Traditionally, data mining techniques are used for constructing user behavior models. The process of user behavior modeling can be presented as KDD-process (Knowledge Discovery in Databases), defined as *extracting nontrivial, previously unknown and potentially useful information from large sets of data* [10]:

J. Filipe, B. Shishkov, and M. Helfert (Eds.): ICSOFT 2006, CCIS 10, pp. 323–332, 2008.

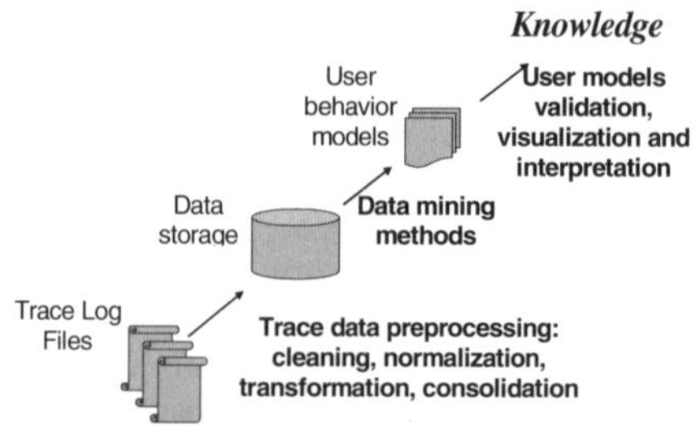


**Fig. 1.** User behavior modeling as KDD process

On the first stage, necessary data is extracted from log-files, transformed into unified representation suitable for analysis, and stored in the data warehouse. Then data mining techniques are applied for building behavior models. Finally, the models are validated and interpreted by an expert. It is necessary to outline such features of the information sources used for user behavior modeling as large volume, heterogeneity and complicated structure of data coming from log-files. But the most significant features are temporal nature of the data and ordering of user actions. The source log-files can be of different levels – from high-level application logs [12] and web access logs [8] to low-level system calls traces [4]. In this paper, we consider the intermediate level, in particular the database access logs. User behavior modeling on this level has not been well studied yet and it was considered mainly in the context of optimization of database server settings [2]. Though, from our point of view, user behavior models built on database level can be very useful in other tasks as well, because nowadays many modern software systems use relational SQL databases as information storage and all important user actions leave a trace in the database access log.

The paper is organized as follows. In Section 2 we give the formal problem statement of probabilistic user behavior modeling. In Section 3 we present our new approach based on *classification method of autoregressive type* and specially designed *empirical feature map* of data from structured log-files into *finite dimensional metric space*. This mapping allows taking into account both *time and frequency* of user actions. Section 4 is devoted to experiments and comparative analysis on real-world data. In the final section we formulate main results and contributions of our research.

## 2 Problem Definition

Traditional probabilistic statement of the user behavior modeling problem is the following [8]. Precedent information on the activity of a user $U$ is given in the form of ordered sequences of actions $H(U) = (A_1, A_2,..., A_N)$. Model of user behavior is defined as the following probabilistic function:

$$P(A^{next} \mid S(U), H(U)). \tag{1}$$

It defines the probability that next user action will be $A^{next}$ under conditions that current user activity is described by the sequence of actions $S(U) = (A_{S1}, A_{S2},..., A_{SK})$ and historical activity of the user is defined by *H(U)*.

Practically all existing methods of constructing models (1) are based on the following propositions. Any user action can be coded as a symbol from some finite alphabet $A_i \in \Omega$. And a training set $H_{train}(U) = \{H_1(U),...,H_L(U)\}$ is formed from the $H(U)$, where $H_i(U) \subset H(U)$ are all available subsequences. Usually $H_i(U)$ are selected consequently by a sliding window method, but sometimes application-oriented methods are used (e.g. sequences may correspond to user sessions).

Probabilistic models (1) can be applied for solving practical tasks of *next action prediction*; *detecting anomalies* (unexpected user actions); and discovering *patterns* and *frequent episodes* of user activity. For the last problem, it is difficult to indicate universal performance evaluation measure since representation of the patterns and frequent episodes depend on the used data mining technique. For the first two problems general performance evaluation measures do exist. Forecasting of next user action $A^{next}$ is performed according to the formula derived form (1) with the use of Bayes rule:

$$A^{next} = \arg\max_{A \in \Omega} P(A \mid S(U), H(U)). \tag{2}$$

In this case, the performance evaluation measure is *hit ratio* that is a proportion of correctly predicted actions to total number of actions. For the anomaly detection, a threshold cutting (confidence level) $\alpha$ should be specified. Then user action $A^{next}$ is considered to be anomalous if:

$$A^{next} \notin \{A \in \Omega \mid P(A \mid S(U), H(U) > \alpha). \tag{3}$$

Precision of anomaly detection is estimated by standard coefficients [6]: *detection rate* and *false positive rate*. They depend on the threshold value and that is why the final comparison is performed with the help of ROC-curves [9] representing the mutual dependence of these coefficients.

The most popular traditional data mining techniques applied for constructing probabilistic model (1) are association rules [7], sequential models [8] and autoregressive classification methods [3]. Although these methods are widely used and demonstrate acceptable results in many practical applications they all can be criticized for calculation complexity; using a priori set critical parameters; and either poor accuracy with good model interpretation or, on the contrary, high accuracy with non-interpreted models. In addition to these disadvantages, almost all these methods rely on *order of actions* and *ignore time* between actions. We think that in the case of database logs analysis it is significant defect for the following reasons. There exists tendency that recent actions have more influence to next possible action then those happened long time ago. Besides, there might be situations, when single db login is used by several different persons simultaneously (for example, most public web systems do not provide individual db logins for their users). In such case the sequences of actions in log files will be *mixed up* that will break the order of actions. The only thing to do here is including time feature in the model.

## 3 Our Approach

Before we turn to the problem of constructing function (1) we need to define the structure of database access log in the form of sequences of actions $A_i \in \Omega$. Most of database access logs consist of records of similar structure:

$$\langle user\ id, event, sql, time, other\ \ features \rangle,$$

where *user id* is user login; *event* is a type of event (e.g., start or finish of a query execution); *sql* is SQL text of a query; *time* is a timestamp; *other features* can be divided into *execution group* that includes numerical characteristics of query execution (e.g. number of read/write operations, duration, etc); and *identification group* with discrete characteristics of query such as identifiers of client process, server's process, user aliases, etc. Thus the problem is to map such structure into a finite alphabet. We suggest the following procedure.

**DB Access Log Pre-processing Procedure:**
**Step 1.** "Uninteresting" attributes reduction.
**Step 2.** Numeric attributes descritization.
**Step 3.** Extracting templates (skeletons) from SQL statement.
**Step 4.** Mapping discrete attributes combination to finite alphabet $\Omega$.

On the first step we exclude attributes that are not interesting for analyzing. For example, db server *process id*, as a rule, is not interesting for the model. On the second step the rest numerical attributes are *discretizied* by some unsupervised discretization algorithm. In particular, we use equal frequency interval method with small (3-10) number of intervals. On the next step SQL statement text is processed. We extract its so called *skeleton* or, in other words, *template*, that presents the query syntax with removed user parameters. We use the approach similar to [14]. SQL statement is converted into the sequence of tokens, where each token has either *keyword* type (for SQL language keywords) or *name* type (for db related names, i.e. table names, fields, stored procedure, etc.). Let us clarify this idea on the example. Assume we are given the following query:

```
SELECT FROM USERS WHERE NAME='Bob' AND CITY='London'
```

We convert it to the sequence of tokens:

```
(SELECT,keyword)(FROM,keyword)(USERS,
name)(WHERE,keyword)(NAME,keyword)(AND, keyword)(CITY,name)
```

Then each unique template gets its unique identifier. Thus, before the fourth step the initial log file record has the form of vector of discrete attributes, where each attribute is either discrete attribute of the initial record or SQL template identifier, or interval id of discretizied initial numeric attribute. Records with the same SQL template, the same discrete attributes and close numeric attributes have the same representation, i.e. the same combination of resulting discrete attributes. Such representation of the *similar records* identifies the *possible action*, to which a unique symbol from the alphabet $\Omega$ is assigned. In this way the alphabet $\Omega$ determines the set of *all possible user actions*. At the first glance the suggested procedure can be criticized for the possibility of unbounded growth of the size of the alphabet $\Omega$. In practice, for production systems being in stable exploitation, it is found that the growth comes to stop quickly enough, just for few hundreds. Besides, number of different possible actions can be reduced by grouping them, using clustering or frequent episodes or expert's domain knowledge.

After applying this procedure for mapping db access logs structures into the alphabet $\Omega$ we can use traditional data mining methods based on association rules, sequential models or autoregressive classification. However, as we outlined before, these methods do not take time feature into account, only the order. To avoid this problem we propose novel approach. Its main idea is constructing *empirical feature map* $\varphi$ that *explicitly* maps an *arbitrary* sequence of symbols from $\Omega$ with timestamps into a *finite-dimension metric space H*. First of all, we need to extend the representation of user actions by adding time labels to them. Then each action from *S(U)* or *H(U)* is described by the pair $(A, tm) \in \Omega \times Time$. Let us formulate the basic assumptions for $\varphi$ mapping:

- *recently* performed actions have more influence on the upcoming action $A^{next}$ than actions performed long time ago;

- *requently* performed actions have more influence on the upcoming action $A^{next}$ than actions performed rarely.

Appropriate background for constructing such mapping comes from the theory of *potential functions* [1]. We assume that any possible action $A_i \in \Omega$ has its own *potential* at any moment *t*. This potential is being reduced proportionally to the time passed from the moment when the action was performed. The exact form of this reduction is given by a priori chosen potential function $Pf : Time \times Time \to \Re$. If the sequence contains the same actions in different times, in accordance to the potential function theory, their potentials are summed up. In this manner, we define the mapping of the sequence (of an arbitrary length) of user actions with timestamps into the real vector space of dimensionality $L = |\Omega|$:

$$\varphi(H(U), t_0) = \left( \sum_{(A,tm) \in H(U), t_0 > tm} Pf(t_0, tm) \right)_{A \in \Omega} . \tag{4}$$

According to (4) the internal "state of a user activity" at any moment *t* is described by the set of $L = |\Omega|$ potentials $\varphi_A(H(U), t_n)$. It allows considering both time and frequency features of previous actions.

Functions from RBF class are convenient for use as potential function *Pf* (4). Potential functions of this type depend only on time interval between actions in series and do not depend on exact time moments. In our experiments we use exponential function $Pf(x, y) = \exp(-\sigma\|x - y\|)$, where parameter $\sigma$ controls the speed of the past actions influence vanishing, i.e. how quickly potentials go down. Besides, such RBF can be efficiently calculated for continuous sequence using the recursive formula (where $\varphi_A(0) = 0$):

$$\varphi_A(t_n) = \begin{cases} \varphi_A(t_{n-1}) * e^{(-\sigma\|t_n - t_{n-1}\|)}, A \neq A_n \\ \varphi_A(t_{n-1}) * e^{(-\sigma\|t_n - t_{n-1}\|)} + 1, A = A_n \end{cases}. \tag{5}$$

The feature mapping function (4) allows a sequence of actions to be presented as a feature vector from $L$-dimensional real vector space. At any moment the "state of the user activity" is unambiguously described by the given vector. Therefore, it is naturally to use the approach based on autoregressive classification methods for constructing user behavior models. In such case a training set is represented as the set of pairs $\left\langle (\varphi_A(t))_{A\in\Omega}, A^t \right\rangle \in \Re^{|\Omega|} \times \Omega$:

$$H_{train}(U) = \{< (\varphi_A(t))_{A\in\Omega}, A^t >\}_t \tag{6}$$

Then learning algorithm is used to construct a *multi-class probabilistic classifier* of the form: $F_{H(U)} : \Re^{|\Omega|} \to \Omega$ that estimates probabilities (1) for any given state $(\varphi_A(t))_{A\in\Omega}$:

$$P(A^t \mid S(U), H(U)) = P(F_{H(U)}((\varphi_A(t))_{A\in\Omega}) = A^t) \tag{7}$$

Since almost all probabilistic multi-class classification method can be applied, when the input space is finite-dimensional real vector space, we concentrate our attention on the two main criteria –accuracy of prediction and understandability of the obtained model for a human expert. From our point of view, decision trees [5], [11] have the best balance between accuracy and interpretation power among all classification methods. Tree based methods partition the input feature space into a set of rectangular regions $R_1, R_2, ..., R_n$, and fit a simple model in each one. Usually this simple model is a class probability distribution. Applying a standard algorithm, e.g. CART [5] or C4.5 [11] to the training set (6) we come to the model that can be represented as a tree, where each terminal node $m$ is connected with a region $R_m$ described by the following predicate system:

$$\begin{aligned} &\texttt{IF } (C_{Ai}^{low} < \varphi_{Ai}(t) < C_{Ai}^{upper}) \texttt{ AND } ... \\ &\quad (C_{Aj}^{low} < \varphi_{Aj}(t) < C_{Aj}^{upper}) \texttt{ AND } ... \\ &\texttt{THEN } (\varphi_A(t))_{A\in\Omega} \in R_m \end{aligned} \tag{8}$$

Here $C_{Aj}^{low}$ and $C_{Aj}^{upper}$ are constants bounding possible value of the potential for action $A_j \in \Omega$ at the moment $t$. The distribution of class probabilities is associated with each region $R_m$. For each possible action $A^t \in \Omega$ we take probability (7) as a ratio of samples presented in the $R_m$ and having class $A^t$ ($Count_{A^t}(R_m)$) to the total number of samples in $R_m$ ($Count(R_m)$):

$$P(A^t \mid (\varphi_A(t))_{A\in\Omega} \in R_m) = \frac{Count_{A^t}(R_m)}{Count(R_m)} \tag{9}$$

Class probabilities (9) are considered as estimates (7), and the whole procedure looks as follows.

```
User Behavior Modeling Procedure:

Preparation process:
Step 1: For any given db trace find Ω and prepare historical
data H(U) using the proposed log translation procedure.
Step 2: Choose potential function type and parameters for fea-
ture map (4).
Step 3: Convert H(U) into training set (6) using feature map
(4).
Training process:
Calculate regions (8) and class probabilities (9) using decision
tree algorithm (e.g. CART or C4.5).
Prediction process:
At any moment t for any current user actions sequence do the
following:
Step 1: Translate the sequence into S(U) using the proposed log
translation procedure.
Step 2: Calculate potentials (φ_A(t))_{A∈Ω} using (4).
Step 3: For each (φ_A(t))_{A∈Ω} use (8) to find the target region R_m
and use (9) to estimate probabilities (7) that define model (1).
```

It should be noticed that proposed model has simple and meaningful interpretation for a human expert. It can be visualized as a decision tree with distributions of possible actions in terminal nodes. Its semantics is described by a system of rules in the form: "*IF at the moment t potentials of the previous user actions are in specified ranges THEN next user action would be A with probability P*".

## 4 Experiments

In this section the results of experimental performance evaluation are presented. The goals of experiments are to check how traditional data mining methods (sequential patterns and association rules) work on real-world data with our proposed SQL-trace translating procedure and to compare performance of existing methods to our novel method, based on time-dependent feature mapping and decision tree learning algorithm. We consider two scenarios: "next action prediction" and "anomaly detection".

Below we denote our method as Pf-DT that stands for "Potential function feature space with Decision Tree". We use recursive exponential RBF (5) as a potential function in the feature map (4), time is calculated in milliseconds, $\sigma = 1000$. In our method, we use C4.5 learning algorithm with probabilistic cutting threshold [11]. As competitors we tried Expectation-Maximization based sequence clustering algorithm (Seq-EM) and Apriori association rules mining algorithm (A-Rules). Both algorithms are implemented in MS 2005 SSAS [13].

We run experiments on real-world data, collected from MS SQL Server trace logs and generated by real-world banking intranet application. The task of the application is registering, evaluating and processing consumer credit requests. An operator enters and processes customer's requests in the system. Several real persons usually work simultaneously under the same operator's login. We collected traces of operators' activity in one branch of the bank during two days, one day – for training, another for testing. There are about 30000 SQL queries per day. Applying SQL trace transformation procedure we consider only SQL query text, execution time, duration and number of read/write operations in a query. As a result we obtain the alphabet size $L = |\Omega| = 65$.

The first series of experiments was for "next action prediction" scenario. To study how the size of the training set affects the model precision we prepared three training sets of different sizes: 2 hours, 4 hours and 8 hours (the whole working day) of activity. The testing dataset is 8 hours of activity in another day. Training time of all algorithms in these experiments was nearly the same, about one minute or less. The experimental performance results (*hit ratio*) are presented in the table below:

**Table 1.** "Next action prediction" experiments

| Experiment Settings | Algorithm | hit ratio |
|---|---|---|
| Training: 8h (33856 records) | Pf-DT | 85.76% |
| Testing: 8h (28060 records) | Seq-EM | 59.72% |
| No anomalies | A-Rules | 42.47% |
| Training: 4h (16180 records) | Pf-DT | 79.77% |
| Testing: 8h (28060 records) | Seq-EM | 43.72% |
| No anomalies | A-Rules | 41.65% |
| Training: 2h (4039 records) | Pf-DT | 51.91% |
| Testing: 8h (28060 records) | Seq-EM | 21.07% |
| No anomalies | A-Rules | 8.6% |

In this scenario our method dramatically outperforms its competitors. Another thing is that accuracy of all algorithms growths with the size of the training set, though the difference between 4 and 8 hours is not significant. It means that user activity in the investigated application is very stable and we do not need large datasets to train.

The second series of experiments is devoted to the investigation of the problem of anomaly detection. To estimate the ability of the algorithms to discover anomalies we have added to the testing dataset 10% of randomly generated anomalous actions (possible actions but in a random places). We also tried 1% and 5% but the results turned out to be very similar to 10%, that is why (and because of space limitation) we leave only results for 10%. They are presented on ROC curve chart below:

However, unlike other methods, our method reached the detection rate of 100% the corresponding false positive rate is too big (about 7%). In the area of smaller false positive rates Seq-EM and even A-Rules outperformed our method. Outstanding performance in the "next action prediction" task and average results in anomaly detection mean that proposed method very precisely guesses the most expected action, but not

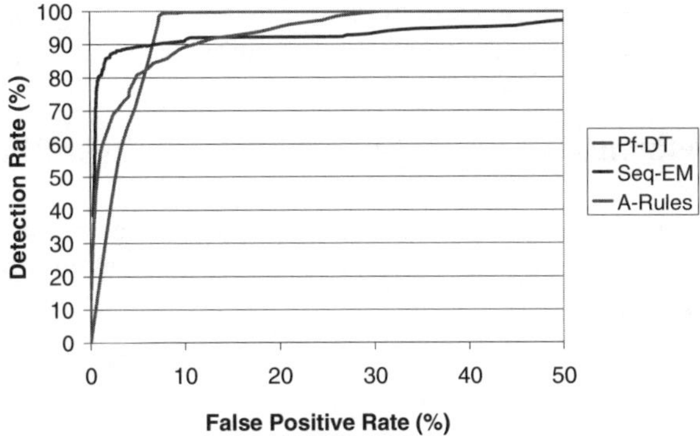


**Fig. 2.** ROC curve for anomaly detection task with 10% anomalies in the testing set (31912 records)

enough accurately estimates the set of all expected actions (that Seq-EM and A-Rules do). It means that the mechanism of probabilities estimation used in the decision tree algorithm (9) is not perfect for the anomaly detection task. In the future research we will check the anomaly detection ability of the proposed approach with other probabilistic multi-class classification algorithms, e.g. with kernel methods [5], and we hope to obtain outperforming results in this scenario as well.

## 5 Conclusions

The main contributions of this paper can be summarized as following:

1. New type of data source for user behavior modeling has been considered. This is the database access log consisting of traces of SQL queries executed by users. It is promising information source because the major part of modern software systems use relational databases as information storage, and usually all critical user actions leave a trace in database access logs.

2. Simple but effective procedure for translating SQL traces structures into a finite alphabet of symbols has been proposed. It allows analyzing database access log data with traditional data mining techniques such as sequential mining and association rules mining methods.

3. Novel method for mining probabilistic user behavior models has been formulated. Unlike other existing data mining methods it incorporates time feature in the user model. The empirical feature map, motivated by potential functions theory, has been proposed for that. Combining this feature map with decision tree algorithm we obtain new method with following advantages: it is precise enough; it takes into account time intervals between user actions; it gives understandable for a human expert interpretation of generated behavior models in the form of "IF…THEN" rules.

4. Experimental performance evaluation on real-world data has been conducted. It has demonstrated that database access logs can be successfully used for user behavior modeling and reliable models can be constructed. In these experiments, our proposed

method has demonstrated outstanding results in the “next action prediction” scenario and competitive results in “anomaly detection” scenario.

## Acknowledgements

This research is supported by grant of RFFI (Russian Foundation for Basic Research) # 05-01-00744 and by grant of the President of Russian Federation MK-2111.2005.9.

## References

1. Aizerman, M.A., Braverman, E.M., Rozonoer, L.I.: Method of Potential Functions in the Theory of Learning Machines. Nauka, Moscow (in Russian) (1970)
2. Dan, P., Yu, S., Chung, J.-Y.: Characterization of database access pattern for analytic prediction of buffer hit probability. VLDB J. 4(1), 127–154 (1995)
3. Debar, H., Becke, M., Siboni, D.: A neural network component for an intrusion detection system. In: IEEE Symp. on Security and Privacy, pp. 240–250 (1992)
4. Ghosh, A., Schwartzbard, A., Schatz, M.: Learning Program Behavior for Intrusion Detection. In: 11th USENIX Workshop on Intrusion Detection and Network Monitoring, Florida, CA (1999)
5. Hastie, T.: The Elements of Statistical Learning. Springer, New York (2001)
6. Lee, W., Stolfo, S.: Data mining approaches for intrusion detection. In: 7th USENIX Security Symposium (SECURITY 1998) (1998)
7. Liu, B., Hsu, W., Ma, Y.: Integrating classification and association rule mining. In: 4th Int. Conf. on KDD and Data Mining, pp. 80–96 (1998)
8. Manavoglu, E., Pavlov, D., Giles, C.: Probabilistic User Behavior Models. In: IEEE Int. Conf. on Data Mining (ICDM-2003), Melbourne, FL (2003)
9. Maxion, R., Roberts, R.: Proper Use of ROC Curves in Intrusion/Anomaly Detection, Tech. report CS-TR-871, University of Newcastle upon Tyne (2004)
10. Piatetsky-Shapiro, G., Fayyad, U., Smyth, P., Uthurusamy, R.: Advances in Knowledge Discovery and Data Mining. AAAI Press/MIT Press, Menlo Park (1996)
11. Quinlan, J.: Generating production rules from decision trees. In: 10th International Joint Conference on Artificial Intelligence, pp. 304–307 (1987)
12. Sarwar, B., Karypis, G., Konstan, J., Riedl, J.: Item-based Collaborative Filtering Recommendation Algorithms. In: 10th International World Wide Web Conference, pp. 285–295 (2001)
13. Tang, Z.-H., MacLennan, J.: Data Mining with SQL Server 2005. Wiley Publishing, Chichester (2005)
14. Valeur, F., Mutz, D., Vigna, G.: A Learning-Based Approach to the Detection of SQL Attacks. In: IEEE Conf. on Detection of Intrusions and Malware & Vulnerability Assessment, pp. 123–140 (2005)

# PART V
# Knowledge Engineering

# Approximate Reasoning to Learn Classification Rules

Amel Borgi

Research Unit SOIE, ISG, Tunis
National Institute of Applied Science and Technology, INSAT
Centre Urbain Nord de Tunis, BP 676, 1080 Tunis, Tunisia
Amel.Borgi@insat.rnu.tn

**Abstract.** In this paper, we propose an original use of approximate reasoning not only as a mode of inference but also as a means to refine a learning process. This work is done within the framework of the supervised learning method SUCRAGE which is based on automatic generation of classification rules. Production rules whose conclusions are accompanied by belief degrees, are obtained by supervised learning from a training set. These rules are then exploited by a basic inference engine: it fires only the rules with which the new observation to classify matches exactly. To introduce more flexibility, this engine was extended to an approximate inference which allows to fire rules not too far from the new observation. In this paper, we propose to use approximate reasoning to generate new rules with widened premises: thus imprecision of the observations are taken into account and problems due to the discretization of continuous attributes are eased. The objective is then to exploit the new base of rules by a basic inference engine, easier to interpret. The proposed method was implemented and experimental tests were carried out.

**Keywords:** Supervised learning, rules generation, approximate reasoning, inference engine, imprecision and uncertainty management.

## 1 Introduction

Facing the increase of data amount recorded daily, the detection of both structures and specific links between them, the organisation and the search of exploitable knowledge have become a strategic stake for decision making and prediction task. This complex problem of data mining has multiple aspects [10], [17]. We focus on one of them: supervised learning. In [4], [3], we have proposed a learning method from examples situated at the junction of statistical methods and those based on Artificial Intelligence techniques. Our method, SUCRAGE (SUpervised Classification by Rules Automatic GEneration) is based on automatic generation of classification rules. Production rules *IF premise THEN conclusion* are a mode of knowledge representation widely used in learning systems because they ensure the transparency and the easy explanation of the classifier [5], [8]. Indeed, the construction of production rules using the knowledge and the know-how of an expert is a very difficult task. The complexity and cost of such a knowledge acquisition have led to an important development of learning methods used for an automatic knowledge extraction, and in particular for rules extraction [8], [5].

J. Filipe, B. Shishkov, and M. Helfert (Eds.): ICSOFT 2006, CCIS 10, pp. 335–347, 2008.

The learning method SUCRAGE is based on a correlation search among the features of the examples and on discretization of continuous attributes. Rules conclusions are of the form «belonging to a class » and are uncertain. In the classification phase, an inference engine exploits the base of rules to classify new observations and also manages rules uncertainty. This reasoning that we called *basic reasoning* allows to obtain conclusions, when the observed facts match *exactly* rules premises.

In this paper, we are interested in an other reasoning: approximate reasoning [16], [8], [6]. It allows to introduce more flexibility and to overcome problems due to discretization. Such reasoning is closer to human reasoning than the basic one: human inferences do not always require a perfect correspondence between facts or causes to conclude.

In [2], we have proposed a context-oriented approximate reasoning. This reasoning, used as an inference mode, allows to manage imprecise knowledge as well as rules uncertainty: according to distance between observations and premises, it computes a neighborhood degree and associates a final confidence degree to rules conclusions. This model is faithful to the classical scheme of Generalized Modus Ponens [16]. In this paper, we propose to see approximate reasoning under another angle. The originality of our approach lies in the use of approximate reasoning, not only as a mode of inference, but to refine the learning. This reasoning allows to generate new rules and to ease in this way problems due to discretization and imprecision of the observations. The aim is that the new base of rules will then be exploited by a *basic inference* engine more easy to interpret. In our model, approximate reasoning has then no more vocation to be a method of inference allowing to fire certain rules but joins in the process of learning itself. The software SUCRAGE was extended: new rules construction through approximate reasoning was implemented. Applications of the extended version to benchmark problems are reported.

This paper is organized as follows. In section 2, the method SUCRAGE is described. More precisely we describe the learning phase (rules generation) and the classification phase. Only the basic inference engine is presented. In section 3, we present the approximate reasoning used as an inference mode. Section 4 attempts to explain the use of approximate reasoning to generate new classification rules and its contribution to the process of learning. Tests and results obtained by computer simulations with two benchmarks are provided in section 5. Finally, section 6 concludes the study.

## 2 The Supervised Learning Method Sucrage

### 2.1 Rules Generation

In this section, we describe the learning phase of the supervised learning method SUCRAGE. The training set contains examples described by numerical features denoted $X_1$, ..., $X_i$, ..., $X_p$. These examples are labelled by the class to which they belong. The classes are denoted $y_1, y_2, ..., y_C$. The generated rules are of the type:

$A_1$ and $A_2$ and ... and $A_k \longrightarrow y, \alpha$

where

$A_i$: condition of the form $X_j$ is in [a,b],
$X_j$: the $j^{th}$ vector component representing an observation,
[a,b]: interval issued from the discretization of the features variation domain (here, it is the variation domain of the feature $X_j$),
y: a hypothesis about membership in a class,
$\alpha$: a belief degree representing the uncertainty of the conclusion.

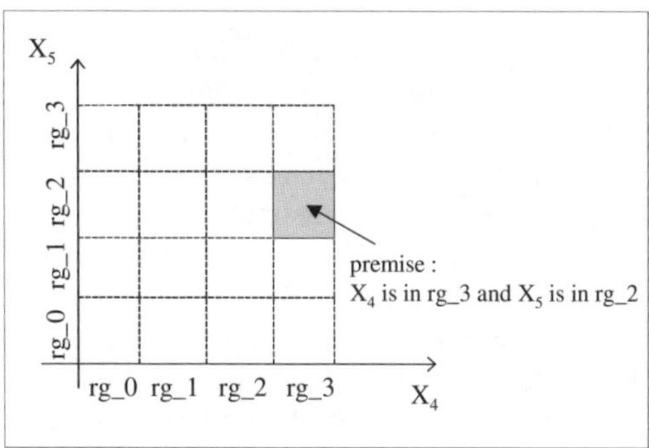


**Fig. 1.** A partition of the correlated features space

Our approach is multi-featured as the features that appear in rules premises are selected in one piece. This selection is realized by linear correlation search among the training set elements [4], [3]. So the first step consists in computing the correlation matrix between the components of the training set vectors. Then to decide which components are correlated, this matrix is thresholded (with a threshold denoted $\theta$). The idea is to detect privileged correlations between the features and to generate the rules according to these correlations. According to Vernazza's approach, we decide to group in the same premises all components that are correlated [14].

Next step in building the rules is feature discretization. Among the non supervised methods of discretization, the simplest one leads to M sub-ranges of equal width. This method called the *regular discretization* is the one we retain for this study. The M obtained sub-ranges are denoted rg_0, rg_1, ..., rg_(M-1), these values are totally ordered.

Once the discretization done, condition parts of rules are then obtained by considering for each correlated components subset, a sub-interval (rg_i) for each component in all possible combinations. Indeed the premises of the rules form a partition of the correlated components space. Figure 1 illustrates such a partition in the case of two correlated features ($X_4$ and $X_5$) and with a subdivision size M=4.

Each premise that we construct leads to the generation of C rules (C: number of classes). The rules conclusions are of the form « belonging to a class » and are not absolutely certain, that's why each conclusion is accompanied by a belief degree $\alpha$. In this paper, we propose to represent the belief degrees by a classical probability estimated on the training set [12], [3], [4].

### 2.2 Basic Inference Engine

The rules were generated for the purpose of a further classification use. In classification phase, the base of rules is exploited to classify new objects that don't belong to the training set. To achieve this goal, our approach consists in using a 0+ order inference engine. The inputs of this engine are the base of rules previously built and a vector representing the object to classify. The inference engine associates then a class to this vector.

We propose two reasoning models. The first one, called *basic reasoning* is presented in this section. The second one, the *approximate reasoning*, will be detailed in section 3. The *basic reasoning* allows the inference engine to fire only the rules with which the new observation components match *exactly*. The engine classifies each new observation using the classical deduction reasoning. It has to manage the rules' uncertainty and take it into account within the inference dynamic. Uncertainty management is done by computations on the belief degrees of the fired rules. Once the rules fired, we have to compute a final belief degree associated with each class. For this we propose to use a triangular co-norm [7]: the final belief degree associated to each class is the result of this co-norm applied on the probabilities of the fired rules that conclude to this considered class. Experimental tests presented in this paper were realized with the Zadeh co-norm (max). Finally the winner class associated with the new observation is the class for which the final belief degree is maximum.

## 3 Approximate Reasoning

Approximate reasoning, in a general way, makes reference to any reasoning which treats imperfect knowledge. This imperfection has multiple facets: for instance the knowledge can be vague, imprecise, or uncertain. In spite of such imperfections, approximate reasoning allows to treat this knowledge and to end in conclusions. In [8], approximate reasoning concerns as well the imprecision and uncertainty representation as their treatment and propagation in a knowledge based system. The term approximate reasoning has however a particular meaning of a word introduced by Zadeh in the field of Fuzzy Logic [16], [15]. In this frame, approximate reasoning corresponds to Generalized Modus Ponens who is an extension of Modus Ponens in fuzzy data. This definition of approximate reasoning is not contradictory to the first one which is more general and concerns all the forms of imperfections.

The approximate reasoning which we introduce is situated in the intersection of these two approaches. We are however more close to "fuzziers" as far as we remain faithful to the Generalized Modus Ponens [16], but we adapt it to a symbolic frame [4], [2], [6]. We propose a model of Approximate Reasoning which allows to associate a final degree of confidence to the conclusions (classes) on the basis of an imprecise correspondance between rules and observations. This reasoning does not fire only the rules the premises of which are exactly verified by the new observation, but also those who are not too much taken away from this observation. Thus, we are in the situation described in figure 2.

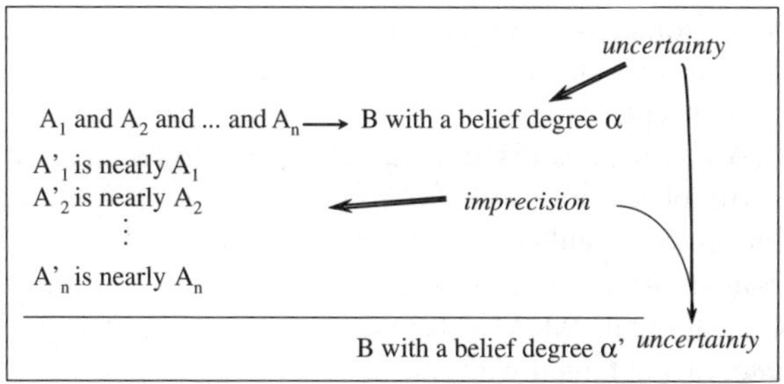


**Fig. 2.** Particular case of Generalized Modus Ponens

The consideration of observations close to rules premises allows to overflow around these premises. More exactly, it allows to extend beyond around the intervals stemming from the discretization and to ease so the problems of borders due to any discretization. So that our approximate reasoning can become operational, it is necessary to formalize first of all the notion of neighborhood. Then, it is necessary to model the approximate inference, that is to determine the degree of the final conclusion (α') of the diagram shown on figure 2.

### 3.1 Proximity between Observation and Premise

In works about approximate reasoning, Zadeh [16] stresses the necessary introduction of a distance in order to define neighbouring facts. In [13], a similarity degree between two objects is introduced. In our case, to define the notion of neighbourhood we have defined two types of measure or distance [4], [2]. A distance that we call local distance will measure the proximity of an observation element to a premise element. These distances will then be aggregated to obtain a global distance between the observation and the whole premise.

**A Local Distance.** We consider, by concern of clearness, the following rule:

$$X_1 \text{ in rg\_r1 and } X_2 \text{ in rg\_r2 and... } X_n \text{ in rg\_rn} \rightarrow y_t, \alpha$$

which groups together in its premise the attributes $X_1$, $X_2$, ...,$X_n$. This rule does not lose in generality: it can be obtained by renaming the attributes.

We note $V=(v_1,v_2,...,v_n)$ the elements of the observation concerned by the premise. To compare V with the following premise: *$X_1$ is in rg_r1* and *$X_2$ is in rg_r2* and ... and *$X_n$ is in rg_rn*, we begin by making local comparisons between $v_1$ and *$X_1$ is in rg_r1*, between $v_2$ and *$X_2$ is in rg_r2* ... So we have to define the local distances $d_1$, $d_2$, ..., $d_n$ of the following schema:

$A_1$ and $A_2$ and ... and $A_n$ →B with a belief degree α
$A'_1$ $d_1$-distant of $A_1$
$A'_2$ $d_2$-distant of $A_2$
...
$A'_n$ $d_n$-distant of $A_n$

---

B with a belief degree α'

More precisely it comes to determine the following distances $d_i$: $v_1$ is $d_1$-distant from rg_r1, …, $v_n$ is $d_n$-distant from rg_rn where the distance is the formal translation of the neighboring concept.

Rule premise ($A_i$) associates discrete values (rg_0, rg_1, ...,rg_(M-1)) to observation components. But observations ($A'_i$ or $v_i$) have numerical values. In order to compare them, we introduce a numerical-symbolic interface [4]. We split each interval rg_k into M sub-intervals of equal range, denoted $\sigma_0$, $\sigma_1$, .... We thus have a finer discretization, and we obtain M*M sub-intervals ($\sigma_0$, $\sigma_1$ ..., $\sigma_{M*M-1}$.). Figure 3 illustrates such sub-intervals obtained with M=3.

We can associate to each numerical value $v_i$ the sub-interval $\sigma_t$ to which it belongs. The distance $d_i$ between $v_i$ and rg_ri is then defined as the number of sub-intervals of type $\sigma$ separating $\sigma_t$ from rg_ri. Of course, $d_i$ is 0 if $v_i$ is in rg_ri. Thus, we obtain the distance vector $D=(d_1,d_2,\ldots,d_n)$ associated to every pair (observation, premise) or (observation, rule).

**A Global Distance.** In order to make approximate inferences, we want to aggregate the different local distances $d_i$. The result of this aggregation is a global distance that we note g-distance, and on which we wish to confer some properties [4]. One property that we impose to that distance is to be very sensitive to little variations of neighboring facts. This global distance that measures distance between approximately equal vectors can be insensitive when facts are very far from each other. This g-distance has to either measure the proximity between two nearby facts, or indicate by a maximal value, that they are not nearby. This is a proximity measure, and not a real distance. This distance is represented by an integer in [0, M-1]. In order to take into account the value dispersion, we do not use tools like min-max functions but we propose an aggregation based on a "dispersion" function $S_D$:

$$S_D: [0..M-1] \longrightarrow \mathbb{N}$$

$$k \longrightarrow S_D(k) = \sum_{i=1}^{n} (d_i - k)^2$$

$S_D(k)$ allows, in a way similar to the variance, to measure the dispersal of the local distances $d_i$ around k. We have then defined a global distance *g-dist* as follows:

$$g\text{-}dist: [0..M\times(M-1)]^n \longrightarrow [0..M-1]$$

$$(d_1,d_2, ...,d_n) \longrightarrow \max[S_D^{-1}(\min_{k=0}^{M-1} S_D(k))]$$

The global distance is presented with more details in [4] and [2]. We have notably proved that the proposed aggregated distance satisfies the above mentioned property.

We can notice that it is possible to have g-distance equal to 0, even if the distance vector is not null. In other words, it is possible to have a global distance equal to 0 for an observation that does not satisfy the considered rule.

### 3.2 Approximate Inference

The use of approximate inference supposes that a meta-knowledge exists in the system and allows it to run. In our case the meta-knowledge gives the possibility to bind imprecision (observation and premise of rule) to uncertainty (conclusion degree). This meta-knowledge has two complementary aspects: the first hypothesis says that a weak difference between observation and premise induces that the conclusion part is not significantly modified. For every rule, a stability area exists around the premise of the rule. The second and stronger hypothesis says that if the distance between observation and premise increases, then uncertainty of the conclusion increases too. A maximal distance must give a maximal uncertainty (in our case, it corresponds to the minimal belief degree, i.e. a probability equal to zero) [4].

The conclusion degree is weakened in accordance with the global distance. In our model, belief degrees (α) associated with rules are numerical, so it is hoped to conserve a numerical final degree (α') for the whole coherence. To compute the final belief degree α'of a conclusion via the approximate reasoning, given the global distance d (symbolic) between the premises and the observation and α the belief degree (numerical) of the conclusion of the fired rule, we propose the following function F :

$$F: \quad [0,1] \times [0..M\text{-}1] \longrightarrow [0,1]$$
$$(\alpha,d) \longrightarrow \alpha.(1-\frac{d}{M-1})$$

This formula includes the two aspects of the meta-knowledge hypothesis mentioned above. It is easy to observe that little imprecisions (in cases where d=0) do not modify uncertainty. On the other hand, a maximal distance (d=M-1) induces a complete uncertainty (α'=0). We note that we find back the basic reasoning in the limit case d=0.

## 4 Approximate Reasoning to Learn New Rules

In this part, we present the use of approximate reasoning not as a mode of inference to exploit rules in classification phase, but as a means to refine the learning. The use of approximate reasoning during the learning phase consists in generating new rules the premises of which are widened. The method consists in generating rules by using the basic approach described in section 2 then to look "around" the rules to verify if we cannot improve them or add better rules. The objective is then to exploit this base of rules thanks to a basic inference engine by hoping to obtain results close to a basic generation of rules exploited by an approximate engine.

For reasons of legibility and simplicity, we shall call the rules generation realized by SUCRAGE in its initial version the *basic generation*. The generation of rules completed by the construction of new rules via approximate reasoning will be called *approximate generation*.

### 4.1 Method with a Constant Number of Rules

This approach can be summarized by: "from an observation situated near the rules which we generated with the basic method of SUCRAGE, we verify if we cannot widen every rule to a rule of better quality". This is made always by using the same whole learning set.

To consider that an observation $O$ is near a rule R, we have to define a *g-threshold*, it is the maximal value authorized by *g-distance(O, R)*.

For every observation $O$ near a rule R (the mother rule) and having the same conclusion (class) as the rule R, we are going to build a new rule (the daughter rule $R_{daughter}$):

- the premise of $R_{daughter}$ is that containing *Premise(R)* and $O$ the most restrictive possible and convex by using the ranges (rg_ri) and the sub-intervals of type σ,
- the class of the conclusion do not change,
- the belief degree of $R_{daughter}$ is recomputed on the whole training set according to the new premise. This new assessment of the belief degree of the daughter rule built through approximate reasoning allows integrating this reasoning into the learning process.

The sentence "Premise containing *Premise(R)* and $O$ the most restrictive possible and convex by using the ranges (rg_ri) and the sub-intervals of type σ" means that to create the new premise, we start from the ancient premise and we add to all the conditions that $O$ does not verify the intervals of type σ which would allow $O$ to verify it. $R_{daughter}$ contains in its premise the same attributes as the mother rule R but with wider values. For instance, as shown in figure 3, in the case of a discretization with M=3, if the given value $O_i \in \sigma_4$ and the condition is $X_i$ *is in rg_*0 then the new condition will be $X_i$ *is in rg_0* $\cup\ \sigma_3 \cup \sigma_4$ (by supposing naturally that the condition of threshold on the global distance is verified).

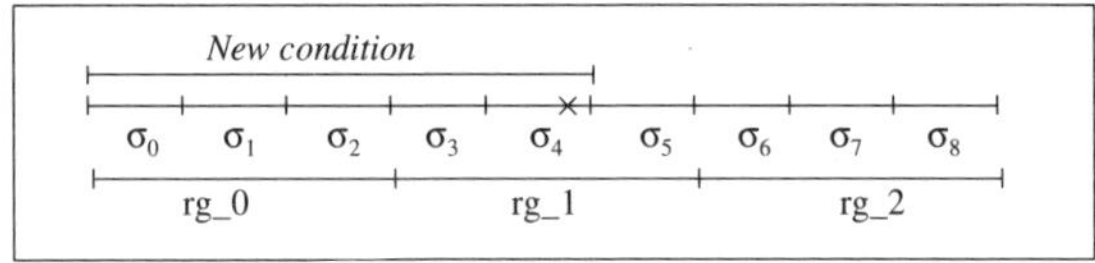


**Fig. 3.** An example of construction of a new rule condition

In the construction procedure of a new rule which we presented, a couple (observation, rule) verifying certain properties gives birth to a new rule. Among the mother rule and all the daughter rules we can generate, only the one who has the strongest belief degree is kept. Thus the initial number of rules does not change.

### 4.2 Method with Addition of Rules

We try here to widen the method of generation of a new base of rules so that the best rule is not the only one kept in the base. For that purpose, we use the "raw force" and we add in the base of rules all the rules that we can generate from each: a rule can then lead to several new rules and either as previously to a unique rule (that of stronger degree). This method allows to create a wide base close to data but this base, because of its size, becomes illegible as for interpretation by an expert. It becomes then necessary to optimize the size of the base of rules [5], [11].

## 5 Tests and Results

The system SUCRAGE that we initially developed allows the generation of rules by the method presented in section 2.1 as well as their exploitation by an inference engine. This engine uses a basic reasoning or approximate one [4], [2]. We completed this system by a module of rules generation via the approximate reasoning. We tested this new application on two learning bases stemming from the server of Irvine's University: those bases are Iris data and Wine data.

To compare the different results, we used the same test methods with the same parameters values for the classification system (size of subdivision M, correlation threshold θ). We used a ten order cross-validation [9]. The obtained results are presented and analyzed in this part.

### 5.1 Results of the Method with a Constant Number of Rules

We present here the tests realized with the method of new rules construction via the approximate reasoning according to the approach with a constant number of rules. The first tests were made with *g-threshold*=0 or *g-threshold*=1 which seem the only reasonable values. Values superior to 2 would throw a search which we could not consider as near the rule. The analysis of the results and the emission of hypotheses to explain them can be made by examining the shape of the generated rules. We distinguish two cases in function of *g-threshold* (0 or 1).

The case *g-threshold*=0 gives results (rates of good classification) almost identical to the basic generation (followed by an exact inference), so they are almost identical to results presented in column "Basic Generator, Basic Inference" of table 1. On the tested data, there are only very few changes between rules generated *basically* and *approximately*. This is mainly due to the following report: it is impossible, for a premise containing a number strictly lower than 3 attributes to have *g-dist*=0. All the rules containing 2 attributes in their premise can not be improved.

Let us focus now on the case *g-threshold*=1. Table 1 presents the rates of good classifications obtained with each of the three possible approaches:

- column « *Approx. Gen.-1, Basic Inference* »: rules were generated by SUCRAGE then new rules were built via approximate reasoning, with a value of *g-threshold=1*. The base of rules is then exploited by a *basic* inference engine.
- column « *Basic Gen., Basic Inference* »: rules were generated by SUCRAGE in its initial version. The rules base is then exploited by a *basic* inference engine.
- column « *Basic Gen., Approx. Inference* »: rules were generated by SUCRAGE in its initial version. The base of rules is then exploited by an approximate inference engine. It is the results of this method that we hope to approach (even improve) by using approximate reasoning to build new rules.

With Iris data, we can see that the results of the approximate generation are close to those obtained with the approximate inference. Moreover, these results are very similar to those obtained with the basic generator followed by basic reasoning. Thus, it is not very interesting in view of the supplementary computations needed.

With the WINE data, the results are very interesting: approximate generation of rules allows improving the case of a basic generation followed by an approximate inference (a single case of identical results). There is also improvement with regard to a basic generation followed by a basic inference.

**Table 1.** Method with a constant number of rules

| Method / Parameter | *Approx. Gen-1, Basic Inference* | *Basic Gen., Basic Inference* | *Basic Gen., Approx. Inference* |
|---|---|---|---|
| | IRIS data | | |
| M=3 θ=0.9 | **98.00** | 97.33 | 97.33 |
| M=3 θ=0.8 | **96.67** | 95.33 | **96.67** |
| M=5 θ=0.9 | 93.33 | **94.00** | **94.00** |
| M=5 θ=0.8 | 90.67 | 90.67 | **93.33** |
| | WINE data | | |
| M=3 θ=0.9 | **88.75** | **88.75** | **88.75** |
| M=3 θ=0.8 | **88.20** | 87.09 | 87.64 |
| M=5 θ=0.9 | **92.68** | 90.92 | 91.50 |
| M=5 θ=0.8 | **93.27** | 92.05 | 92.05 |

For *g-threshold*=1, the observation "around" the rules is not insufficient any more here (case *g-threshold*=0) but can be too much: we sometimes witness the creation of double rules. The observation near a rule can go up to another basic rule which was already generated, it is then the strongest which is going to win. We can have here a loss of information. The algorithm tends then to create an absorption of weak rules by strong rules rather than an extension of the strong rules.

## 5.2 Results of the Method with Addition of Rules

Table 2 presents the results obtained with the second method of new rules generation via the approximate reasoning: this time every new generated rule is added to the initial base of rules. The column "*Approx. Gen. Add., Basic Inference*" of this table contains the results obtained with this approach, the title of the last two columns is unchanged in comparison with table 1. In addition, every cell contains the rate of good classifications followed by the number of rules between brackets (for this method the number of rules takes importance).

The analysis of these results shows that they are very correct at the level of good classifications rate: with the *approximate generator with addition* the rates of good classifications are generally improved or maintained in comparison with the basic generator followed by a basic inference as well as the basic generator followed by an approximate inference. With the WINE data, two cases of light depreciation are to be noted.

**Table 2.** Method with addition of rules

| Method / Parameter | *Approx. Gen. Add., Basic Inference* | *Basic Gen., Basic Inference* | *Basic Gen., Approx. Inference* |
|---|---|---|---|
| IRIS data | | | |
| M=3 θ=0.9 | **97.33** (61.4) | **97.33** (23.5) | **97.33** (23,5) |
| M=3 θ=0.8 | **96.67** (123.6) | 95.33 (21.5) | **96.67** (21.5) |
| M=5 θ=0.9 | **95.33** (119.1) | 94.00 (37.7) | 94.00 (37.7) |
| M=5 θ=0.8 | **94.67** (303.9) | 90.67 (39.7) | 93.33 (39.7) |
| WINE data | | | |
| M=3 θ=0.9 | **90.45** (245.2) | 88.75 (96.7) | 88.75 (96.7) |
| M=3 θ=0.8 | **89.93** (214) | 87.09 (97.9) | 87.64 (97.9) |
| M=5 θ=0.9 | 89.35 (388.2) | 90.92 (152.4) | **91.50** (152.4) |
| M=5 θ=0.8 | 91.57 (343.8) | **92.05** (152.2) | **92.05** (152.2) |

On the other hand, the number of generated rules increases very widely. Moreover, it is evident that we generate many useless rules, even harmful rules entailing a decline of the results. A work to reduce the number of rules becomes here indispensable as well to eliminate the harmful rules that for reasons of legibility of the base of rules [11], [5]. A work was realized in this sense: we used Genetic Algorithms to reduce the size of the base of rules without losing too much performance. This approach tested in the case of basic generation of rules led to very interesting experimental results [1].

## 6 Conclusions

The supervised learning method SUCRAGE allows to generate classification rules then to exploit them by an inference engine which implements a basic reasoning or an approximate reasoning. The originality of our approach lies in the use of approximate reasoning to refine the learning: this reasoning is not only considered any more as a second running mode of the inference engine but is considered as a continuation of the learning phase. Approximate reasoning allows to generate new wider and more general rules. Thus imprecision of the observations are taken into account and problems due to the discretization of continuous attributes are eased. This process of learning refinement allows to adapt and to improve the discretization. The initial discretization is regular, it is not supervised. It becomes, via the approximate reasoning, supervised, as far as the observations are taken into account to estimate their adequacy to rules and as far as the belief degrees of these new rules are then computed on the whole training set. Moreover the interest of this approximate generation is that the new base of rules is then exploited by a basic inference engine, easier to interpret. Thus approximate reasoning complexity is moved from the classification phase (a step that has to be repeated) to the learning phase (a step which is done once). The realized tests lead to satisfactory results as far as they are close to those obtained with a basic generation of rules exploited by an approximate inference engine.

The continuation of the work will focus on the first method of new rules generation (with constant number of rules) to make it closer to what takes place during approximate inference. The search for other forms of g-distance can turn out useful notably to be able to obtain results of generation between the g-threshold value 0 (where we remain too close to the observation) and the g-threshold value 1 (where we go away too many "surroundings" of the observation). The second method, which enriches the base of rules with all the new rules, is penalized by the final size of the obtained base. An interesting perspective is to bend over the manners to reduce the number of rules without losing too much classification performance.

## References


1. Borgi, A.: Différentes méthodes pour optimiser le nombre de règles de classification dans SUCRAGE. In: 3rd Int. Conf. Sciences of Electronic, Technologies of Information and Telecom. SETIT 2005, Tunisia, p. 11 (2005)
2. Borgi, A., Akdag, H.: Apprentissage supervisé et raisonnement approximatif, l'hypothèse des imperfections. Revue d'Intelligence Artificielle, Editions Hermès, 15(1), 55–85 (2001)
3. Borgi, A., Akdag, H.: Knowledge based supervised fuzzy-classification: An application to image processing. Annals of Mathematics and Artificial Intelligence 32, 67–86 (2001)
4. Borgi, A.: Apprentissage supervisé par génération de règles: le système SUCRAGE, Thèse de doctorat (PhD thesis), Université Paris VI (1999)
5. Duch, W., Setiono, R., Zurada, J.M.: Computational Intelligence Methods for Rule-Based Data Understanding. Proceedings of the IEEE 92, 5 (2004)
6. El-Sayed, M., Pacholczyk, D.: Towards a Symbolic Interpretation of Approximate Reasoning. Electronic Notes in Theoretical Computer Science, vol. 82(4), pp. 1–12 (2003)
7. Gupta, M.M., Qi, J.: Connectives (And, Or, Not) and T-Operators in Fuzzy Reasoning. Conditional Inference and Logic for Intelligent Systems, 211–233 (1991)
8. Haton, J.-P., Bouzid, N., Charpillet, F., Haton, M., Lâasri, B., Lâasri, H., Marquis, P., Mondot, T., Napoli, A.: Le raisonnement en intelligence artificielle. InterEditions (1991)
9. Kohavi, R.: A Study of Cross-Validation and Bootstrap for Accuracy Estimation and Model Selection. In: Proc. of the Fourteenth International Joint Conference on Artificial Intelligence, vol. 2 (1995)
10. Michalski, R.S., Ryszard, S.: A theory and methodology of inductive learning. Machine Learning: An Artificial Intelligence Approach I, 83–134 (1983)
11. Nozaki, K., Ishibuchi, H., Tanaka, H.: Selecting Fuzzy Rules with Forgetting in Fuzzy Classification Systems. In: Proc. 3rd IEEE. Conf. Fuzzy Systems, vol. 1 (1994)

12. Pearl, J.: Numerical Uncertainty In Expert Systems. In: Shafer, Pearl (eds.) Readings in Uncertain Reasoning. Morgan Kaufman publishers, California (1990)
13. Ruspini, E.: On the semantics of fuzzy logic. International Journal of Approximate Reasoning 5 (1991)
14. Vernazza, G.: Image Classification By Extended Certainty Factors. Pattern Recognition 26(11), 1683–1694 (1993)
15. Yager, R.R.: Approximate reasoning and conflict resolution. International Journal of Approximate Reasoning 25, 15–42 (2000)
16. Zadeh, L.A.: A Theory of Approximate Reasoning. Machine Intelligence 9, 149–194 (1979)
17. Zhou, Z.H.: Three perspectives of data mining. Artificial Intelligence 143, 139–146 (2003)

# Combining Metaheuristics for the Job Shop Scheduling Problem with Sequence Dependent Setup Times

Miguel A. González, María R. Sierra, Camino R. Vela, Ramiro Varela, and Jorge Puente

Artificial Intelligence Center, Dept. of Computing, University of Oviedo
Campus de Viesques, 33271 Gijón, Spain
raist@telecable.es
{mariasierra,camino,ramiro,puente}@aic.uniovi.es

**Abstract.** The Job Shop Scheduling $(JSS)$ is a hard problem that has interested to researchers in various fields such as Operations Research and Artificial Intelligence during the last decades. Due to its high complexity, only small instances can be solved by exact methods, while instances with a size of practical interest should be solved by means of approximate methods guided by heuristic knowledge. In this paper we confront the Job Shop Scheduling with Sequence Dependent Setup Times $(SDJSS)$. The $SDJSS$ problem models many real situations better than the $JSS$. Our approach consists in extending a genetic algorithm and a local search method that demonstrated to be efficient in solving the $JSS$ problem. We report results from an experimental study showing that the proposed approaches are more efficient than other genetic algorithm proposed in the literature, and that it is quite competitive with some of the state-of-the-art approaches.

**Keywords:** Metaheuristics, Genetic Algorithms, Local Search, Job Shop Scheduling.

## 1 Introduction

The Job Shop Scheduling Problem with Sequence Dependent Setup Times $(SDJSS)$ is a variant of the classic Job Shop Scheduling Problem $(JSS)$ in which a setup operation on a machine is required when the machine switches between two jobs. This way the $SDJSS$ models many real situations better than the $JSS$. The $SDJSS$ has interested to a number of researchers, so we can find a number of approaches in the literature, many of which try to extend solutions that were successful to the classic $JSS$ problem. This is the case, for example, of the branch and bound algorithm proposed by Brucker and Thiele in [1], which is an extension of the well-known algorithms proposed in [2], [3] and [4], and the genetic algorithm proposed by Cheung and Zhou in [5], which is also an extension of a genetic algorithm for the $JSS$. Also, in [6] a neighborhood search with heuristic repairing is proposed that it is an extension of the local search methods for the $JSS$.

In this paper we apply a similar methodological approach and extend a genetic algorithm and a local search method that we have applied previously to the $JSS$ problem. The genetic algorithm was designed by combining ideas taken from the literature such

J. Filipe, B. Shishkov, and M. Helfert (Eds.): ICSOFT 2006, CCIS 10, pp. 348–360, 2008.

as for example the well-known $G\&T$ algorithm proposed by Giffler and Thomson in [7], the codification schema proposed by Bierwirth in [8] and the local search methods developed by various researchers, for example Dell' Amico and Trubian in [9], Nowicki and Smutnicki in [10] or Mattfeld in [11]. In [12] we reported results from an experimental study over a set of selected problems showing that the genetic algorithm is quite competitive with the most efficient methods for the $JSS$ problem.

In order to extend the algorithm to the $SDJSS$ problem, we have firstly extended the decoding algorithm, which is based on the $G\&T$ algorithm. Furthermore, in our local search method, we have adapted the neighborhood structure termed $N_1$ in the literature to obtain a neighborhood that we have termed $N_1^S$.

The experimental study was conducted over the set of 45 problem instances proposed by Cheung and Zhou in [5] and also over the set of 15 instances proposed by Brucker and Thiele in [1]. We have evaluated the genetic algorithm alone and then in conjunction with local search. The results show that the proposed genetic algorithm is more efficient than the genetic algorithm proposed in [5] and that the genetic algorithm combined with local search improves with respect to the raw genetic algorithm when both of them run during similar amount of time. Moreover, the efficiency of the genetic algorithm is at least comparable to the exact approaches proposed in [1] and [13].

The rest of the paper is organized as it follows. In section 2 we formulate the $SDJSS$ problem. In section 3 we outline the genetic algorithm for the $SDJSS$. In section 4 we describe the extended local search method. Section 5 reports results from the experimental study. Finally, in section 6 we summarize the main conclusions.

## 2 Problem Formulation

We start by defining the $JSS$ problem. The classic $JSS$ problem requires scheduling a set of $N$ jobs $J_1, \ldots, J_N$ on a set of $M$ physical resources or machines $R_1, \ldots, R_M$. Each job $J_i$ consists of a set of tasks or operations $\{\theta_{i1}, \ldots, \theta_{iM}\}$ to be sequentially scheduled. Each task $\theta_{il}$ having a single resource requirement, a fixed duration $p\theta_{il}$ and a start time $st\theta_{il}$ whose value should be determined.

The $JSS$ has two binary constraints: precedence constraints and capacity constraints. Precedence constraints, defined by the sequential routings of the tasks within a job, translate into linear inequalities of the type: $st\theta_{il} + p\theta_{il} \leq st\theta_{i(l+1)}$ (i.e. $\theta_{il}$ before $\theta_{i(l+1)}$). Capacity constraints that restrict the use of each resource to only one task at a time translate into disjunctive constraints of the form: $st\theta_{il} + p\theta_{il} \leq st\theta_{jk} \vee st\theta_{jk} + p\theta_{jk} \leq st\theta_{il}$. Where $\theta_{il}$ and $\theta_{jk}$ are operations requiring the same machine. The objective is to come up with a feasible schedule such that the completion time, i.e. the $makespan$, is minimized.

In the sequel a problem instance will be represented by a directed graph $G = (V, A \cup E)$. Each node in the set $V$ represents a operation of the problem, with the exception of the dummy nodes $start$ and $end$, which represent operations with processing time 0. The arcs of the set $A$ are called *conjunctive arcs* and represent precedence constraints and the arcs of set $E$ are called *disjunctive arcs* and represent capacity constraints. Set $E$ is partitioned into subsets $E_i$ with $E = \cup_{i=1,\ldots,M} E_i$. Subset $E_i$ corresponds to resource $R_i$ and includes an arc $(v, w)$ for each pair of operations requiring that

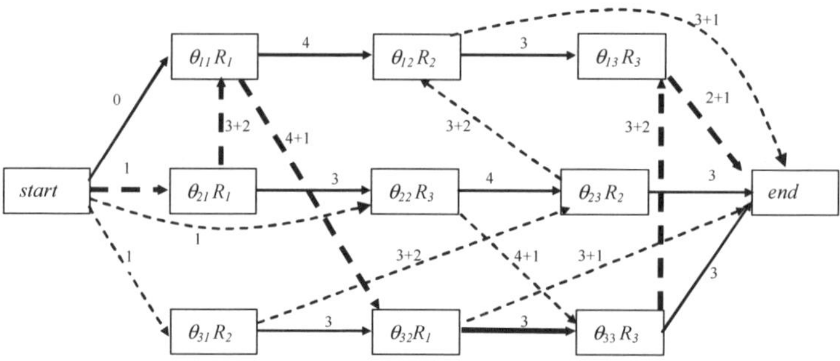


**Fig. 1.** A feasible schedule to a problem with 3 jobs and 3 machines. Bold face arcs show a critical path whose length, i.e. the $makespan$, is 22.

resource. The arcs are weighed with the processing time of the operation at the source node. The dummy operation $start$ is connected to the first operation of each job; and the last operation of each job is connected to the node $end$.

A feasible schedule is represented by an acyclic subgraph $G_s$ of $G$, $G_s = (V, A \cup H)$, where $H = \cup_{i=1..M} H_i$, $H_i$ being a hamiltonian selection of $E_i$. Therefore, finding out a solution can be reduced to discovering compatible hamiltonian selections, i.e. orderings for the operations requiring the same resource or partial schedules, that translate into a solution graph $G_s$ without cycles. The $makespan$ of the schedule is the cost of a *critical path*. A *critical path* is a longest path from node start to node end. The critical path is naturally decomposed into subsequences $B_1, \ldots, B_r$ called $critical\ blocks$. A critical block is a maximal subsequence of operations of a critical path requiring the same machine.

In the $SDJSS$, after an operation $v$ of a job leaves machine $m$ and before entering an operation $w$ of another job on the same machine, a setup operation is required with duration $S^m_{vw}$. The setup operation can be started as soon as operation $v$ leaves the machine $m$, hence possibly in parallel with the operation preceding $w$ in its job sequence. The setup time $S^m_{vw}$ is added to the processing time of operation $v$ to obtain the cost of each disjunctive arc $(v, w)$. $S^m_{0v}$ is the setup time of machine $m$ if $v$ is the first operation scheduled on $m$ and $S^m_{v0}$ is the cleaning time of machine $m$ if $v$ is the last operation scheduled on $m$.

Figure 1 shows a feasible solution to a problem with 3 jobs and 3 machines. Dotted arcs represent the elements of set $E$ included in the solution, while conjunctive arcs are represented by continuous arrows.

## 3 Genetic Algorithm for the SDJSS Problem

The $JSS$ is a paradigm of constraint satisfaction problems and was confronted by many heuristic techniques. In particular genetic algorithms [8],[11],[14],[12] are a promising approach due to their ability to be combined with other techniques such as tabu search and simulated annealing. Moreover genetic algorithms allow for exploiting any kind

**Algorithm 1.** Conventional Genetic Algorithm.

input: a JSS problem $P$
output: a schedule $H$ for problem $P$
1. Generate the Initial Population;
2. Evaluate the Population;
**while** No termination criterion is satisfied **do**
  3. Select chromosomes from the current population;
  4. Apply the Crossover and Mutation operators to the chromosomes selected at step 3. to generate new ones;
  5. Evaluate the chromosomes generated at step 4;
  6. Apply the Acceptation criterion to the set of chromosomes selected at step 3. together with the chromosomes generated at step 4.;
**end while**
7. Return the schedule from the best chromosome evaluated so far;

of heuristic knowledge from the problem domain. In doing so, genetic algorithms are actually competitive with the most efficient methods for $JSS$.

As mentioned above, in this paper we consider a conventional genetic algorithm for tackling the $JSS$ and extend it to the $SDJSS$. This requires mainly the adaptation of the decoding algorithm. Additionally we consider a local search method for the $JSS$ and adapt it to the $SDJSS$.

Algorithm 1 shows the structure of the genetic algorithm we have considered. In the first step the initial population is generated and evaluated. Then the genetic algorithm iterates over a number of steps or generations. In each iteration a new generation is built from the previous one by applying the genetic operators of selection, crossover, mutation and acceptation. In principle, these four operators can be implemented in a variety of ways and are independent each one to the others. However in practice all of them should be chosen considering their effect on the remaining ones in order to get a successful convergence. The approach taken in this work is the following. In the selection phase all chromosomes are grouped into pairs, and then each one of these pairs is mated and mutated accordingly to the corresponding probabilities to obtain two offsprings. Finally a tournament selection is done among each pair of parents and their offsprings.

To codify chromosomes we have chosen permutations with repetition proposed by C. Bierwirth in [8]. In this scheme a chromosome is a permutation of the set of operations, each one being represented by its job number. This way a job number appears within a chromosome as many times as the number of operations of its job. For example, the chromosome (2 1 1 3 2 3 1 2 3) actually represents the permutation of operations $(\theta_{21}\,\theta_{11}\,\theta_{12}\,\theta_{31}\,\theta_{22}\,\theta_{32}\,\theta_{13}\,\theta_{23}\,\theta_{33})$. This permutation should be understood as expressing partial schedules for every set of operations requiring the same machine. This codification presents a number of interesting characteristics; for example, it is easy to evaluate with different algorithms and allows for efficient genetic operators. In [15] this codification is compared with other permutation based codifications and demonstrated to be the best one for the $JSS$ problem over a set of 12 selected problem instances of common use. For chromosome mating we have considered the *Generalized Order*

*Crossover* $(GOX)$ that works as it is shown in the following example. Let us consider that the two following chromosomes are selected as parents for crossover

$$\text{Parent1 } (1\ 2\ 3\ \underline{3\ 2\ 1\ 1}\ 3\ 2)\ \ \text{Parent2 } (3\ \underline{3}\ 2\ 3\ 1\ \underline{1\ 2}\ 2\ \underline{1})$$

Firstly, a substring is selected from Parent1 and inserted in the Offspring at the same position as in this parent. Then the remaining positions of the Offspring are completed with genes from Parent2 after having removed the genes selected from Parent1. If the selected substring from Parent1 is the one marked with underlined characters, the resulting Offspring is

$$\text{Offspring } (3\ 2\ 3\ 3\ 2\ 1\ 1\ 1\ 2).$$

By doing so, $GOX$ preserves the order and position of the selected substring from Parent1 and the relative order of the remaining genes from Parent2. The mutation operator simply selects and swaps two genes at random. In practice the mutation would not actually be necessary due to the $GOX$ operator has an implicit mutation effect. For example the second 3 from Parent1 is now the third one in the Offspring.

### 3.1 Decoding Algorithm

As decoding algorithm we have chosen the well-known $G\&T$ algorithm proposed by Giffler and Thomson in [7] for the $JSS$ and then we have made a natural extension for the $SDJSS$. The $G\&T$ algorithm is an active schedule builder. A schedule is active if one operation must be delayed when you want another one to start earlier. Active schedules are good in average and, what is most important, it can be proved that the space of active schedules contains at least an optimal one, that is, the set of active schedules is *dominant*. For these reasons it is worth to restrict the search to this space. Moreover, the $G\&T$ algorithm is complete for the $JSS$ problem. Algorithm 2 shows the $G\&T$ algorithm for the $JSS$.

In order to adapt the $G\&T$ algorithm for the $SDJSS$ we consider an extension termed $EG\&T$. $EG\&T$ can be derived from the algorithm $EGTA1$ developed by Ovacik and Uzsoy in [16], by simply taking into account the setup times in Algorithm 2. So, the step 4 of Algorithm 2 is exchanged by

4. Remove from $B$ every operation $\theta$ that $st\theta \geq st\theta' + p\theta' + S^R_{\theta'\theta}$ for any $\theta' \in B$;

In Algorithm 2, $st\theta$ refers to the maximum completion time of the last scheduled operation on the machine required by operation $\theta$ and the preceding operation to $\theta$ in its job. Hence the algorithm can be adapted to the $SDJSS$ problem by considering $st\theta$ as the maximum completion time of the preceding operation in the job and the completion time of the last scheduled operation in the machine plus the corresponding setup time. It is easy to demonstrate that $EG\&T$ is not complete. In [17] two more extensions of the $G\&T$ schedule generation scheme are proposed, one of them is not complete either, and the other is complete but it is very time consuming due to it needs to do backtracking. In any case, the lack of completeness of a decoding algorithm is not a serious problem in the framework of $GAs$ due to a $GA$ itself is not complete. Moreover, the local search schema outlined in the next section gives to any chromosome the chance of being reached, so in any way the lack of completeness of the decoding algorithm is compensated.

**Algorithm 2.** The decoding Giffler and Thomson algorithm for the $JSS$ problem.

input: a chromosome $C$ and a problem $P$
output: the schedule $H$ represented by chromosome $C$ for problem $P$
1. $A$ = set containing the first operation of each job;
**while** $A \neq \emptyset$ **do**
  2. Determine the operation $\theta' \in A$ with the earliest completion time if scheduled in the current state, that is $st\theta' + p\theta' \leq st\theta + p\theta, \forall\theta \in A$;
  3. Let $R$ be the machine required by $\theta'$, and $B$ the subset of $A$ whose operations require $R$;
  4. Remove from $B$ every operation that cannot start at a time earlier than $st\theta' + p\theta'$;
  5. Select $\theta^* \in B$ so that it is the leftmost operation of $B$ in the chromosome sequence;
  6. Schedule $\theta^*$ as early as possible to build the partial schedule corresponding to the next state;
  7. Remove $\theta^*$ from $A$ and insert the succeeding operation of $\theta^*$ in set $A$ if $\theta^*$ is not the last operation of its job;
**end while**
8. return the built schedule;

## 4 Local Search

Conventional genetic algorithms, like the one described in the previous section, often produce moderate results. However meaningful improvements can be obtained by means of hybridization with other methods. One of such techniques is local search, in this case the genetic algorithm is called a memetic algorithm. Hybridization of a genetic algorithm with local search is carried out by applying the local search algorithm to every chromosome just after this chromosome is generated, instead of simply applying the Algorithm 2 as it is done in the simple genetic algorithm. Algorithm 3 shows the typical strategy of a local search.

Roughly speaking local search is implemented by defining a neighborhood of each point in the search space as the set of chromosomes reachable by a given transformation rule. Then a chromosome is replaced in the population by one of its neighbors, if any of them satisfies the acceptation criterion. The local search from a given point completes either after a number of iterations or when no neighbor satisfies the acceptation criterion.

In this paper we consider the neighborhood structure proposed by Nowicki and Smutnicki in [10], which is termed $N_1$ by D. Mattfeld in [11], for the $JSS$. As other strategies, $N_1$ relies on the concepts of critical path and critical block. It considers every critical block of a critical path and made a number of moves on the operations of each block. After a move inside a block, the feasibility must be tested. Since an exact procedure is computationally prohibitive, the feasibility is estimated by an approximate algorithm proposed by Dell' Amico and Trubian in [9]. This estimation ensures feasibility at the expense of omitting a few feasible solutions. In [11] the transformation rules of $N_1$ are defined as follows.

**Definition 1.** *($N_1$) Given a schedule $H$ with partial schedules $H_i$ for each machine $R_i$, $1 \leq i \leq M$, the neighborhood $N_1(H)$ consist of all schedules derived from $H$ by reversing one arc $(v, w)$ of the critical path with $(v, w) \in H_i$. At least one of $v$ and $w$ is*

**Algorithm 3.** The Local Search Algorithm.

input: a chromosome $C$ and a $JSS$ problem $P$
output: a (hopefully) improved chromosome
1. Evaluate chromosome $C$ (Algorithm 2) to obtain schedule $H$;
**while** No termination criterion is satisfied **do**
  2. Generate the neighborhood of $H$ with some method $N$, $N(H)$;
  3. Select $H' \in N(H)$ with the selection criterion;
  4. Replace $H$ by $H'$ if the acceptation criterion holds;
**end while**
5. Rebuild chromosome $C$ from schedule $H$;
6. return chromosome $C$;

*either the first or the last member of a block. For the first block only $v$ and $w$ at the end of the block are considered whereas for the last block only $v$ and $w$ at the beginning of the block must be checked.*

The selection strategy of a neighbor and the acceptation criterion are based on a $makespan$ estimation, which is done in constant time as it is also described in [9], instead of calculating the exact $makespan$ of each neighbor. The estimation provides a lower bound of the $makespan$. The selected neighbor is the one with the lowest $makespan$ estimation whenever this value is lower than the $makespan$ of the current chromosome. Notice that this strategy is not steepest descendent because the exact $makespan$ of selected neighbor is not always better than the $makespan$ of the current solution. We have done this choice in the classic $JSS$ problem due to it produces better results than a strict steepest descent gradient method. [12].

The Algorithm stops either after a number of iterations or when the estimated $makespan$ of selected neighbor is larger than the $makespan$ of the current chromosome.

This neighborhood relies on the fact that, for the $JSS$ problem, reversing an arc of the critical path always maintains feasibility. Moreover, the only possibility to obtain some improvement by reversing an arc is that the reversed arc is either the first or the last of a critical block.

However, things are not the same for $SDJSS$ problem due to the differences in the setup times. As can we see in [6], feasibility is not guaranteed when reversing an arc of the critical path, and reversing an arc inside a block could lead to an improving schedule. The following results give sufficient conditions of no-improving when an arc is reversed in a solution $H$ of the $SDJSS$ problem. In the setup times the machine is omitted for simplicity due to all of them refers to the same machine.

**Theorem 1.** *Let $H$ be a schedule and $(v, w)$ an arc that is not in a critical block. Then reversing the arc $(v, w)$ does not produce any improvement even if the resulting schedule is feasible.*

**Theorem 2.** *Let $H$ be a schedule and $(v, w)$ an arc inside a critical block, that is there exist arcs $(x, v)$ and $(w, y)$ belonging to the same block. Even if the schedule $H'$*

*obtained from $H$ by reversing the arc $(v, w)$ is feasible, $H'$ is not better than $H$ if the following condition holds*

$$S_{xw} + S_{wv} + S_{vy} \geq S_{xv} + S_{vw} + S_{wy} \quad (1)$$

**Theorem 3.** *Let $H$ be a schedule and $(v, w)$ an arc in a critical path so that $v$ is the first operation of the first critical block and $z$ is the successor of $w$ in the critical path and $M_w = M_z$. Even if reversing the arc $(v, w)$ leaves to a feasible schedule, there is no improvement if the following condition holds*

$$S_{0w} + S_{wv} + S_{vz} \geq S_{0v} + S_{vw} + S_{wz} \quad (2)$$

*Analogous, we can formulate a similar result if $w$ is the last operation of the last critical block.*

Hence we can finally define the neighborhood strategy for the $SDJSS$ problem as it follows

**Definition 2.** $(N_1^S)$ *Given a schedule $H$, the neighborhood $N_1^S(H)$ consist of all schedules derived from $H$ by reversing one arc $(v, w)$ of the critical path provided that none of the conditions given in previous theorems 1, 2 and 3 hold.*

### 4.1 Feasibility Checking

Regarding feasibility, for the $SDJSS$ it is always required to check it after reversing an arc. As usual, we assume that the triangular inequality holds, what is quite reasonable in actual production plans, that is for any operations $u$,$v$ and $w$ requiring the same machine

$$S_{uw} \leq S_{uv} + S_{vw} \quad (3)$$

Then the following is a necessary condition for no-feasibility after reversing the arc $(v, w)$.

**Theorem 4.** *Let $H$ be a schedule and $(v, w)$ an arc in a critical path, $PJ_w$ the operation preceding $w$ in its job and $SJ_v$ the successor of $v$ in its job. Then if reversing the arc $(v, w)$ produces a cycle in the solution graph, the following condition holds*

$$stPJ_w > stSJ_v + duSJ_v + S_{min} \quad (4)$$

*where*

$$S_{min} = min\{S_{kl}/(k, l) \in E, J_k = J_v\}$$

*and $J_k$ is the job of operation $k$.*

Therefore the feasibility estimation is efficient at the cost of discarding some feasible neighbor.

### 4.2 Makespan Estimation

For $makespan$ estimation after reversing an arc, we have also extended the method proposed by Taillard in [18] for the $JSS$. This method was used also by Dell'Amico and Trubian in [9] and by Mattfeld in [11]. This method requires calculating *heads* and *tails*. The head $r_v$ of an operation $v$ is the cost of the longest path from node *start* to node $v$ in the solution graph, i.e. is the value of $stv$. The tail $q_v$ is defined so as the value $q_v + p_v$ is the cost of the longest path from node $v$ to node *end*.

For every node $v$, the value $r_v + p_v + q_v$ is the length of the longest path from node *start* to node *end* trough node $v$, and hence it is a lower bound of the $makespan$. Moreover, it is the $makespan$ if node $v$ belongs to the critical path. So, we can get a lower bound of the new schedule by calculating $r_v + p_v + q_v$ after reversing $(v, w)$.

Let us denote by $PM_v$ and $SM_v$ the predecessor and successor nodes of $v$ respectively on the machine sequence in a schedule. Let nodes $x$ and $z$ be $PM_v$ and $SM_w$ respectively in schedule $H$. Let us note that in $H'$ nodes $x$ and $z$ are $PM_w$ and $SM_v$ respectively. Then the new heads and tails of operations $v$ and $w$ after reversing the arc $(v, w)$ can be calculated as the following

$$r'_w = max(r_x + px + S_{xw}, r_{PJ_w} + pPJ_w)$$

$$r'_v = max(r'_w + pw + S_{wv}, r_{PJ_v} + pPJ_v)$$

$$q'_v = max(q_z + pz + S_{vz}, q_{SJ_v} + pSJ_v)$$

$$q'_w = max(q'_v + pv + S_{vw}, q_{SJ_w} + pSJ_w)$$

From these new values of heads and tails the $makespan$ of $H'$ can be estimated by

$$C'_{max} = max(r'_v + pv + q'_v, r'_w + pw + q'_w)$$

which is actually a lower bound of the new $makespan$. This way, we can get an efficient $makespan$ estimation of schedule $H'$ at the risk of discarding some improving schedule.

## 5 Experimental Study

For experimental study we have used the set of problems proposed by Cheung and Zhou in [5] and also the benchmark instances taken from Brucker and Thiele [1]. The first one is a set of 45 instances with sizes (given by the number of jobs and number of machines $N \times M$) $10 \times 10$, $10 \times 20$ and $20 \times 20$, which is organized into 3 types. Instances of type 1 have processing times and setup times uniformly distributed in (10,50); instances of type 2 have processing times in (10,50) and setup times in (50,99); and instances of type 3 have processing times in (50,99) and setup times in (10,50). Table 1 shows the results from the genetic algorithm termed $GA\_SPTS$ reported in [5]. The data are grouped for sizes and types and values reported are averaged for each group. This algorithm was coded in FORTRAN and run on PC 486/66. The computation time with problem sizes $10 \times 10$, $10 \times 20$ and $20 \times 20$ are about 16, 30 and 70 minutes respectively. Each

**Table 1.** Results from the $GA\_SPTS$

| ZRD Instance | Size $N \times M$ | Type | Best | Avg | StDev |
|---|---|---|---|---|---|
| 1-5 | $10 \times 10$ | 1 | 835,4 | 864,2 | 21,46 |
| 6-10 | $10 \times 10$ | 2 | 1323,0 | 1349,6 | 21,00 |
| 11-15 | $10 \times 10$ | 3 | 1524,6 | 1556,0 | 35,44 |
| 16-20 | $20 \times 10$ | 1 | 1339,4 | 1377,0 | 25,32 |
| 21-25 | $20 \times 10$ | 2 | 2327,2 | 2375,8 | 46,26 |
| 26-30 | $20 \times 10$ | 3 | 2426,6 | 2526,2 | 75,90 |
| 31-35 | $20 \times 20$ | 1 | 1787,4 | 1849,4 | 57,78 |
| 36-40 | $20 \times 20$ | 2 | 2859,4 | 2982,0 | 93,92 |
| 41-45 | $20 \times 20$ | 3 | 3197,8 | 3309,6 | 121,52 |

**Table 2.** Results from the $GA\_EG\&T$

| ZRD Instances | Size $N \times M$ | Type | Best | Avg | StDev |
|---|---|---|---|---|---|
| 1-5 | $10 \times 10$ | 1 | 785,0 | 803,0 | 8,76 |
| 6-10 | $10 \times 10$ | 2 | 1282,0 | 1300,2 | 9,82 |
| 11-15 | $10 \times 10$ | 3 | 1434,6 | 1455,4 | 12,87 |
| 16-20 | $20 \times 10$ | 1 | 1285,8 | 1323,0 | 15,38 |
| 21-25 | $20 \times 10$ | 2 | 2229,6 | 2278,2 | 22,24 |
| 26-30 | $20 \times 10$ | 3 | 2330,4 | 2385,8 | 23,91 |
| 31-35 | $20 \times 20$ | 1 | 1631,6 | 1680,4 | 17,99 |
| 36-40 | $20 \times 20$ | 2 | 2678,0 | 2727,8 | 23,60 |
| 41-45 | $20 \times 20$ | 3 | 3052,0 | 3119,6 | 29,33 |

**Table 3.** Results from the $GA\_EG\&T\_LS$

| ZRD Instances | Size $N \times M$ | Type | Best | Avg | StDev |
|---|---|---|---|---|---|
| 1-5 | $10 \times 10$ | 1 | 778,6 | 788,5 | 6,70 |
| 6-10 | $10 \times 10$ | 2 | 1270,0 | 1290,4 | 9,16 |
| 11-15 | $10 \times 10$ | 3 | 1433,8 | 1439,8 | 6,71 |
| 16-20 | $20 \times 10$ | 1 | 1230,2 | 1255,5 | 12,74 |
| 21-25 | $20 \times 10$ | 2 | 2178,4 | 2216,8 | 18,61 |
| 26-30 | $20 \times 10$ | 3 | 2235,2 | 2274,0 | 19,32 |
| 31-35 | $20 \times 20$ | 1 | 1590,0 | 1619,8 | 15,90 |
| 36-40 | $20 \times 20$ | 2 | 2610,2 | 2668,0 | 27,48 |
| 41-45 | $20 \times 20$ | 3 | 2926,0 | 2982,2 | 26,32 |

algorithm run was stopped at the end of the 2000th generation and tried 10 times for each instance.

Tables 2 and 3 reports the results reached by the genetic algorithm alone and the genetic algorithm with local search, termed $GA\_EG\&T$ and $GA\_EG\&T\_LS$ respectively, proposed in this work. In the first case thc genetic algorithm was parameterized

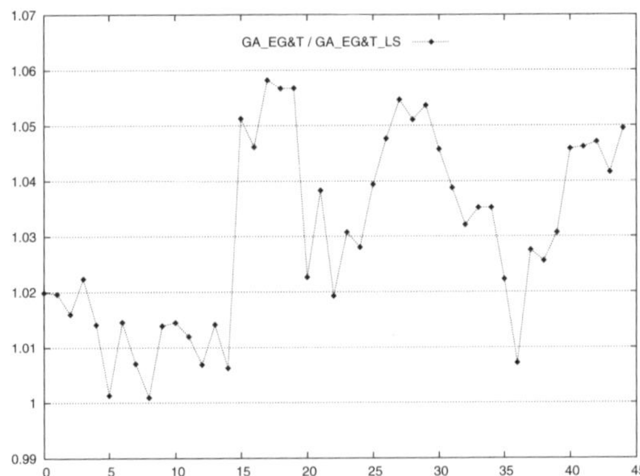


**Fig. 2.** Comparison of the raw genetic algorithm with the memetic algorithm. The graphic shows for each problem the quotient of the mean makespan of the best solutions reached in all 30 trials by the raw GA and the GA with local search.

with a population of 100 chromosomes, a number of 140 generations, crossover probability of 0.7, and mutation probability of 0.2. For the experiments combining the genetic algorithm with local search, we have parameterized the genetic algorithms with 50 chromosomes in the population and 50 generations in order to have similar running times.

The rest of the parameters remain as in previous experiments. The genetic algorithm was run 30 times and reported the values of the best solution reached, the average of the best solutions of the 30 runs and the standard deviation. The machine was a Pentium IV at 1.7 Ghz. and the computation time varied from about 1 sec. for the smaller instances to about 10 sec. for the larger ones. As we can observe both algorithms improved the results obtained by the $GA\_SPTS$. Moreover algorithm $GA\_EG\&T\_LS$ has outperformed $GA\_EG\&T$. Figure 2 shows the relative improvement of $GA\_EG\&T\_LS$ over $GA\_EG\&T$ in all problems. The improvement is clear in almost all cases. Regarding the benchmark from Brucker and Thiele [1], these instances are defined from the classical $JSS$ instances, proposed by Lawrence [19], by introducing setup times. There are 15 instances named $t2\_ps01$ to $t2\_ps15$. Instances $t2\_ps01$ to $t2\_ps05$ are of type $10 \times 5$ (small instances). Instances $t2\_ps06$ to $t2\_ps10$ are of type $15 \times 5$ (medium instances). Instances $t2\_ps11$ to $t2\_ps15$ are of type $20 \times 5$ (large instances). Table 4 shows results from two state-of-the-art methods: the branch and bound algorithms proposed by Brucker and Thiele [1] (denoted as $BT96$) and Artigues et al. in [13] (denoted as $ABF04$). In the results reported in [1] and [13] the target machine was Sun 4/20 station and Pentium IV at 2.0 GHz. in both cases the time limit for the experiments was 7200 sec. In this case, our memetic algorithm was parameterized as the following: population size = 100 for small and medium instances and 200 for larger instances, and the number of generations has been 100 for small instances, 200 for medium instances, and 400 for larger instances. The rest of the parameters remain as in previous experiments. We run the algorithm 30 times for each instance, and the computation time for the larger instances was 30 sec. for each run, i.e. 900 sec. of running time for each instance.

**Table 4.** Comparison between $BT96$, $ABF04$ and $GA\_EG\&T\_LS$

| Problem Instance | Size $N \times M$ | $BT96$ | $ABF04$ | $GA\_EG\&T\_LS$ |
|---|---|---|---|---|
| $t2\_ps01$ | $10 \times 5$ | **798** | **798** | **798** |
| $t2\_ps02$ | $10 \times 5$ | **784** | **784** | **784** |
| $t2\_ps03$ | $10 \times 5$ | **749** | **749** | **749** |
| $t2\_ps04$ | $10 \times 5$ | **730** | **730** | **730** |
| $t2\_ps05$ | $10 \times 5$ | **691** | **691** | 693 |
| $t2\_ps06$ | $15 \times 5$ | 1056 | 1026 | 1026 |
| $t2\_ps07$ | $15 \times 5$ | 1087 | **970** | **970** |
| $t2\_ps08$ | $15 \times 5$ | 1096 | 1002 | 975 |
| $t2\_ps09$ | $15 \times 5$ | 1119 | 1060 | 1060 |
| $t2\_ps05$ | $15 \times 5$ | 1058 | **1018** | **1018** |
| $t2\_ps06$ | $20 \times 5$ | 1658 | - | 1450 |
| $t2\_ps07$ | $20 \times 5$ | 1528 | 1319 | 1347 |
| $t2\_ps08$ | $20 \times 5$ | 1549 | 1439 | 1431 |
| $t2\_ps09$ | $20 \times 5$ | 1592 | - | 1532 |
| $t2\_ps05$ | $20 \times 5$ | 1744 | - | 1523 |

values in bold are optimal

As we can observe, $GA\_EG\&T\_LS$ is able to reach optimal solutions for the smaller instances, as $BT96$ and $ABF04$, with only one exception. For the medium and large instances reaches solutions that are better or equal than $ABF04$ and much better that $BT06$. Unfortunately, for the larger instances, results from only two instances are reported in [13].

## 6 Conclusions

In this work we have confronted the Job Shop Scheduling Problem with Sequence Dependent Setup Times by means of a genetic algorithm hybridized with local search. As other approaches reported in the literature, we have extended a solution developed for the classic $JSS$ problem. We have reported results from an experimental study on the benchmark proposed in [5] showing that the proposed genetic algorithms produce better results than the genetic algorithm proposed in [5], mainly when these algorithms are hybridized with local search. Here it is important to remark that the running conditions of both genetic algorithms are not strictly comparable. Also we have experimented with the benchmark proposed by Brucker and Thiele in [1], and compare our memetic algorithm with two state-of-the-art exact branch and bound approaches due to Brucker and Thiele [1] and Artigues et al. in [13] respectively. In this case the results shown that our approach is quite competitive.

As future work we plan to look for new extensions of the $G\&T$ algorithm in order to obtain a complete decoding algorithm and more efficient operators. Also we will try to extend other local search algorithms and neighborhoods that have been proved to be very efficient for the $JSS$ problem.

## Acknowledgements

We would like to thank Waiman Cheung and Hong Zhou, and Christian Artigues for facilitating us the benchmarks used in the experimental study. This research has been supported by FEDER-MCYT under contract TIC2003-04153 and by FICYT under grant BP04-021.

## References

1. Brucker, P., Thiele, O.: A branch and bound method for the general-job shop problem with sequence-dependent setup times. Operations Research Spektrum 18, 145–161 (1996)
2. Brucker, P., Jurisch, B., Sievers, B.: A branch and bound algorithm for the job-shop scheduling problem. Discrete Applied Mathematics 49, 107–127 (1994)
3. Brucker, P.: Scheduling Algorithms, 4th edn. Springer-Verlag, Heidelberg (2004)
4. Carlier, J., Pinson, E.: Adjustment of heads and tails for the job-shop problem. European Journal of Operational Research 78, 146–161 (1994)
5. Cheung, W., Zhou, H.: Using genetic algorithms and heuristics for job shop scheduling with sequence-dependent setup times. Annals of Operational Research 107, 65–81 (2001)
6. Zoghby, J., Barnes, J.W., Hasenbein, J.J.: Modeling the re-entrant job shop scheduling problem with setup for metaheuristic searches. European Journal of Operational Research 167, 336–348 (2005)
7. Giffler, B., Thomson, G.: Algorithms for solving production scheduling problems. Operations Reseach 8, 487–503 (1960)
8. Bierwirth, C.: A generalized permutation approach to jobshop scheduling with genetic algorithms. OR Spectrum 17, 87–92 (1995)
9. Dell' Amico, M., Trubian, M.: Applying tabu search to the job-shop scheduling problem. Annals of Operational Research 41, 231–252 (1993)
10. Nowicki, E., Smutnicki, C.: A fast taboo search algorithm for the job shop problem. Management Science 42, 797–813 (1996)
11. Mattfeld, D.C.: Evolutionary Search and the Job Shop. Investigations on Genetic Algorithms for Production Scheduling. Springer-Verlag, Heidelberg (1995)
12. González, M., Sierra, M., Vela, C., Varela, R.: Genetic Algorithms Hybridized with Greedy Algorithms and Local Search over the Spaces of Active and Semi-active Schedules. In: Marín, R., Onaindía, E., Bugarín, A., Santos, J. (eds.) CAEPIA 2005. LNCS (LNAI), vol. 4177, pp. 231–240. Springer, Heidelberg (2006)
13. Artigues, C., Belmokhtar, S., Feillet, D.: A New Exact Algorithm for the Job shop Problem with Sequence Dependent Setup Times. In: Régin, J.-C., Rueher, M. (eds.) CPAIOR 2004. LNCS, vol. 3011, pp. 96–109. Springer, Heidelberg (2004)
14. Varela,R.,Vela,C.R.,Puente,J.,Gómez,A.:Aknowledge-basedevolutionarystrategyforscheduling problems with bottlenecks. European Journal of Operational Research 145, 57–71 (2003)
15. Varela, R., Serrano, D., Sierra, M.: New Codification Schemas for Scheduling with Genetic Algorithms. In: Mira, J., Álvarez, J.R. (eds.) IWINAC 2005. LNCS, vol. 3562, pp. 11–20. Springer, Heidelberg (2005)
16. Ovacik, I., Uzsoy, R.: Exploiting shop floors status information to schedule complex jobs. Operations Research Letters 14, 251–256 (1993)
17. Artigues, C., Lopez, P., Ayache, P.D.: Schedule generation schemes for the job shop problem with sequence-dependent setup times: Dominance properties and computational analysis. Annals of Operational Research 138, 21–52 (2005)
18. Taillard, E.D.: Parallel taboo search techniques for the job shop scheduling problem. ORSA Journal of Computing 6, 108–117 (1993)
19. Lawrence, S.: Resource constrained project scheduling: an experimental investigation of heuristic scheduling techniques (supplement). Technical report, Graduate School of Industrial Administration, Carnegie Mellon University (1984)

# A Description Clustering Data Mining Technique for Heterogeneous Data

Alejandro García López[1], Rafael Berlanga[2], and Roxana Danger[2]

[1] European Laboratory for Nuclear Research (CERN), Geneva, Switzerland
Alejandro.Garcia.Lopez@cern.ch
[2] Depto. de Lenguajes y Sistemas Informáticos. Universitat Jaume I, Castellón, Spain
berlanga@lsi.uji.es, roxana.danger@alumail.uji.es

**Abstract.** In this work we present a formal framework for mining complex objects, being those characterized by a set of heterogeneous attributes and their corresponding values. First we introduce several Data Mining techniques available in the literature to extract association rules. We show as well some of the drawbacks of these techniques and how our proposed solution is going to tackle them. Then demonstrate how applying a clustering algorithm as a pre-processing step on the data allow us to find groups of attributes and objects that provide us with a richer starting point for the Data Mining process. Then define the formal framework, its decision functions and its interesting measurement rules, as well as a newly designed Data Mining algorithms specifically tuned for our objectives. We also show the type of knowledge to be extracted in the form of a set of association rules. Finally state our conclusions and propose the future work.

**Keywords:** Complex objects, association rules, clustering, data mining.

## 1 Introduction

The problem of mining complex objects, as we understand it, is that of extracting useful information out of multidimensional heterogeneous data. To fully comprehend this concept we need therefore to define what we mean by *extracting useful information* and *multidimensional heterogeneous data*.

When we talk about *multidimensional heterogeneous data*, we are referring to collections of attributes of different types (boolean, categorical, numerical, etc.) which are represented in an structured way. This structured representation would normally be based on a relational schema, although we could also think of, for example, a collection of XML documents.

On the other hand, what we mean by *extracting useful information* is mainly the discovery of frequent and approximate underlying patterns (Association Rules, ARs), which can help users to undertake a number of decision taking tasks. Examples of these are: summarizing a data collection, finding interesting relations amongst its attributes, finding certain trends, etc.

This kind of association rules can be applied to a wide range of applications. Our main motivating application consists of mining large log repositories that contain data about the performance of a GRID infrastructure for ALICE experiments at CERN.

J. Filipe, B. Shishkov, and M. Helfert (Eds.): ICSOFT 2006, CCIS 10, pp. 361–373, 2008.

Stored data records include heterogeneous attributes involving different data types (e.g. location of a node, average serving time, number of processes, etc.) In this context, users can be interested on finding frequent patterns amongst these attributes in order to plan properly the distribution of tasks over the GRID.

The definition of ARs was first stated in [1], referring to binary attributes. Basically it is defined as follows. Let $I = I_1, I_2, ..., I_m$ be a set of binary attributes, called items. Let $T$ be a database of transactions. Each transaction $t$ is represented as a binary vector, with $t[k] = 1$ if $t$ bought the item $I_k$, and $t[k] = 0$ otherwise. Let $X$ be a set of some items in $I$. We say that a transaction $t$ satisfies $X$ if for all items $I_k$ in $X$, $t[k] = 1$. An AR is then, an implication of the form $X \Rightarrow I_j$, where $X$ is a set of some items in $I$, and $I_j$ is a single item in $I$ that is not present in $X$. An example of this type of rule is: "90% of transactions that purchased bread and butter also purchased milk". The antecedent of this rule consists of bread and butter and the consequent consists of milk alone.

In [2] where the concept of Quantitative Association Rules (QARs) is first shown, the authors deal with the fact that the vast majority of relational databases, either based on scientific or business information are not filled with binary data-types (as requested by the classical ARs) but with a much richer range of data-types both numerical and categorical.

A first approach to tackle this problem consists of mapping the QARs problem into the *boolean* ARs problem. The key idea is that if all attributes are categorical or the quantitative attributes have only a few values, this mapping is straightforward. However, this approach generates problems as if the intervals are too large, some rules may not have the required *minimum confidence* and if they are too small, some rules may not have the required *minimum support*. We could also think of the strategy of considering all possible continuous ranges over the values of the quantitative attribute to cover the partitioned intervals (to solve the *minimum confidence* problem) and increase the number of intervals (solving the problem of *minimum support*). Unfortunately two new problems arise: First, if a quantitative attribute has *n* values (or intervals), there are on average $O(n^2)$ ranges that include a specific value or interval, fact that blows up the execution time and second, if a value (or interval) of a quantitative attribute has *minimum support*, so will any range containing this value/interval, therefore, the number of rules increases dramatically.

The approach taken by [2] is different. Considering ranges over adjacent values/ intervals of quantitative attributes to avoid the *minimum support* problem. To mitigate the problem of the excess of execution time, they restricted the extent to which adjacent values/intervals may be combined by introducing a user-specified *maximum support* parameter; they stop combining intervals if their combined *support* exceeds this value. They introduce as well a *partial completeness measure* in order to be able to decide whether to partition a quantitative attribute or not and how many partitions should there be, in case it's been decided to partition at all. To address the problem of the appearance of too many rules, they propose an *interest measure* based on the deviation from the expectation that helps to prune out the uninteresting rules (extension of the *interest measure* already proposed in [3]). Finally an algorithm to extract QARs is presented, sharing the same idea of the algorithm for finding ARs over binary data given in [4] but adapting the implementation to the computational details of how candidates are generated and how their *supports* are now counted.

In [5], the authors pointed out the pitfalls of the equi-depth method (interest measure based on deviation), and presented several guiding principles for quantitative attribute partitioning. They apply clustering methods to determine sets of dense values in a single attribute or over a set of attributes that have to be treated as a whole. But although they took distance among data into account, they did not take the relations among other attributes into account by clustering a quantitative attribute or a set of quantitative attributes alone. Based on this, [6] improved the method to take into account the relations amongst attributes.

Another improvement in the mining of quantitative data is the inclusion of Fuzzy Sets to solve the *sharp boundary problem* [7]. An element belongs to a set category with a membership value, but it can as well belong to the neighbouring ones.

In [8] a mixed approach based on the *quantitative approach* introduced by [2], the hash-based technique from the Direct Hashing and Pruning (DHP) algorithm [9] and the methodology for generating ARs from the *apriori* algorithm [4] was proposed. The experimental results prove that this approach precisely reflects the information hidden in the data-sets, and on top of it, as the data-set increases, it scales-up linearly in terms of processing time and memory usage.

On the other hand, the work realised by Aumann et al. in [10], proposes a new definition for QARs. An example of this rule would be: $sex = female \Rightarrow Wage : mean = \$7.90\ p/hr$ (overall mean wage = $\$9.02$). This form of QAR, unlike others doesn't require the discretisation of attributes with real number domains as a pre-processing step. Instead it uses the statistical theory and data-driven algorithms to process the data and find regularities that lead to the discovery of ARs. A step forward in this kind of rules was given by [11]. They provide variations of the algorithm proposed in [10] enhancing it by using heuristic strategies and advanced database indexing. The whole methodology is completed with the proposition of post-processing techniques with the use of similarity and significance measures.

The motivation of this work is to tackle some of the drawbacks of the previous techniques. Most of them require the translation of the original database so that each non-binary attribute can be regarded as a discrete set of binary variables over which the existing data mining algorithms can be applied to. This approach can be sometimes unsatisfactory due to the following reasons: the translated database can be larger than the original one, the transformation of the quantitative data could not correspond to the intended semantics of the attributes. Moreover, current approaches do not deal with heterogeneous attributes but define ad-hoc solutions for particular data types (mainly numerical ones). As a consequence, they do not provide a common data mining framework where different representations, interesting measures and value clustering techniques can be properly combined.

### 1.1 Overview of Our Proposal

In this article, we extend the work introduced in [12] by applying clustering techniques in two steps of the mining process. A schematic view of the overall process can be seen in Figure 1. First, clustering is applied to the attribute domains, so that each object can

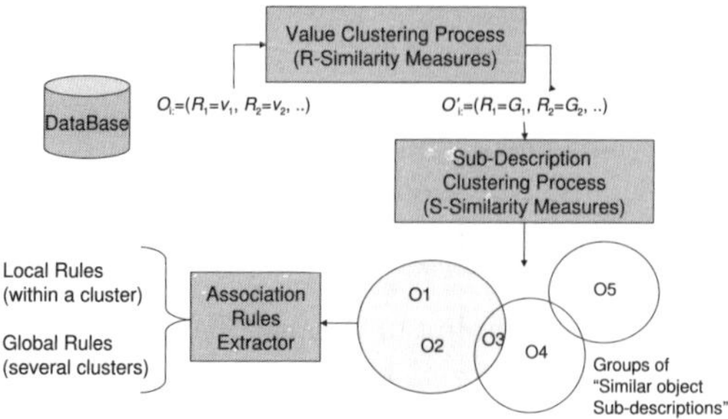


**Fig. 1.** Overview of our proposal

be expressed as a set of pairs $\langle attribute, cluster \rangle$ instead of $\langle attribute, value \rangle$. This new representation allows users to define the most appropriate technique to discretize numeric domains or to abstract categorical values. We name *object sub-description* to the characterisation of an object through value clusters. The second step consists of clustering *object sub-descriptions* in order to find frequent patterns between their features.

Finally, we propose an algorithm capable of obtaining the frequent itemsets from the found object subdescription clusters. We distinguish two kind of ARs, namely: inter and intra-cluster. The former relate attributes of different clusters, whereas the latter relate attributes locally defined in a cluster. Both kind of ARs provide different levels of details to users, which can mine a selected cluster involving a restricted set of attributes (i.e. local analysis) or the whole set of clusters (i.e. global analysis).

The paper is organised as follows: in the next section, we introduce the necessary concepts of the proposed framework. Then, in Section 3 we explain how we include clustering in our mining process. In Section 4 we describe a data-mining algorithm that finds frequent object sub-descriptions, and in Section 5 we describe the preliminary experimental results. Finally, in Section 6 we give our conclusions and we outline the future work.

## 2 Formal Definitions

In the proposed framework, a data collection consists of a set of objects, $\Omega = o_1, o_2, ..., o_n$, which are described by a set of features $R = R_1, R_2, ..., R_m$. We will denote with $D_i$ the domain of the i-th feature, which can be of any data type.

We will apply the clustering algorithm to the attributes' domains in order to find groups (clusters) of close values and use them instead of the original values. Thus each object will be no longer characterised by it's attributes' values but by the clusters to which these values belong. We will denote the set of clusters in the domain ($D_i$) of a given attribute $i$ as $\Pi_i = G_{i,1}, ..., G_{i,r}$, being $r \geq 1$ and $G_{i,r}$ the r-th cluster in the domain of the i-th attribute.

On the other hand, we will apply a second clustering step to the object sub - descriptions in order to generate groups of objects that will help us in reducing the final number of rules. We will denote the set of clusters in $\Omega$ as $\Theta = OG_1, ..., OG_t$, being $t \geq 1 \leq n$ and $OG_i$ the i-th cluster in the objects' domain.

In order to compare two attribute-clusters, each feature $R_i$ has associated a *comparison criterion*, $C_i(x, y)$, which indicates whether the pair of clusters, $x, y \in \Pi_i$, must be considered equal or not. This comparison criterion can include specifications for the case of invalid and missing values in order to deal with incomplete information.

The simplest comparison criterion is the strict equality, which can be applied to any domain:

$$C(x,y) = \begin{cases} 1 & \textit{if } x = y \\ 0 & \textit{otherwise} \end{cases}$$

Another interesting criteria can use the centroid of each domain cluster. For example, being $c_{i,r}$ the centroid of the r-th cluster over the i-th attribute the comparison function looks as follows:

If $x \in G_{a,1}$ and $y \in G_{a,2}$ then

$$C(x,y) = \begin{cases} 1 & \textit{if } |c_{a,1} - c_{a-2}| \leq \epsilon \\ 0 & \textit{otherwise} \end{cases}$$

Which expresses the fact that two clusters are considered equal if their centroids differ from each other in at most a given threshold $\epsilon$.

Since the mining process is intended to discover the combinations of object features and object clusters that frequently co-occur, it is necessary to manage the different object projections. Thus, a *subdescription* of an object $o$ for a subset of features $S \subseteq R$, denoted as $I|_S(o)$, is the projection of $o$ over the feature set $S$. In this context, we denote $o[r]$ the value of the object $o$ for the feature $r$.

Moreover, we assume that there exists a *similarity function* between two object subdescriptions, which allow us to decide whether two objects $o_i$ and $o_j$ must be considered equal or not by the mining process. All the similarity functions are binary, that is, they return either 0 (not equal) or 1 (equal).

The simplest similarity function is the following one:

$$Sim(I|_S(o), I|_S(o')) = \begin{cases} 1 & \textit{if } \forall r \in S, C(o[r], o'[r]) = 1) \\ 0 & \textit{otherwise} \end{cases}$$

which expresses the strict equality by considering the comparison criterion of each of the subdescription features.

Alternatively, the following similarity function states that two sub-descriptions are considered equal if they have at least $\epsilon$ features belonging to the same cluster:

$$Sim(I|_S(o), I|_S(o')) = \begin{cases} 1 & \textit{if } |\{r \in S | C(o[r], o'[r]) = 1\}| \geq \epsilon \\ 0 & \textit{otherwise} \end{cases}$$

In order to compare object-clusters, we can take one *representative* object of each cluster. In our approach, such a representative corresponds to the object with maximum connectivity according to the adopted similarity function. This is because we use a clustering algorithm that generates star-shaped clusters.

Analogously to the traditional Data Mining works, we also provide definitions of *support* and ARs, but applied to this new context.

We define the *support* of a *subdescription* $v = I|_S(o)$, denoted with $Sup(v)$, based in the work by [12], as the percentage of objects in $\Omega$ whose sub-descriptions are similar to $v$, that is:

$Sup(v) = \frac{|\{o' \in \Omega | Sim(I|_S(o'), v) = 1\}|}{|\Omega|}$

We say that a pair of *sub-descriptions* $v_1 = I|_{R_1}(o)$ and $v_2 = I|_{R_2}(o)$, with $R_1 \bigcap R_2 = \emptyset$ and $R_1, R_2 \subset R$, are associated through the AR $v_1 \Rightarrow v_2(s, c)$, if $Sup(v') \geq s$ and $\frac{Sup(v')}{Sup(v_1)} \geq c$, where $v' = I|_{R_1 \cup R_2}(o)$. The values of $s$ and $c$ are called *support* and *confidence* of the rule respectively.

The problem of computing the AR for complex objects consists of finding all the AR of the *subdescriptions* of $\Omega$ whose *support* and *confidence* satisfy the user-specified thresholds.

It must be pointed out that the previous definitions subsume the traditional concept of AR, therefore, if we use strict equality in both the comparison criterion and the similarity function, we obtain the classical definition of AR.

Besides, we can include other comparison criteria such as the interval-based partitions, for quantitative data, and the *is-a* relationship of the concept taxonomies, in order to represent other kinds of ARs [3] [13] [14].

The idea that different items have different levels of interest for the user, as suggested in [15], can be also incorporated in this framework by assigning a weight to each variable in the similarity function. Moreover, when the variables' data is fuzzy, it is perfectly admissible to use as a comparison criterion the membership of the values to the same fuzzy set.

## 3 Finding Interesting Subdescriptions

In a previous step to that of finding the interesting ARs we will pre-process the data by means of clustering algorithms in order to find the groups that will be the base of our mining process.

The objective of this pre-processing step is that of identifying clusters in the domain of the attributes that will characterise the objects we will use to extract intra-cluster rules, and identifying clusters in the domain of the recently discovered object subdescriptions in order to extract inter-cluster rules. Mainly, in the first clustering step, we would be translating the Database from a combination of $< attribute, value >$ into a combination of $< attribute, cluster >$. The second clustering process, on the other hand, would have as a result the creation of groups of objects, that will be the base of our mining process and the input of the association rules extraction algorithm that will be presented in the next section.

The algorithm chosen for this process is the Star Clustering Algorithm introduced in [16], and modified to be order independent in [17]. The main reason for choosing it is that the Star-based representation of the objects subdescriptions seems a good way of representing the support of each subdescription (i.e. the number of objects that are similar to it, also called, satellites, as we will see later). Briefly, the star-shaped graphs capture the most supported subdescriptions w.r.t. the defined similarity function.

This algorithm approximates the minimal dominant set of the $\beta - similarity$ graph. The minimal dominant set [18] is the smallest set composed of graph's vertexes that contains every vertex in the graph, or at least if a vertex is not contained, it has a neighbour that does. The members of the minimal dominant set are called *stars* and their neighbours *satellites*.

A star-shaped sub-graph of $l + 1$ vertexes consists of a star an $l$ satellites. Each sub-graph forms a group and the *stars* are the objects with the biggest connectivity. If an object is isolated in the graph it is considered as well a *star*.

The basic steps of this algorithm are the following ones:

- Obtain the $\beta - similarity$ graph.
- Calculate the degree of every vertex.
- While there's still ungrouped sub-vertexes do:
  - Take the ungrouped vertex with the highest degree.
  - Build a group with it an its neighbours.

Figure 2 shows the star-shaped graph of a cluster of object subdescriptions. The complexity of the algorithm is in $O(n^2)$, being $n$ the number of processed objects.

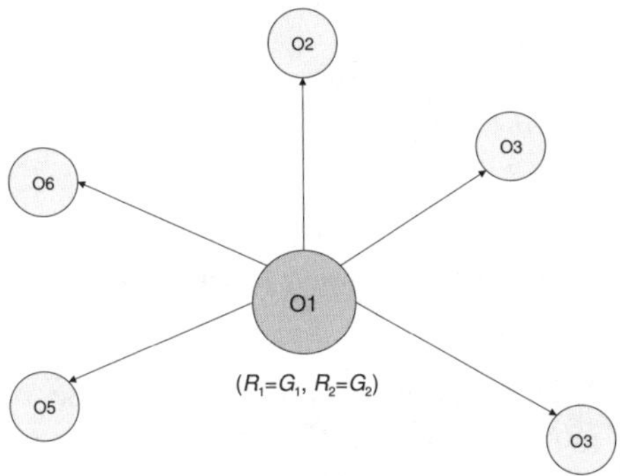


**Fig. 2.** Example of star-based object cluster

## 4 Extracting Association Rules

In this section we present an algorithm (see Figure 3) for computing the frequent subdescriptions for a collection of complex objects. This algorithm is inspired in the original algorithm of [4]. However, it also uses the strategy of the Partition algorithm [3] to compute the *support* of object subdescriptions. The implementation used is the one appearing in the R Statistical Package (http://www.r-project.org/), on its module *arules* introduced in [19]

It is worth mentioning that in this work an item-set is a *subdescription*, and its *support* is the number of objects in the database that are similar to it.

The algorithm works as follows: first, it finds all the frequent itemsets in side each cluster, so that the intra-cluster association rules can be extracted. Then, once this rules

FreqItemSets_ComplexObjects($\Omega$, CriterionComps, SimilFuncs, MinSupp, FreqSets)

Input: $\Omega = \{o_1, o_2, ..., o_n\}$, a set of complex objects.
CriterionComps: array of comparison's functions. The i-th component in the array corresponds to the comparison criterion for the i-th feature.
SimilFuncs: Dictionary of similarity functions, such that the key that corresponds to the similarity function for the subdescription $S' = \{K_{i_1}, ..., K_{i_s}\}$ is the own set $S'$.
MinSupp: Minimal support to consider a subdescription as frequent.

Output: FreqSets: Set of dictionaries that maintains for each size and combination of features (with at least one frequent cluster) the frequent sub-descriptions in $\Omega$ and the index of the objects that are similar to each one of these sub-descriptions.

Method:

$F_1 = SetFreqClusters(\Omega, CriterionComps)$
$k = 2$
while $F_{k-1} \neq \emptyset$ do
  $SetCandidatesVars = \{\{f_i, f_j\} / f_i, f_j \in F_{k-1}.keys(), |f_i \cup f_j| = k\}$
  for each pair of features $\{f_i, f_j\}$ in $SetCandidatesVars$ do
    for each $O \in \Omega$ do
      if $I|_{f_i}(O) \in F_{k-1}[f_i].keys()$ and $I|_{f_j}(O) \in F_{k-1}[f_j].keys()$ then
        $IndexSimObjs = F_{k-1}[f_i][I|_{f_i}(O)] \cap F_{k-1}[f_j][I|_{f_j}(O)]$
        $SimObjs = \{\}$
        for $O_k \in \Omega$, $k \in IndexSimObjs$ do
          if $SimilFuncs[f_i \cup f_j](I|_{f_i \cup f_j}(O), I|_{f_i \cup f_j}(O_k)) = 1$ then
            $SimObjs = SimObjs \cup \{k\}$
        if $|SimObjs| \geq MinSupp$ then
          $F_k[f_i \cup f_j][I|_{f_i \cup f_j}(O)] = SimObjs$
  $FreqSets = FreqSets \cup \{F_k\}$
  $k = k + 1$

**Fig. 3.** Data Mining Algorithm

have been devised, it will start combining clusters (if a number of conditions occur) by combining their sets of intra-rules, in order to extract inter-cluster rules.

The conditions to combine to clusters of objects are: First, that the overlap between both clusters (number of objects present on both clusters) is greater than the minimum support threshold and second, that there is a minimum number of complementary variables to assure that the combination of clusters does not result in a redundant set of frequent item-sets/association rules.

To illustrate the process, let's think of a clustering process resulting in 3 object clusters, $OG_1, OG_2$ and $OG_3$ (resulting from the second clustering process). We would apply the apriori algorithm to extract the frequent itemsets from within them, being those denoted by $S_1, S_2$ and $S_3$. This 3 sets are the rules at the first level. Then, if the aforementioned conditions are matched, we would start combining the sets of itemsets two by two in order to find the 2nd level rules: $S_1 - S_2, S_1 - S_3$ and $S_2 - S_3$ pruning out the items present on both sides, and keeping only those who add new information. Finally we would combine the three of them, in the same way, provided the conditions of overlapping and complementary variables would be complied, and there would be some item adding new information to them.

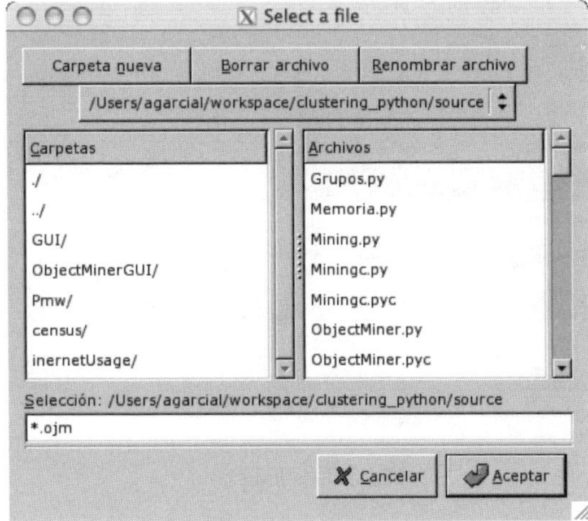


**Fig. 4.** Select File Dialog

It's important to take into account that in order to guarantee the monotonic construction of the frequent itemsets, it is necessary that the similarity functions satisfy the following condition: if two objects are different with respect to a subdescription $S_1$, they are also different with respect to any other subdescription $S_2$, such that $S_1 \subset S_2$ [12].

## 5 Preliminary Results

To validate our approach we have developed ObjectMiner, an association rules extraction tool that implements our method giving the user some extra functionalities like defining its own similarity measures (Figures 6 and 7) or choosing the clustering algorithm to be applied (Figures 4 and 5). The data on the application's snapshots is from our training Database the CIA World Fact-book Database.

Executing the model over the aforementioned Database, we prove that we are able to generate rules in two levels. From all 175 Countries, 3 clusters are generated at the first level. This number may vary depending on the similarity measure used for clusters (that is, for telling if an object belongs in one cluster or not).

The attributes of the Database subject to study were:

- WaterArea: total water area in $km^2$
- LandArea: total land area in $km^2$
- Climate: text describing the type of climate
- Terrain: text describing the type of terrain
- Natural Resources: text explaining the natural resources
- LandUseArableLand: percentage of arable Land.
- LandUsePermanentCrops: percentage of land destined to permanent crops
- LandUseOther: percentage of land destined to other things
- IrrigatedLand: percentage of irrigated land

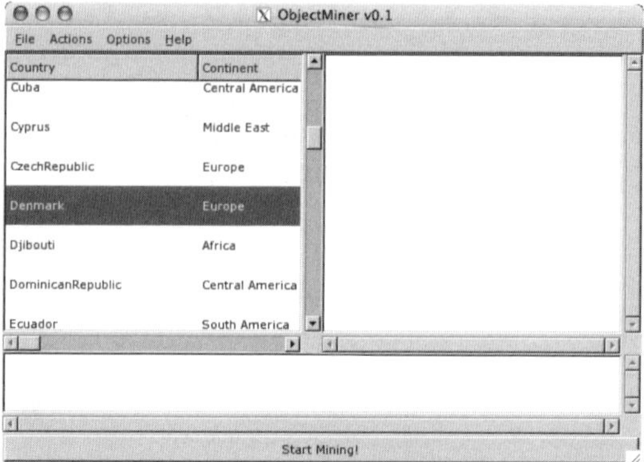


**Fig. 5.** Loaded Data

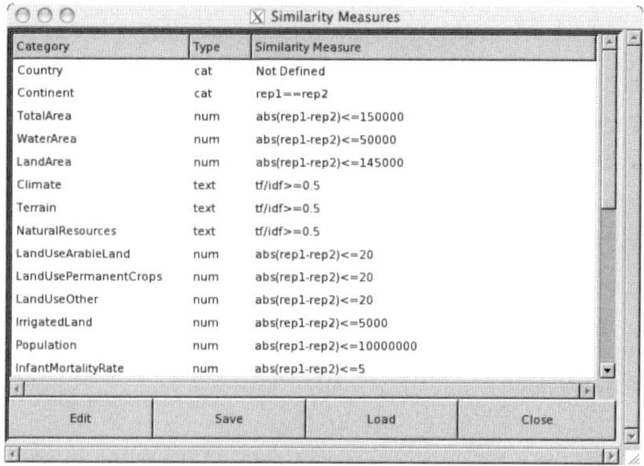


**Fig. 6.** Similarity Measures

- Population: number of inhabitants
- InfantMortalityRate: infant mortality rate
- LifeExpectancyAtBirthTotal: life expectancy at birth of the whole population in years
- LifeExpectancyAtBirthFemale: life expectancy at birth for females in years
- LifeExpectancyAtBirthMale: life expectancy at birth for males in years
- Religions: percentage of the different religions represented in the country.
- GDPRealGrowthRate: gross domestic product growth rate
- GDPAgriculture: percentage of the gross domestic product coming from agriculture
- GDPIndustry: percentage of the gross domestic product coming from industry
- GDPServices: percentage of the gross domestic product coming from services
- PopulationBelowPovertyLine: percentage of population considered to be poor
- Industries: description of the main industries of the country
- ElectricityProduction: electricity production in kw/hour
- ExternalDebt: external debt in dollars

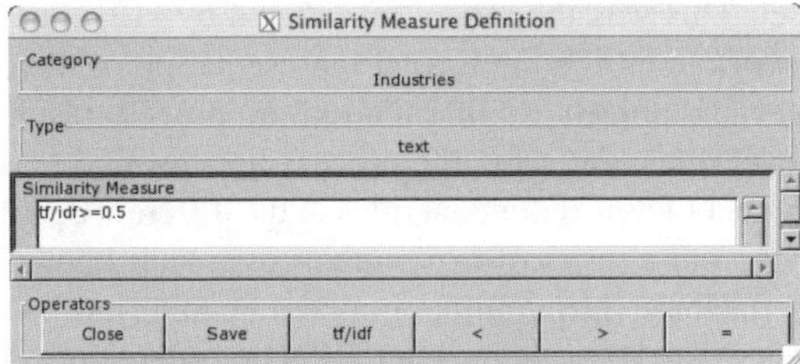


**Fig. 7.** Textual Similarity Measure

Examples of the similarity measures used are:

- Numerical attributes, as Water Area: $abs(rep1 - rep2) <= 50000$. That means, that two objects will fall within the same cluster if their Water Area surface is different by less than 50000 $Km^2$.
- Textual attributes, as Climate: $tf/idf >= 0.5$, meaning that two objects will belong in the same cluster if the tf/idf vector of their climate description differs by 0.5 units.

Examples of the rules generated with our approach in the different levels of detail available are the following:

- GDPServices $\sim 55$ and IrrigatedLand $\sim 75620$ then InfantMortalityRate $\sim 0.14476$
- LandUseArableLand $\sim 20$ and InfantMortalityRate $\sim 0.14476$ and WaterArea $\sim 470131$ then LandArea $\sim 7617960$
- LandUseArableLand $\sim 20$ and InfantMortalityRate $\sim 0.14476$ and LandUseOther $\sim 80$ and LandUsePermanentCrops $\sim 10$ then WaterArea $\sim 470131$

Statistics of the execution:

- Total number of attribute clusters: 1124
- Total number of object clusters: 3 (1st Level), 2(2nd Level) and 1(3rd Level)
- Total number of rules: $\sim 2230$
- Execution Time: $\sim 20$ minutes (on an Intel Dual Core $2.16Ghz$ machine)

Further testing is necessary with real-life Databases in order to determine if the method scales when the size of the problem grows dramatically. The future tests will be performed over the Internet Usage Database, the Census Income Database and the MonAlisa Repository Database.

## 6 Conclusions and Future Work

This paper presents a general framework for mining complex objects stemming from any of the existing data models (e.g. relational, object-oriented and semi-structured data models). The mining process is guided by the semantics associated to each object description feature (attributes), which are stated by the users by selecting the appropriate representation model. This was the model introduced by [12]. Furthermore, we have extended the framework to enrich the formal representation of the objects using clusters of both attributes and objects, so that the mining process results in an acceptable number of higher level rules. We show the process finishes in a reasonable amount of time producing a number of non-redundant rules in the different detail levels. The future work includes carrying out a series of experiments over well-known databases and the Monalisa repository database (http://alimonitor.cern.ch:8889), which is the Grid monitoring database for the ALICE experiment at CERN, in order to prove that the proposed method is generating the expected results. Our future research ideas include as well the introduction of modifications on the way the similarity measures are defined, and the reduction of the first mining step to the centroid of each cluster instead of its complete collection of documents.

## References


1. Agrawal, R., Imielinski, T., Swami, A.N.: Mining association rules between sets of items in large databases. In: Buneman, P., Jajodia, S. (eds.) Proceedings of the 1993 ACM SIGMOD International Conference on Management of Data, Washington, D.C., pp. 207–216 (1993)
2. Srikant, R., Agrawal, R.: Mining quantitative association rules in large relational tables. In: Jagadish, H.V., Mumick, I.S. (eds.) Proceedings of the 1996 ACM SIGMOD International Conference on Management of Data, Montreal, Quebec, Canada, pp. 1–12 (1996)
3. Srikant, R., Agrawal, R.: Mining generalized association rules, vol. 13, pp. 161–180 (1997)
4. Agrawal, R., Srikant, R.: Fast algorithms for mining association rules. In: Bocca, J.B., Jarke, M., Zaniolo, C. (eds.) Proc. 20th Int. Conf. Very Large Data Bases, VLDB, pp. 487–499. Morgan Kaufmann, San Francisco (1994)
5. Miller, R.J., Yang, Y.: Association rules over interval data, pp. 452–461 (1997)
6. Tong, Q., Yan, B., Zhou, Y.: Mining quantitative association rules on overlapped intervals. In: Li, X., Wang, S., Dong, Z.Y. (eds.) ADMA 2005. LNCS (LNAI), vol. 3584, pp. 43–50. Springer, Heidelberg (2005)
7. Kuok, C.M., Fu, A.W.C., Wong, M.H.: Mining fuzzy association rules in databases. SIGMOD Record 27, 41–46 (1998)
8. Dong, L., Tjortjis, C.: Experiences of using a quantitative approach for mining association rules. In: Liu, J., Cheung, Y.-m., Yin, H. (eds.) IDEAL 2003. LNCS, vol. 2690, pp. 693–700. Springer, Heidelberg (2003)
9. Park, J.S., Chen, M.-S., Yu, P.S.: An effective hash based algorithm for mining association rules. In: Carey, M.J., Schneider, D.A. (eds.) Proceedings of the 1995 ACM SIGMOD International Conference on Management of Data, San Jose, California, pp. 175–186 (1995)
10. Aumann, Y., Lindell, Y.: A statistical theory for quantitative association rules. In: KDD, pp. 261–270 (1999)
11. Okoniewski, M., Gancarz, L., Gawrysiak, P.: Mining multi-dimensional quantitative associations. In: Bartenstein, O., Geske, U., Hannebauer, M., Yoshie, O. (eds.) INAP 2001. LNCS (LNAI), vol. 2543, pp. 265–275. Springer, Heidelberg (2003)
12. Danger, R., Ruiz-Shulcloper, J., Berlanga, R.: Objectminer: A new approach for mining complex objects. In: ICEIS, (2), pp. 42–47 (2004)
13. Zhing, Z., Lu, Y., Zhang, B.: An effective partitioning-combining algorithm for discovering quantitative association rules. In: Proc. of the First Pacific-Asia Conference on Knowledge Discovery and Data Mining (1997)

14. Hipp, J., Myka, A., Wirth, R., Güntzer, U.: A new algorithm for faster mining of generalized association rules. In: Żytkow, J.M. (ed.) PKDD 1998. LNCS, vol. 1510, pp. 74–82. Springer, Heidelberg (1998)
15. Gyenesei, A.: Mining weighted association rules for fuzzy quantitative items. In: Zighed, A.D.A., Komorowski, J., Żytkow, J.M. (eds.) PKDD 2000. LNCS (LNAI), vol. 1910, pp. 416–423. Springer, Heidelberg (2000)
16. Aslam, J.A., Pelekhov, K., Rus, D.: Static and dynamic information organization with star clusters. In: CIKM, pp. 208–217 (1998)
17. Gil-García, R., Badía-Contelles, J.M., Pons-Porrata, A.: Extended star clustering algorithm. In: Sanfeliu, A., Ruiz-Shulcloper, J. (eds.) CIARP 2003. LNCS, vol. 2905, pp. 480–487. Springer, Heidelberg (2003)
18. Kann, V.: A compendium of NP optimization problems. In: Complexity and Approximation. Springer, Heidelberg (1999)
19. Hahsler, M., Grün, B., Hornik, K.: `arules` — A computational environment for mining association rules and frequent item sets. Journal of Statistical Software 14, 1–25 (2005)

# A Pattern Selection Algorithm in Kernel PCA Applications

Ruixin Yang, John Tan, and Menas Kafatos

Center for Earth Observing and Space Research (CEOSR)
College of Science, George Mason University, VA 22030, U.S.A.
{ryang,jtan,mkafatos}@gmu.edu

**Abstract.** Principal Component Analysis (PCA) has been extensively used in different fields including earth science for spatial pattern identification. However, the intrinsic linear feature associated with standard PCA prevents scientists from detecting nonlinear structures. Kernel-based principal component analysis (KPCA), a recently emerging technique, provides a new approach for exploring and identifying nonlinear patterns in scientific data. In this paper, we recast KPCA in the commonly used PCA notation for earth science communities and demonstrate how to apply the KPCA technique into the analysis of earth science data sets. In such applications, a large number of principal components should be retained for studying the spatial patterns, while the variance cannot be quantitatively transferred from the feature space back into the input space. Therefore, we propose a KPCA pattern selection algorithm based on correlations with a given geophysical phenomenon. We demonstrate the algorithm with two widely used data sets in geophysical communities, namely the Normalized Difference Vegetation Index (NDVI) and the Southern Oscillation Index (SOI). The results indicate the new KPCA algorithm can reveal more significant details in spatial patterns than standard PCA.

**Keywords:** Data mining, knowledge acquisition, large-scale, dimension reduction.

## 1 Introduction

Principal Component Analysis (PCA) has been extensively used in different fields since introduced by Pearson in 1902 (as cited in [1]). This data decomposition procedure is known by various names in different disciplines such as Karhunen-Loève Transformation (KLT) in digital signal processing [2], Proper Orthogonal Decomposition (POD) in studies of turbulence coherent structure with nonlinear dynamical systems [3], and Empirical Orthogonal Function (EOF) for one variable data [4] or Singular Value Decomposition (SVD) for multiple variables [5] applied to earth science, in particular for climate studies.

In principle, PCA is a linear procedure to transform data for various purposes including dimension reduction (factor analysis), separation of variables, coherent structure identification, data compression (approximation), feature extraction, etc. Consequently, PCA results can be viewed and explained from various perspectives. One way to interpret the PCA results is to consider the PCA procedure projecting the original high

J. Filipe, B. Shishkov, and M. Helfert (Eds.): ICSOFT 2006, CCIS 10, pp. 374–387, 2008.

dimensional data into a new coordinate system. In the new system, the space spanned by the first few principal axes captures most of the information (variances) of the original data [6]. Another point of view, commonly used in earth science, is to consider the PCA results of spatio-temporal data as a decomposition between the spatial components and temporal components. Once again, the approximation defined by the first few principal components gives the smallest total mean-square error compared to any other expansions with the same number of items [7].

In earth science applications, the spatial components from the PCA decomposition are recognized as representative patterns because the spatial components are orthogonal to each other. Correspondingly, the uncorrelated time series (temporal components) are often used to study the relationships between the corresponding spatial patterns and a predetermined phenomenon such as the well-known El Niño, characterized by abnormally warm sea surface temperature (SST) over the eastern Pacific Ocean. Through this procedure, patterns can be associated with natural phenomena. One example of such association is found between Normalized Difference Vegetation Index (NDVI) patterns and ENSO (El Niño Southern Oscillation) by directly using spatial components of PCA [8]. Another well-known spatial pattern is obtained by regressing the leading principal time series from the sea-level-pressure (SLP) to the surface air temperature (SAT) field [9].

Although PCA is broadly used in many disciplines as well as in earth science data analysis, the intrinsic linear feature prevents this method from identifying nonlinear structures. This may be necessary as many geophysical phenomena are intrinsically nonlinear. As a consequence, many efforts have been made to extend PCA to grasp nonlinear relationships in data sets such as the principal curve theory [10] and the neutral network-based PCA [11,12], which is limited to low dimensional data or needs standard PCA for preprocessing. More recently, as the kernel method has been receiving growing attention in various communities, another nonlinear PCA implementation as a kernel eigenvalue problem has emerged [13].

The kernel-based principal component analysis (KPCA) actually is implemented via a standard PCA in feature space, which is related to the original input space by a nonlinear implicit mapping [13]. KPCA has been recently applied to earth science data to explore nonlinear low dimensional structures [14,15]. Ideally, the intrinsic nonlinear low dimensional structures in the data can be uncovered by using just a few nonlinear principal components. However, the dimension numbers of the feature space are usually much larger than the dimension numbers of the input space. Moreover, the variance cannot be quantitatively transferred from the feature space back into the input space. Consequently, the numbers of principal components which contribute meaningful amounts of variances are much larger than we commonly encounter in standard PCA results. Therefore, we need a mechanism to select nonlinear principal patterns and to construct the representative patterns. In this paper, we present the KPCA algorithms in language used in climate studies and propose a new KPCA pattern selection algorithm based on correlation with a natural phenomenon for KPCA applications to earth science data.

To the best of our knowledge, this work is the first and only effort on using KPCA in climate studies for knowledge acquisition. Therefore, in the following section, we first describe the PCA algorithm and then the KPCA algorithm in language comparable to the standard PCA applications in climate studies. Next, we present the newly proposed KPCA pattern selection algorithm. Then we briefly discuss the earth science data used for this work in Section 3 and describe the results in Section 4. In Section 5, we first discuss in-depth our understanding of the KPCA concepts, and finally present conclusions.

The contribution of this work includes two main points: 1) The emerging KPCA technique is described in the notation of PCA applications commonly used in earth science communities, and this work is the first KPCA application to earth science data; 2) A new spatial pattern selection algorithm based on correlation scores is developed here to overcome the problems of KPCA applications in earth science data sets, the overwhelming numbers of components and the lack of quantitative variance description.

## 2 Algorithm

KPCA emerged only recently from the well-known kernel theory, parallel to other kernel-based algorithms such as support vector machine (SVM) classification [16]. In order to compare the similarities and the differences between standard PCA and KPCA, in this section, we first describe the commonly used standard PCA algorithms applied to earth science data analysis and then the KPCA algorithm for the same applications. Then, we discuss the limitation and new issues of KPCA such as pattern selection. Finally, we describe the new KPCA pattern selection algorithm.

### 2.1 PCA Algorithm

We follow the notations and procedures common to earth science communities to describe the standard PCA algorithm and a variant of its implementation by dual matrix [7].

Suppose that we have a spatio-temporal data set, $\psi(\boldsymbol{x}_m, t)$, where $\boldsymbol{x}_m$ represents the given geolocation with $1 \leq m \leq M$, and $t$, the time, which is actually discretized at $t_n(1 \leq n \leq N)$. The purpose of the PCA as a data decomposition procedure is to separate the data into spatial parts $\phi(\boldsymbol{x}_m)$ and temporal parts $a(t)$ such that

$$\psi(\boldsymbol{x}_m, t) = \sum_{i=1}^{M} a_i(t)\phi_i(\boldsymbol{x}_m). \tag{1}$$

In other words, the original spatio-temporal data sets with $N$ time snaps of spatial field values of dimension $M$ can be represented by $M$ number of spatial patterns. The contribution of those patterns to the original data is weighted by the corresponding temporal function $a(t)$. To uniquely determine the solution satisfying Equation (1), we place the spatial orthogonality condition on $\phi(\boldsymbol{x}_m)$ and the uncorrelated time variability condition on $a(t)$.

The above conditions result in an eigenvalue problem of covariance matrix $C$ with $\lambda$ being the eigenvalues and $\phi$ the corresponding eigenvectors. We can construct a data matrix $D$ with

$$D = \begin{pmatrix} \psi_1(t_1) & \psi_1(t_2) & ... & \psi_1(t_N) \\ \psi_2(t_1) & \psi_2(t_2) & ... & \psi_2(t_N) \\ ... & ... & ... & ... \\ \psi_M(t_1) & \psi_M(t_2) & ... & \psi_M(t_N) \end{pmatrix}, \tag{2}$$

where $\psi_m(t_n) \equiv \psi(\boldsymbol{x}_m, t_n)$. Each row of matrix $D$ is corresponding to one time series of the given physical values at a given location, and each column is a point in an $M$-dimensional space spanned by all locations, corresponding to a temporal snap. With the data matrix, the covariance matrix can be written as

$$C = fac * DD' \tag{3}$$

where the apostrophe symbol denotes the transpose operation. In the equation above, $fac = 1/N$.

Since the size of matrix $D$ is $M \times N$, the size of matrix $C$ is $M \times M$. Each eigenvector of $C$ with $M$ components represents a spatial pattern, that is, $\phi(\boldsymbol{x}_m)$. The corresponding time series (N values as a vector, $\boldsymbol{a}$) associated with a spatial pattern represented by $\overrightarrow{\phi_j}$ can be obtained by the projection of the data matrix onto the spatial pattern in the form of $\overrightarrow{a_j}' = \overrightarrow{\phi_j}' D$.

The advantage of the notations and procedures above, as normally defined in the earth science communities, is that they shed light on interpretation of the PCA results. Based on the matrix theory, the trace of covariance matrix $C$ is equal to the sum of all eigenvalues. That is, $trace(C) \equiv \sum_{i=1}^{M} c_{ii} = \sum_{i=1}^{M} \lambda_i$. From Equation (2) and Equation (3), we have

$$c_{ii} = \frac{1}{N} \sum_{n=1}^{N} [\psi_i(t_n)]^2, \tag{4}$$

which is the the variance of the data at location $\boldsymbol{x}_i$ if we consider that the original data are centered against the temporal average (anomaly data). Therefore, the trace of $C$ is the total variance of the data, and the anomaly data are decomposed into spatial patterns with corresponding time series for the contribution weights. The eigenvalues measure the contribution to the total variance by the corresponding principal components.

Computationally, solving an eigenvalue problem of an $M \times M$ matrix of $DD'$ form is not always optimal. When $N < M$, the eigenvalue problem of matrix $DD'$ can be easily converted into an eigenvalue problem of a dual matrix, $D'D$, of size $N \times N$ because the ranks and eigenvalues of $DD'$ and $D'D$ are the same [1]. Actually, the rank of the covariance matrix, $r_C$, is equal to or smaller than $min(M, N)$. The summation in Equation (1) and other corresponding equations should be from 1 to $r_C$ instead of $M$.

The element of the dual matrix of the covariance matrix:

$$S = fac * D'D \tag{5}$$

is not simply the covariance between two time series. Instead,

$$s_{ij} = \frac{1}{N} \sum_{m=1}^{M} [\psi_m(t_i)\psi_m(t_j)], \tag{6}$$

can be considered as an inner product between two vectors which are denoted by columns in the data matrix $D$ and are of $M$ components. One should note that spatial averaging does make sense in earth science applications for one variable, unlike in traditional factor analysis. Nevertheless, the data values are centered against the temporal averages at each location. Due to this fact, the dual matrix $S$ cannot be called a covariance matrix in strict meanings.

Since the matrix $S$ is of size $N \times N$, the eigenvectors are not corresponding to the spatial patterns. Actually, they are corresponding to the temporal principal components, or the time series, $a(t)$. To obtain the corresponding spatial patterns, we need to project the original data (matrix $D$) onto the principal time series by

$$\overrightarrow{\phi_j} = D\overrightarrow{a_j} \tag{7}$$

with an eigenvalue dependent scaling.

## 2.2 KPCA Algorithms

In simple words, KPCA is the implementation of linear PCA in feature space [13]. With the same notation as we used in the previous section for the spatio-temporal data, we can recast the KPCA concept and algorithm as follows.

As in the case with dual matrix, we consider each snap of spatial field with $M$ points as a vector of $M$ components. Then, the original data can be considered as $N$ $M$-dimensional vectors or $N$ points in an $M$-dimensional space. Suppose that there is a map transforming a point from the input space (the space for original data) into a feature space, then we have

$$\Phi : R^M \rightarrow F; \boldsymbol{\psi} \mapsto \boldsymbol{X}. \tag{8}$$

Assume the dimension of the feature space is $M_F$, one vector in the input space, $\overrightarrow{\psi}_k$, is transformed into

$$\overrightarrow{X}_k \equiv \overrightarrow{\Phi(\overrightarrow{\psi}_k)} = \left(\Phi_1(\overrightarrow{\psi_k}), \Phi_2(\overrightarrow{\psi_k}), ..., \Phi_{M_F}(\overrightarrow{\psi_k})\right). \tag{9}$$

Similar to the data matrix in input space, we can denote the data matrix in the feature space as

$$D_\Phi = \begin{pmatrix} \Phi_1(\overrightarrow{\psi_1}) & \Phi_1(\overrightarrow{\psi_2}) & ... & \Phi_1(\overrightarrow{\psi_N}) \\ \Phi_2(\overrightarrow{\psi_1}) & \Phi_2(\overrightarrow{\psi_2}) & ... & \Phi_2(\overrightarrow{\psi_N}) \\ ... & ... & ... & ... \\ \Phi_{M_F}(\overrightarrow{\psi_1}) & \Phi_{M_F}(\overrightarrow{\psi_2}) & ... & \Phi_{M_F}(\overrightarrow{\psi_N}) \end{pmatrix}. \tag{10}$$

Unlike the standard PCA case, where we can actually solve an eigenvalue problem for either $DD'$ or $D'D$ depending on the spatial dimension size and the number of observations (temporal size), we can only define

$$K = fac * D_\Phi{}' D_\Phi \tag{11}$$

for the eigenvalue problem in the feature space. This limitation comes from the so called "kernel trick" used for evaluating the elements of matrix $K$.

Comparing to the definition of $s_{ij}$ for the standard PCA case, we will have the element of matrix $K$ as

$$k_{ij} = fac * \left(D_{\Phi}' D_{\Phi}\right)_{ij} = \frac{1}{N}(\overrightarrow{\Phi(\overrightarrow{\psi_i})} \bullet \overrightarrow{\Phi(\overrightarrow{\psi_j})}). \tag{12}$$

The key of the kernel theory is that we do not need to explicitly compute the inner product. Instead, we define a kernel function for this product such that

$$k(\boldsymbol{x}, \boldsymbol{y}) = \left(\overrightarrow{\Phi(\boldsymbol{x})} \bullet \overrightarrow{\Phi(\boldsymbol{y})}\right). \tag{13}$$

Through the "kernel trick," we do not need to know either the mapping function $\Phi$ or the dimension size of the feature space, $M_F$, in all computations.

The main computation step in the KPCA is to solve the eigenvalue problem with $K\boldsymbol{\alpha} = \lambda\boldsymbol{\alpha}$. The eigenvalues still can be used to estimate the variance but only in the feature space. The eigenvector, as the case with dual matrix $S$ in the standard PCA case, is playing a role of a time series. For the spatial patterns in the feature space, another projection, similar to that described in Equation (7),

$$\boldsymbol{v} = \sum_{m=1}^{M} \alpha_i \overrightarrow{\Phi(\overrightarrow{\psi_i})}. \tag{14}$$

is needed. In practice, we do not need to compute $\boldsymbol{v}$ either. What we are more interested in is the spatial patterns we can obtain from the KPCA process. Therefore, we need to map back the structures represented by $\boldsymbol{v}$ in the feature space into the input space. Since the mapping from the input space to the feature space is nonlinear and implicit, it is not expected that the reverse mapping is simple or even unique. Fortunately, a preimage (data in the input space) reconstruction algorithm based on certain optimal condition has been developed already [17]. In this process, all needed computations related to the mapping can be performed via the kernel function, and the algorithm is used in this work.

### 2.3 KPCA Pattern Selection Algorithm

Kernel functions are the key part in KPCA applications. There are many functions that can be used as kernel as long as certain conditions are satisfied [13]. Examples of kernel functions include polynomial kernels and Gaussian kernels [18]. When the kernel function is nonlinear as we intend to choose, the dimension in the feature space is usually much higher than the dimension in the input space [13]. In special situations, the number of the dimensions could be infinite as in the case presented by the Gaussian kernel [17]. The higher dimension in feature space is the desired feature for machine learning applications such as classification because data are more separated in the feature space and special characters are more easily identified. However, for spatial pattern extraction in earth science applications, the higher dimensionality introduces new challenges because we cannot simply pick one or a few spatial patterns associated with the largest eigenvalues.

Moreover, in standard PCA, the principal directions represented by the spatial patterns can be considered as the results of rotation of the original coordinate system. Therefore, the total variance of the centered data is conserved under the coordinate system rotation. As a result, significant spatial patterns are selected based on the contribution of variance by the corresponding patterns to the total variance. This can simply be calculated by the eigenvalues as discussed in Section 2.1. In the KPCA results, the mapping between the input space and the feature space is nonlinear. Therefore, the variance is not conserved from input space into the feature space. Consequently, although the eigenvalues still can be used to estimate the variance contribution in feature space, the variance distribution in the feature space cannot be quantitatively transferred back into variance distribution in the input space.

The introduction of higher dimensions in KPCA, that is, a large number of principal components and the difficulty to quantitatively describe the variance contribution in the input space by each component require a new mechanism for identifying the significant spatial patterns. A new pattern selection algorithm is developed [14] to overcome these problems as described below.

In standard PCA applications for earth science data analysis, the temporal components are usually correlated with a time series representing a given natural phenomenon. And the corresponding spatial pattern is claimed to be related to the phenomenon if the correlation coefficient is significantly different from zero. In KPCA, we cannot easily identify such spatial patterns, but we generally have more temporal components, as discussed in Section 2.2. The eigenvectors $\alpha$, i.e., the KPCA temporal components, can be used to select KPCA components which can enhance the correlation to the given phenomenon.

After we perform the KPCA process on a particular set of data, we utilize an algorithm to obtain a reduced set of temporal components in the pattern selection procedure. Although the variance in feature space does not represent the variance in the input space, we can still use the eigenvalues as a qualitative measurement to filter KPCA components which may contribute to the spatial patterns in the input space. We are interested in the significant KPCA components which are associated with, say, $99.9\%$ variance in feature space as measured by the corresponding eigenvalues, and treat other components associated with very small eigenvalues as components coming from various noises. The algorithm sorts the temporal components in descending order according to their correlation score with a given phenomenon. Then linear combinations of them are tested from the highest score to the lowest, only the combinations that increase the correlation with the signal of interest are retained. The steps for combining the temporal components are:

- The correlation score of a vector **V** with the signal of interest is denoted as corr(**V**).
- Sort the normalized PC's according to the correlation score $\rightarrow PC_1, PC_2, \ldots, PC_p$.
- Save the current vector with the highest correlation score $\rightarrow \mathbf{V} := PC_1$.
- Save the current high correlation score as cc $\rightarrow cc := corr(PC_1)$.
- Maintain a list of the combination of sorted PC's $\rightarrow List\{\ \}$.

Where List.Add$\{1\}$ results in List$\{1\}$, List.Add$\{2\}$ results in List$\{1, 2\}$ , etc... Loop over the possible combinations of PC's that can increase the correlation score. If the

```
V := PC_1
cc := corr(PC_1)
List.Add(1)
FOR i := 2 TO p
    IF corr(V + PC_i) > cc THEN
       V := V + PC_i
       cc := corr(V + PC_i)
       List.Add(i)
    END IF
END FOR
```

**Fig. 1.** Pseudo-code of the pattern selection procedure

score is increased, then keep the combination of PC's. The pseudo-code for the new pattern selection algorithm is given in Figure 1. In the pseuso-code and in the list above, $p$ is the number of KPCA components we are interested in after de-noise.

The spatial patterns in input space are computed based on the preimage algorithm with all selected components [17].

## 3 Data

A gridded global monthly Normalized Difference Vegetation Index (NDVI) data set was chosen to implement the KPCA. NDVI is possibly the most widely used data product from earth observing satellites. The NDVI value is essentially a measure of the vegetation greenness [19]. As vegetation gains chlorophyll and becomes greener, the NDVI value increases. On the other hand, as vegetation loses chlorophyll, the value decreases.

The NDVI data used here were obtained from the NASA data web site [20]. The data are of $1^0 \times 1^0$ latitude-longitude spatial resolution with global coverage, and monthly temporal resolution with temporal coverage from January 1982 to December 2001. Since PCA analysis usually needs data without gaps, only data points with valid NDVI data in the whole period are chosen in the analysis. Therefore, we worked on global NDVI data for the 1982-1992 period only. Before using the data with PCA or KPCA, the NDVI data are deseasonalized by subtracting the climatological values from the original data. For that reason, the analysis is actually on NDVI anomalies.

In implementations, each point (location) in the physical coordinate system (the globe in latitude-longitude coordinates) is treated as one dimension, and time another dimension. Consequently, the data sets are represented in matrix format, and each column represents one month and each row element in the column represents a grid point value. In other words, all the latitude-by-longitude grid points for each month will be unrolled into one column of the data matrix. Therefore, the rows in each column represent a spatial location in latitude and longitude and each column represents a point in time as shown in the data matrix of Equation (2).

As a relationship between NDVI PCA patterns and El Niño Southern Oscillation (ENSO) was found [8], we pick ENSO as the natural phenomenon for implementing the pattern selection algorithm. El Niño refers to a massive warming of the coastal waters of the eastern tropical Pacific. The Southern Oscillation refers to the fluctuations of atmospheric pressure in eastern and western Pacific [21], and its amplitude is described by a normalized sea level pressure difference between Tahiti and Darwin, also called Southern Oscillation Index (SOI) [22]. Because El Niño is highly correlated with one phase of the southern oscillation, the phenomenon is usually called El Niño Southern Oscillation (ENSO). ENSO is the largest known global climate variability on interannual timescales, and the SOI is one of the representative signals of ENSO. The SOI represents a climatic anomaly that has significant global socio-economic impacts including flooding and drought pattern modification. The SOI data used here were obtained from NOAA National Weather Service, Climate Prediction Center [23].

## 4 Results

The standard linear PCA is first used to the spatio-temporal NDVI anomaly data. As a widely used procedure, we correlate the principal temporal components with the SOI time series and find that the correlation is strongest between the fourth component (the component corresponding to the fourth largest eigenvalue) and SOI. The correlation coefficient is $0.43$, and this component contributes $3.8\%$ of the total variance. The corresponding simple spatial pattern is displayed in Figure 2.

In the KPCA analysis, with trials of several kernels for the best results, we choose the Gaussian kernel,

$$k(\boldsymbol{x}, \boldsymbol{y}) = exp\left(-\frac{\| \boldsymbol{x} - \boldsymbol{y} \|^2}{2\sigma^2}\right). \tag{15}$$

for the demonstration. We then use the pattern selection algorithms described in Section 2.3 to obtain a combined spatial pattern. In order to attain a high correlation, the free parameter $\sigma$ in the Gaussian kernel had to be adjusted. Using the data set's standard deviation for $\sigma$ in the Gaussian kernel did not produce the best results. It is possible that the kernel under-fits the data with that $\sigma$. A $\sigma$ being equal to $26\%$ of the standard deviation of the NDVI data set resulted in the correlation score with SOI of $r = 0.68$.

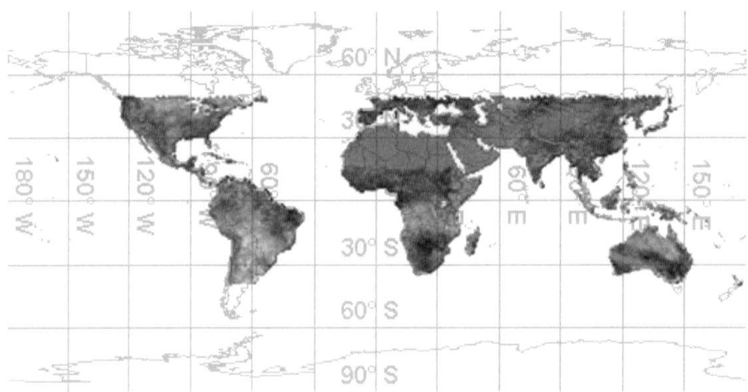


**Fig. 2.** Simple NDVI spatial pattern of the fourth spatial component from standard PCA. The gray scale denotes the anomaly values. The darkest is corresponding to the highly positive anomaly values.

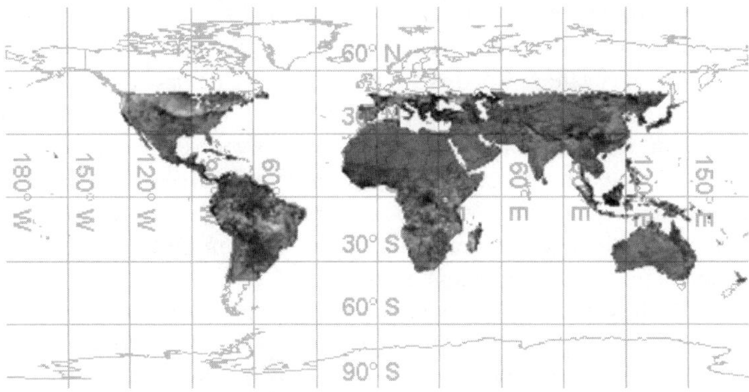


**Fig. 3.** Combined NDVI spatial pattern from KPCA results based on Gaussian kernel. The gray scale is the same as that in Figure 2.

Twenty (20) of the 131 eigenvectors were used, and those are about 15% of the significant KPCA components. The corresponding combined spatial pattern with those selected components is presented in Figure 3.

For comparison, the same pattern selection algorithm is also applied to the standard PCA results. In this case, 28 of 120 eigenvectors are selected for enhancing the correlation set initially by the fourth component. The resulting correlation coefficient is $r = 0.56$. The corresponding combined spatial pattern based on the 28 selected PCA components is demonstrated in Figure 4. Apparently, the pattern selection algorithm is more efficient and effective with the KPCA application than with the standard PCA application because we achieve higher correlation scores with fewer components in the KPCA case than in the standard PCA case.

By comparing Figure 4 and Figure 3 against Figure 2, we can find that the combined patterns from either the standard PCA or KPCA components show higher-resolution

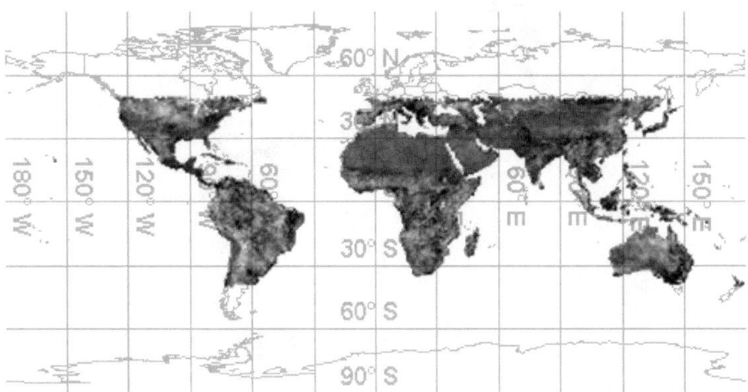


**Fig. 4.** Combined NDVI spatial pattern from standard PCA results with the same pattern selection algorithm as for KPCA. The gray scale is the same as that in Figure 2.

structure than the simple pattern presented by a single PCA component. This result is not unexpected because PCA components contain high resolution information in components with low eigenvalues. In other words, the first principal component associated with the largest eigenvalue catches large scale features of the data. The key point is that with standard principal component analysis, we can only pick one component to be associated with a given phenomenon through a correlation analysis. Once the component is identified, we cannot associate other components to the same phenomenon. The pattern selection algorithm described in this paper provides a mechanism to select multiple principal components with one phenomenon.

To explore the difference for information extraction from the combined patterns and the simple pattern, we display a world drought map for the 1982-1983 El Niño episode in Figure 5 [24] because the NDVI dataset used here spans the 1982-1992 period. Please note that the correlation selection in our case is based on a positive correlation coefficient while the values of SOI associated with El Niño are negative. Therefore, in the spatial patterns based on an NDVI anomaly (Figres 2–4), positive values are actually associated with a negative NDVI anomaly due to ENSO, which in return, is associated with the drought patterns in Figure 5.

The simple PCA pattern (Figure 2) does capture drought patterns, but in large scale only, such as droughts in the Amazon area, southern Africa, and Australia in the 1982-1983 period. However, the shapes and sizes of the drought patterns are difficult to compare with the simple PCA pattern. In contrast, the combined patterns from the selection algorithm applications on standard PCA and KPCA capture the details such as the curvature in the drought patterns in the continental US for the 1983 drought. The combined KPCA pattern also shows good agreement on the drought patterns in western Africa around Ivory Coast. The drought pattern in Malaysia and Borneo Island (around 112E longitude near the Equator) in the South & East Asia region is evident in the combined patterns from both standard PCA and KPCA, but they are not exhibited in the simple

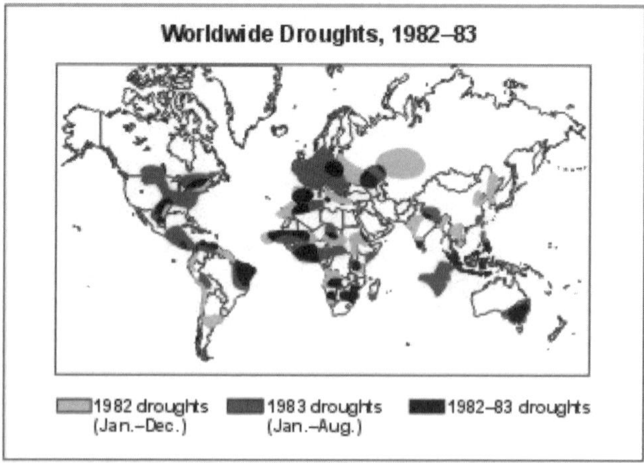


**Fig. 5.** World drought pattern during the 1982-1983 El Niño episode (from the web site of National Drought Mitigation Center [24])

PCA pattern. Another apparent improvement from the combined KPCA spatial pattern is that the drought in Europe is more accurately identified in contrast to the simple PCA pattern.

## 5 Discussion and Conclusions

From a data decomposition perspective, PCA as well as KPCA are data adaptive methods. That means that the bases for the decomposition are not chosen *a priori,* but are constructed from the data. In the standard linear PCA case, the orthogonality condition on the spatial patterns and the uncorrelated condition on temporal components guarantee the uniqueness of the decomposition. Additional freedoms introduced by the implicit nonlinear transformation make the uniqueness condition invalid, and the KPCA results depend on the nonlinear structure implicitly described by the kernel. As a result, different kernels should be tested before significant results can be discovered because the underlying nonlinear structure can only be picked up by a kernel with a similar structure.

In a broad sense, principal component analysis describes the co-variability among multivariate observations. The definition of the co-variability between two observations actually determines the core structure one may expect from the result. The most commonly used definition is covariance or correlation between points defined in either object space or variable space [6]. In KPCA application, if we do not consider the process as a mapping from input space into the feature space, we can treat the "kernel trick" as another definition of the pair-wise co-variability. However, this definition of the co-variability can only be implemented on data points defined for each observation. That is, the KPCA is applied to object space only. This results in the eigenvalue problem for KPCA being always on a matrix of size $N \times N$, even when $M$, the number of variables or geolocations in earth science applications is smaller than $N$. Since the mapping function $\Phi$ is never determined in the procedure, the computationally efficient SVD procedure cannot be used either, because the data matrix in feature space, $D_{\Phi}$, is not known.

The pair-wise co-variability is actually a measure of the pair-wise proximities. Therefore, KPCA can be understood in a broad sense as a general means to discover "distance" or "similarity (dissimilarity)" based structure. That is why most dimension reduction algorithms such as Multidimensional Scaling (MDS) [25], Locally Linear Embedding (LLE) [26], and Isomap [27] can be related to KPCA algorithm [28].

In conclusion, the KPCA algorithm is recast in the notation of PCA commonly used in earth science communities and is used for NDVI data. To overcome the problems of KPCA applications in earth sciences, namely the overwhelming numbers of components and lack of quantitative variance description, a new spatial pattern selection algorithm based on correlation scores is proposed here. This selection mechanism works both on standard PCA and KPCA, and both give superior results compared to the traditional simple PCA pattern. In the implementation example with NDVI data and the comparison with the global drought patterns during the 1982-1983 El Niño episode, the combined patterns show much better agreement with the drought patterns on details such as locations and shapes.

## References


1. Von Storch, H., Zwiers, F.W.: Statistical Analysis in Climate Research. Cambridge University Press, Cambridge (1999)
2. Haddad, R.A., Parsons, T.W.: Digital Signal Processing: Theory, Applications, and Hardware. Computer Science Press (1991)
3. Holmes, P., Lumley, J.L., Berkooz, G.: Turbulence, Coherent Structures, Dynamical Systems and Symmetry. Cambridge University Press, Cambridge (1996)
4. Lorenz, E.N.: Empirical orthogonal functions and statistical weather prediction. In: Final Report, Statistical Forecasting Project, 1959. Massachusetts Institute of Technology, Dept. of Meteorology, pp. 29–78 (1959)
5. Wallace, J.M., Smith, C., Bretherton, C.S.: Singular Value Decomposition of Wintertime Sea Surface Temperature and 500-mb Height Anomalies. Journal of Climate 5, 561–576 (1992)
6. Krzanowski, W.J.: Principles of Multivariate Analysis: A User's Perspective. Oxford University Press, Oxford (1988)
7. Emery, W.J., Thomson, R.E.: Data Analysis Methods in Physical Oceanography. Elsevier, Amsterdam (2001)
8. Li, Z., Kafatos, M.: Interannual Variability of Vegetation in the United States and Its Relation to El Niño/Southern Oscillation. Remote Sensing of Environment 71, 239–247 (2000)
9. Thompson, D.W.J., Wallace, J.M.: Annular Modes in the Extratropical Circulation. Part I: Month-to-Month Variability. Journal of Climate 13, 1000–1016 (2000)
10. Hastie, T., Stuetzle, W.: Principal Curves. Journal of the American Statistical Association 84, 502–516 (1989)
11. Kramer, M.A.: Nonlinear Principal Component Analysis Using Autoassociative Neural Networks. AIChE J. 37(2), 233–243 (1991)
12. Monahan, A.H.: Nonlinear Principal Component Analysis: Tropical Indo–Pacific Sea Surface Temperature and Sea Level Pressure. Journal of Climate 14, 219–233 (2001)
13. Schölkopf, B., Smola, A., Müller, K.R.: Nonlinear Component Analysis as a Kernel Eigenvalue Problem. Neural Computation 10, 1299–1319 (1998)
14. Tan, J.: Applications of Kernel PCA Methods to Geophysical Data. George Mason University, PhD Thesis (2005)
15. Tan, J., Yang, R., Kafatos, M.: Kernel PCA Analysis for Remote Sensing Data. In: 18th Conference on Climate Variability and Change, Altanta, GA, CD-ROM, Paper P1.5. American Meteorological Society (2006)
16. Schölkopf, B., Burges, C.J.C., Smola, J.: Advances in Kernel Methods: Support Vector Learning. MIT Press, Cambridge (1999)
17. Mika, S., Schölkopf, B., Smola, A., Müller, K.R., Scholz, M., Rätsch, G.: Kernel PCA and De-noising in Feature Spaces. In: Kearns, M.S., Solla, S.A., Cohn, D.A. (eds.) Advances in Neural Information Processing Systems, vol. 11, pp. 536–542. MIT Press, Cambridge (1999)
18. Schölkopf, B., Mika, S., Burges, C., Knirsch, P., Müller, K.R., Rätsch, G., Smola, A.: Input Space vs. Feature Space in Kernel-Based Methods. IEEE Transactions on Neural Networks 10, 1000–1017 (1999)
19. Cracknell, A.P.: The Advanced Very High Resolution Radiometer. Taylor & Francis Inc, Abington (1997)
20. GES DISC (NASA Goddard Earth Sciences (GES) Data and Information Services Center (DISC)): Pathfinder AVHRR Land Data (2006), (Last accessed on February 9, 2006), `ftp://disc1.gsfc.nasa.gov/data/avhrr/Readme.pal`
21. Philander, S.G.: El Niño, La Niña, and the Southern Oscillation. Academic Press, London (1990)

22. Ropelewski, C.F., Jones, P.D.: An Extension of the Tahiti–Darwin Southern Oscillation Index. Monthly Weather Review 115, 2161–2165 (1987)
23. CPC (Climate Predication Center/NOAA): (STAND TAHITI - STAND DARWIN) SEA LEVEL PRESS ANOMALY (2006), (Last accessed on February 5, 2006), http://www.cpc.ncep.noaa.gov/data/indices/soi
24. NDMC (National Drought Mitigation Center): What is Drought? (2006), (Last accessed on February 8, 2006), http://www.drought.unl.edu/whatis/elnino.htm
25. Cox, T.F., Cox, M.A.: Multidimensional Scaling. Chapman and Hall, Boca Raton (2000)
26. Roweis, S., Saul, L.: Nonlinear Dimensionality Reduction by Locally Linear Embedding. Science 290, 2323–2326 (2000)
27. Tenenbaum, J.B., de Silva, V., Langford, J.: A Global Geometric Framework for Nonlinear Dimensionality Reduction. Science 290, 2319–2323 (2000)
28. Ham, J., Lee, D., Mika, S., Schölkopf, B.: Kernel View of the Dimensionality Reduction of Manifolds. In: Proceedings of the 21st International Conference on Machine Learning (2004)

# Author Index

Printing: Mercedes-Druck, Berlin
Binding: Stein + Lehmann, Berlin